# CLINICAL EMG FOR SURFACE RECORDINGS: VOLUME 2

Jeffrey R. Cram, Ph.D., and Associates

Clinical Resources
Nevada City, California

First Printing, 1990
Second Printing, 1990
Third Printing, 1991
Fourth Printing, 1991



Made in the United States of America

Published By:

Clinical Resources
14618 Tyler Foote Road
Nevada City, California 95959
1-800-669-7187

ISBN: 1-878701-00-2

# CONTRIBUTORS

Jeffrey R. Cram, Ph.D.
Owner and Director
Clinical Resources
Nevada City, CA

C.C. Stuart Donaldson, Ph.D.
Tarus Consultants
Calgary, Alberta
Canada

Mary Donaldson, M.Ed.
Calgary, Alberta
Canada

Dennis L. Ettare, M.A.
Owner and Clinical Director
Biofeedback Associates of CA
San Jose, CA

Ross C. Ettare, M.D.
Pacific Institute of
Cancer Research
Los Angeles, CA

Barbara J. Headley, M.S., P.T.
Director
Pain Evaluation &
Rehabilitation Center
St. Paul, MN

Leonard Hudzinski, Ph.D.
Dept of Psychiatry
Ochsner Clinic,
New Orleans, LA

Robert Kall, M.Ed.
Owner Future Health
Ben Salem, PA

Glenn Kasman, M.S., P.T.
Virginia Mason Medical Center
Seattle, WA

Gary Lawrence, B.A.
Dept of Psychiatry
Ochsner Clinic
New Orleans, LA

John Perry, Ph.D.
Owner Biotechnologies
Bryn Mawr, PA

Leslie Talcott, M.S., R.N.
Director
Perineometry Research
Institute
Bryn Mawr, PA

Robert Tosh, D.C.
Private Practitioner
Quakertown, PA

Will Taylor, M.D.
Occupational Health
Consultant
Falmouth, ME

Edgar Wilson, M.D.
Colorado Center for
Behavioral Health
Boulder, CO

# *FOREWORD*

In the scholarly world of books and authors, one becomes somewhat saturated with seeing and reading new book after new book. A natural ennui results. But when I saw the proofs of this fine book, my fatigue vanished and I voraciously read. It is truly unique and so a pleasure for me to write this foreword for it. Jeffrey Cram and his associates deserve not only commendation and congratulations, we owe them also a great vote of thanks.

For many advanced readers experienced in EMG and/or EMG Biofeedback, the book is an essential to expand their horizons. For the novice, it is an excellent part of the training required to keep up to date on these fields. Scientific, but not dry and obscure, the book will be a happy companion for all who become acquainted with it. In short, it is a fine book that takes a prominent place among the top books on the subject.

John V. Basmajian, MD, FRCPC, FACA,
FACRM (Australia), FABMR, FSBM

Professor Emeritus, Medicine and
Anatomy, McMaster University

# PREFACE

This volume has grown out of my sincere desire to revise and update the prior volume (usually referred to as the "Gray Manual"), while strengthening the neuromuscular perspective by asking clinicians in the field to lend their expertise and broaden the base of our understanding.

My chapter on Muscle Scanning now includes a firmer, broader basis of understanding of this technique. The more recent studies on reliability, validity, specificity and sensitivity have been included, along with more case presentations and a new "imaging" technique to "see" the findings. The graphics for dynamic assessment have been greatly improved, and hopefully will aid in differentiating what is normal from what is abnormal.

I have asked several biofeedback practitioners who deal with pain to share their knowledge. The chapter by Donaldson and Donaldson lays down some very basic axioms for dynamic assessments which we should all memorize. Will Taylor's chapter encourages us to think of the neuromuscular system from a "holistic" perspective. He wants us to remember that the dynamic forces of excitation and inhibition are at work as muscles push and pull bones through space and time. Dennis Ettare has created some very nice and simple dynamic rituals for the world of EMG feedback training which we can all benefit from. If you want to begin generalization training for EMG oriented work, this chapter will put you into action.

The treatment of the neuromuscular system would not be complete without input from physical therapy. Barbara Headley's chapter clearly looks at the interaction of the neuromuscular system through the concepts of excitation, inhibition, posture, activity and trigger points. Her work is the only work I am aware of which describes the relationship between trigger points and surface EMG. Glenn Kasman provides a very comprehensive view of evaluation and treatment, breaking each into its various tasks and components. If you want to know all of the potential variables which might affect treatment, and how to approach these with biofeedback while using all of the "tricks of the trade," you will find this chapter enlightening.

It has always been my opinion that both static and dynamic EMG monitoring would become the staple of knowledge for the Chiropractic profession. Dr. Tosh presents a conceptual framework for surface EMG which should aid chiropractic (and alike) with their diagnostic and treatment considerations.

Hudzinski and Lawrence have taken their many years of study and experience in treating TMJ and Headache related disorders at the

Oschner Clinic, and provided the practitioner with all of the diagnostic and treatment components which they have found important and useful.

The treatment of incontinence associated with a disturbance of the muscle of the pelvic floor is a frontier of EMG biofeedback. Perry and Talcott are some of the original pioneers of this area, and share their information in a sensitive fashion.

A book on the neuromuscular system would not be complete without due consideration being given to the muscles which move skin instead of bones. Rob Kall's chapter reviews how these muscles of E-motion may be used to encourage a positive affective state.

The last chapter by Dr. Ed Wilson is a very special one for me. One which should be copied and given out to every chronic pain patient you treat. It is written to and for the patient, and addresses all of the issues they might encounter in their long journey towards a new lifestyle.

I hope that you enjoy the contents of this book as much as I enjoyed working with the authors and putting their materials together.

Jeffrey R. Cram, Ph.D.
Seattle, Washington
Spring, 1990

# *ACKNOWLEDGEMENTS*

The work in this book would not exist if it were not for some interesting and supportive people who I encountered down the path of life. One such person is Jan Hoover, the developer of the EMG (biofeedback) instruments which I cut my teeth on. He encouraged me to think about "what an instrument could do." He also allowed me to write technical manuals for his instruments, letting the mildly dyslexic child within me find a rewarding route of expression. It was also his "accidental" production of the scanning electrode system (the IP-5 for the M-57 EMG) which caught my attention. I call it accidental, in that I don't believe he truly understood the impact such a simple device could have on our understanding of muscles, and upon the professional community.

The "gift" of the muscle scanning device came while I was working at the Swedish Hospital Pain Clinic. Dr. Charles Freeman played a major role by allowing me to follow my instincts, using EMG monitoring and feedback in a fashion other than the traditional relaxation based training. Dr. Steven Wolf introduced me to the concepts of dynamic EMG evaluation and training, which dovetailed wonderfully with the static evaluation of the scan. With these two tools in hand, the rest is history.

There are several people who actually made the production of this volume possible. First of all, my wife and children, who tolerated the time and energy which was directed towards this work. Without the perseverance of David Skeels the book would have never made it to press. He spent many hours inputing and massaging the manuscript into its final form. Stephanie Edmonds was responsible for reading the manuscript word for word, dotting the i's and crossing the t's.

Last, but not least, are the authors who have worked hard to make their contributions. All of these contributors are clinicians, who were not under the traditional publish or perish dilemma. Their contributions come straight from the heart.

# *TABLE OF CONTENTS*

# CHAPTER 1

# *EMG Muscle Scanning and Diagnostic Manual for Surface Recordings*

Jeffrey R. Cram, Ph.D.

---

## *TABLE OF CONTENTS*

# *LIST OF FIGURES*

# *LIST OF TABLES*

# *INTRODUCTION*

## DIAGNOSTIC USES OF SURFACE EMG

Surface electromyography may be used in many ways. Diagnostically, it may be used to monitor the "emotional" tone of particular muscle groups and therefore the general arousal level of the individual patient. Since muscles primarily move bones, EMG may also be measured during movement to quantify some of its normal or abnormal aspects. Since muscles also provide the basis of postural support, surface EMG may also be described in terms of posture. Splinting and guarding patterns associated with pain are another feature contributing to the source EMG.

Perhaps the most interesting diagnostic feature of surface EMG has to do with muscle resting length. A muscle may shorten and feel hard to the touch with both contraction or contracture. With contraction, the muscle's hardness is associated with increased EMG activity. However, with contracture, a muscle at rest feels hard to the touch and is electrically silent. A physician may palpate the neck or shoulders of a patient and notice what he or she perceives as hard/tense muscles. With this finding, a physician may prescribe medications (muscle relaxants) or refer the patient for relaxation (based biofeedback) training. When the biofeedback practitioner scans or attaches electrodes on these muscles, the EMG values make the decisive diagnosis. If the values are high, relaxation is in order. If the values are low, stretching is called for.

Once one understands that the resting length of a muscle can be altered through rhythmic exercise or restrictions of lengthening, it becomes possible to understand how a hard muscle may be electrically silent. The large and hard muscles of body builders, for example, are electrically silent in their resting state. They look and feel hard due to the building of muscle bulk, not increased EMG activity. Likewise, shortening of a muscle occurs when one does not engage in activities which stretch or lengthen the muscle. The hamstrings may shorten from a job which requires a lot of sitting. When this happens, touching the toes becomes difficult. Similarly, someone who has a painful area may immobilize it in order to avoid pain (protective guarding). If this is done for a long enough length of time, the muscles will shorten, feel hard to the touch and yet be electrically silent in their resting state. If one discovers a hard muscle which is electrically silent (has a low level of EMG activity), one should conclude that the patient is in need of mobilizing the muscle through exercise rather than suggesting an EMG biofeedback relaxation program. In essence, "Diagnosis should lead to treatment".

## THE TREATMENT POTENTIAL OF SURFACE EMG (BIOFEEDBACK)

EMG biofeedback is the process of monitoring and displaying muscle activity to an individual patient. The muscle activity is converted to either a visual or auditory display which can be easily understood. This information provides a "feedback loop" to the patient about the "state" of their muscle activity. Since we are primarily blind to the sensations of our muscles, unless they hurt, the EMG biofeedback allows the patient to become aware of an otherwise unconscious process. By using this information, the patient learns to better control and regulate the number and rate of motor unit activity. This may involve learning to alter the output from the cortex, the basal ganglia, the cerebellum, the limbic system and/or the spinal reflexes. The major goal of EMG biofeedback is to teach the patient to activate the right number of motor units for the task at hand. To reach this goal, the patient might receive feedback about: 1) Attempts to relax a given muscle or region; 2) Attempts at discriminating specific levels of muscle activity; 3) Attempts to tense or contract a given muscle or region; or 4) How alterations in posture effect the pattern of muscle activity.

EMG biofeedback is commonly used in conjunction with more general relaxation techniques. Here, quieting the mind and emotions are thought to be as important as quieting the muscles. In fact, the best known of these techniques is the general or "systemic" relaxation model. Here the electrodes are typically placed on the face, shoulders or arms and the patient is encouraged to lower his or her general level of arousal through lowering the resting tone of particular emotionally laden muscle groups. It may also be used as an adjunct to stress management or psychotherapy where emotional displays in the muscular system can be monitored and altered "on-line", so to speak.

Surface EMG recordings are used in several different types of treatment protocols. The work of Dr. Perry, presented in one of the later chapters, introduces the use of EMG in the treatment of incontinence through the re-education of the pelvic floor muscles. (biofeedback aided Kegal exercises, if you would). Surface EMG recordings are being used more and more in the retraining of the muscles in pain related disorders. The "postural or dynamic model", as it has been dubbed, has been gaining more and more momentum over the last five years.

As one begins working with the neuromuscular system, one's effectiveness in treatment seems to exponentially increase with one's understanding of this system.

## SECTION 1. OVERVIEW OF THE NEUROMUSCULAR SYSTEM

### THE SENSORY AND MOTOR ASPECTS OF THE SYSTEM.

The neuromuscular system is very rich, complex and subtle. Its many levels of sensory information (muscle spindles, golgi tendon organs, ruffini terminals and free nerve endings) informs the CNS of the instantaneous status of each bone, muscle group and muscle fiber. This sensory input alters muscle tension and action. It does this by strongly modulating the two motor systems. The gamma motor system controls the sensitivity of the stretch receptor and the alpha motor system controls the large muscle fibers which ultimately move the bones. Interestingly enough, the primary function of the skeletal muscles is to provide postural support. Once posture is assumed, movement, a secondary function, can take place.

Posture and movement may be acquired or refined. We learn by integrating our spinal reflexes to walk, to write, etc.. If injury to the muscle or the supporting soft tissue occurs, posture and movement patterns may be altered by reflexively bound muscle splinting. When this is maintained in a longer rhythm, learning occurs and a protective guarding pattern emerges to prevent pain. If the range of motion of the affected muscle group is consciously or unconsciously restricted for a prolonged length of time, a structural change in its resting length will occur further altering movement and posture.

On yet another level, the neuromuscular system is intimately involved in the displays of emotions. While facial displays of emotions are universally recognized, postural displays are less likely to be registered in a conscious fashion. Again, these emotional displays alter movement and posture. All of these aspects should enter into our diagnostic evaluation of this system.

The sensory side of the neuromuscular system is quite fascinating. The information which comes from our joints emanates from the ruffini nucleus. This CNS goniometer is sensitive to two degrees of arc and never turns off (habituates). It is also one of the few aspects of the neuromuscular system which projects to the cortex. It is for this reason that patients seem to be more aware of their bone or limb position than their muscle tension. It is easier to teach patients about their muscle function by asking them to move a limb or region in this direction or that than by asking them to tense this muscle or that. One of the easiest ways to help people release tension in their shoulders, for example, is to request that they drop or lower their shoulders. Using this positional, joint related language is vastly superior compared to using words such as "relax or let go".

One other sensory element projects to the cortex. This is the free nerve ending or nociceptor. This is the nerve ending which is responsible for the sensation of pain. These primarily surround the arteries and arterioles supplying the muscle as well as connective tissue near the muscle. Output from these nerve endings is caused by mechanical pressure (swelling or edema) or by changes in the biochemical environment of the muscle. These biochemical changes are towards a stronger acid condition caused by the build up of muscle metabolites (lactic acid). When pain occurs, it is due solely to these two events.

In general, an individual is more likely to be aware that a muscle hurts before they are aware that it is tense. This is because there are no "tension receptors" in the muscle. The threshold of nocioception however, may be lowered by a diminished supply of oxygen to the muscle (ischemia) commonly associated with prolonged tonic muscle contractions. Severe or prolonged reductions in oxygen will lead to an increase in the release of bradykinin, prostaglandin and serotonin which sensitize the free nerve endings (Ulmer, 1977; Robard, 1975). This lowered pain threshold may now allow the internal mechanical pressure associated with moderate muscle contractions (or tension) to lead to pain.

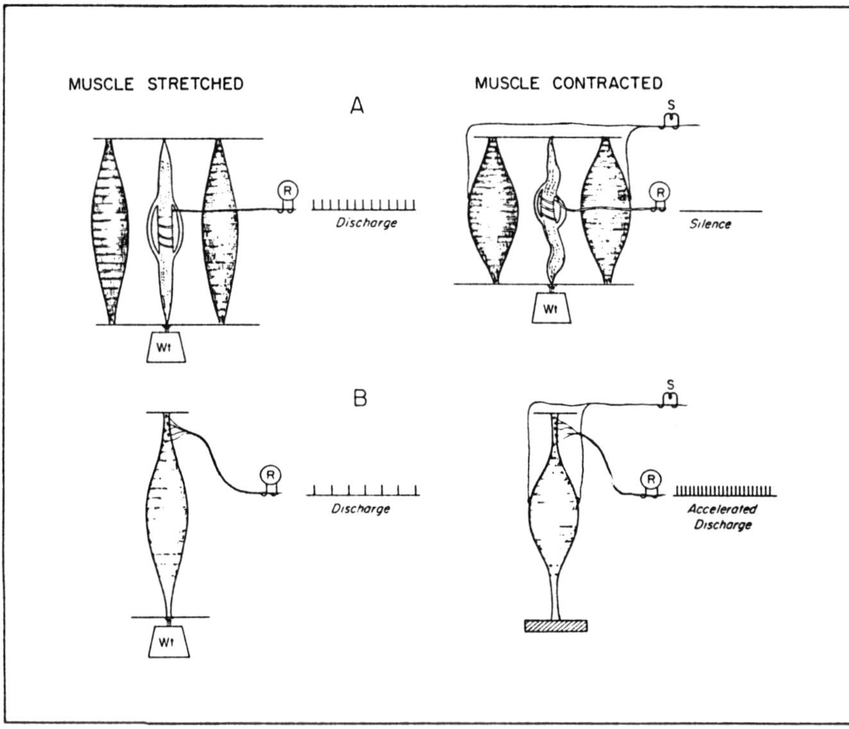


**FIGURE 1.** THE MUSCLE SPINDLE (A) AND GOLGI TENDON ORGAN (B) REPRINTED WITH PERMISSION FROM RUCH, T. & PATTON, H., PHYSIOLOGY AND BIOPHYSICS

The most similar sensory endings to a "tension receptor" are the muscle spindle and the golgi tendon organ. Simply put, the muscle spindle is our stretch receptor. Our stretch receptors provide the CNS with information concerning the instantaneous length and velocity of every muscle movement. They play a vital role in setting our muscle tone, and guiding our movements through time and space against gravity. Whenever a muscle spindle is placed on stretch, the muscle which houses the stretch receptor contracts. A common but very artificial example of this occurs when the physician tests the reflexes by hitting the patella tendon. When the reflexes are intact a tap to this tendon stretches the quadriceps muscle leading to its contraction and thus the movement of the lower leg. A real life example would occur when you are standing and lean forward. The calf muscles (primarily soleus) are placed on stretch as you go forward off your center of gravity and they reflexively contract to prevent you from falling forward. The sensitivity of the stretch receptor is set by the gamma motor system which we shall discuss later.

If the stretch receptor turns on its muscle of origin, the golgi tendon organ turns it off. The golgi tendon organ resides in the tendon which connects the muscle tissue to the bone. It is active only during muscle contraction, not stretching. As the muscle contracts more and more, the golgi tendon organ gets more and more excited. Its primitive function is to stop the muscle from contracting so hard that it tears itself from its attachments. Its subtler function is to inform the CNS of the strength of the contraction and ultimately when to stop the contraction. For the piano player, it guides the loudness of the notes played by providing information concerning the strength of the contraction used to strike the key.

At a spinal cord or segmental level, the lower motor interneurons are barraged by the push and pull outputs of the muscle spindles and golgi tendon organs. If the excitatory potentials are strong enough the lower motor neuron fires. The output of the lower motor neuron, along with its axons and the muscle fibers it innervates, is called a "motor unit". It is also known as the alpha motor system; the system that makes the bones move. It is this system that we measure with our EMG sensors and amplifiers.

There is however, a second motor system known as the gamma motor system. This motor system innervates and activates the small muscles upon which the muscle spindle or stretch receptor is located. The more active the gamma motor system, the more sensitive the stretch receptors and the greater the general muscle tone. The less active the gamma motor system, the less active the stretch receptor and the more deeply relaxed the muscle will be. It was the gamma motor system which the Progressive Muscle Relaxation training procedures of Dr. Edmond Jacobson (1924) were ultimately directed. While he may have initiated his training at an alpha motor level with gentle tense and release motions, by the end of training it was common for the patella reflex (see above) of the patient to be absent while in the relaxed state.

What controls or commands the gamma motor system to fire? Certainly it receives some of its excitatory potentials from the brain and brain stem. The cerebellum, for example, is well known for its use of the gamma motor system in the coordination of posture. The gamma motor system is also activated at a segmental, spinal cord level. Interestingly enough, nocioception is known to excite the gamma motor system at this level. If this nocioception comes from an injured joint or muscle it increases gamma motor activity. This, in turn, activates muscles crossing that joint, thus providing the basis for muscular "splinting". As this occurs, the injured area loses it range of motion as a protective, healing measure. The gamma motor system is also amenable to learning and may provide part of the basis for a protective guarding pattern if it were to emerge. Unfortunately, we cannot directly monitor the gamma motor system with our EMG instruments. We can only indirectly note its effect by monitoring alpha motor activity.

## PHYSIOLOGICAL BASIS OF MUSCLE ACTIVITY

The source for the EMG signal is the alpha motor system and its motor units. As you might recall, a motor unit is comprised of the lower motor neuron, its axon and the muscle fibers which it innervates. When the nerve action potential travels down the axon and reaches the neuromuscular junction it releases acetylcholine (ACH). This chemical event begins the process of depolarization of the muscle fibers of the motor unit. The energy of the depolarization is released along the direction of the muscle fiber, and it is this energy which creates the muscle action potential or MAP. It is the MAP which comprises the source EMG signal.

When MAPs are studied using needle EMG electrodes, the recording electrode is inserted into the muscle tissue. The size of the recording electrode is very small and specific, literally the size of the head of a pin. Typically a single motor unit is studied at multiple muscle sites along what is known as myotomes. Myotomes are innervation patterns for the motor system, just as dermatomes are innervation patterns for the sensory aspects of the skin. When more than one motor unit is voluntarily recruited for diagnostic purposes the phenomena is called an "interference pattern". This term reflects the bias of the invasive electromyographer. The primary emphasis is the single motor unit.

Needle EMG studies are conducted primarily by physiatrists and neurologists to study the integrity of the nerve which feeds a muscle. If there is a peripheral nerve injury or if the nerve is being impinged by a bulging disk (radiculopahy), the motor unit activity associated with this trauma will be identifiable from the wave form of single motor unit activity. Only on a rare occasion does the needle EMG study show a "myopathies" or something wrong with the integrity of the muscle tissue itself. It is very important to differentiate between needle EMG studies and surface EMG studies. In fact, there is a movement in physical medicine to put a small "i" in front of the "EMG", to indicate an integrated or surface EMG. This would appear as "iEMG".

While the needle EMG studies a single motor unit, the surface EMG studies populations of motor units. The energy from the individual MAPs of the motor units travels through the body tissues summating at the skin's surface along the lines of energy from which it (they) emanated. The amplitude of this energy in incredibly small and is expressed in microvolts, a millionth of a volt. The amplitude of the surface EMG recording is determined by many factors, some of which are listed below: 1) The number of motor units firing; 2) The rate of these firings; 3) The proximity of these motor units to the sensors or electrodes; 4) The orientation of the recording electrodes relative to the muscle fibers; 5) The distance between the recording electrodes; 6) The filter band width which you use, and 7) The obesity of the individual.

The energy of surface EMG also has a power spectrum ranging somewhere between 20 to 10,000 Hz. The shape of this energy spectrum varies slightly from muscle group to muscle group, but it is safe to say that close to 80% of the signal resides somewhere between 30 to 80 Hz. Some believe that energy above 300 Hz is primarily associated with vibration or noise from the skin tissue. Obviously, how you "view" this spectrum of energy is determined by the filters you select on your EMG amplifier. These instrumentation related issues will be discussed next.

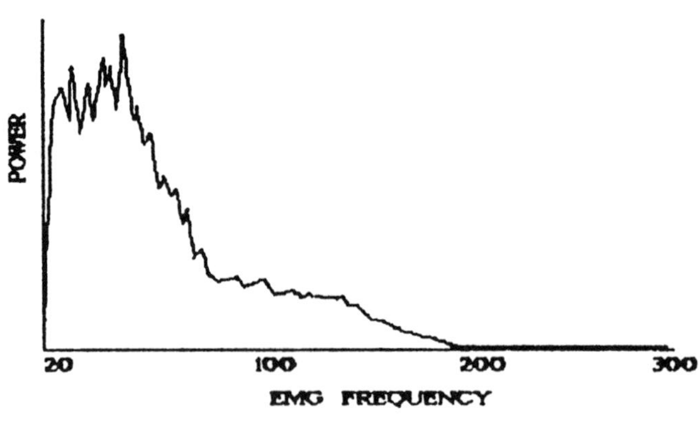


**FIGURE 2.** THE EMG POWER SPECTRUM

## SECTION 2. INSTRUMENT RELATED ISSUES

The instruments which are used to amplify and monitor the muscle action potentials have changed considerably over the course of the last 25 year. Most of the changes have occurred in the way the EMG signal is amplified. The first major change involved the use of differential amplification and common mode rejection. Essentially, the differential amplifier rejects from further amplification any signal which is common to the two "active" or recording electrodes. The vast majority of what is thrown away is external noise, such as 60 cycle. This has allowed EMG recordings to emerge out from behind copper shielded cages and rooms.

The next major improvement occurred with a large increase in what is known as "input impedance" of the amplifier. The input impedance of an amplifier functionally represents the amount of impedance (resistance) at the electrode site which can be tolerated by the pre-amplifiers. If the impedance between the electrode and the skin is 10 to 100 times greater than the input impedance of the amplifier, the amplifier will become so noisy as to invalidate the incoming EMG signal. In the 1950's and 60's, the input impedance of amplifiers was quite low, 10,000 Ohms, thus requiring vigorous abrading of the skin prior to the application of the electrode. The general rule of thumb was that the impedance at the electrode site needed to be below 10,000 ohms. The impedance issue was considered so important that almost all of the EMG instruments included an ohm meter, or "electrode test" feature, as part of the instrument itself. Since the early 1980's, the new generation of amplifiers have taken input impedance up into the Megohm (or even Gigohm) range. This increased the level of input impedance which functionally increased the level of impedance which could be tolerated at the electrode site. Interestingly enough, it was this small change in the amplifiers which made the muscle scanning procedure possible. The future will continue to bring changes in instrumentation. On-line spectral analysis of the EMG signal and digital/computerized filtering of the EMG signal are just around the corner.

### THE IMPORTANCE OF FILTER BAND WIDTH AND ITS SELECTION

Since the EMG spectrum can span 20-10,000 Hz, the filter band width of your EMG instrument will determine which parts of the EMG spectrum you will monitor or allow to pass on for further amplification.

Some instruments have only one filter width, while others have several to choose from.

The bandpass filter most commonly used in biofeedback instruments is very unique to biofeedback industry. Some 20 years ago, an engineer for the largest biofeedback manufacturer decided that the biofeedback practitioners needed help. Presumably, using the rationale that the biofeedback practitioner might not be able to readily discriminate the live EMG signal from 60 cycle noise and biological artifact, this engineer created an optional 100 - 200 Hz (or narrow) bandpass filter. This filter vigorously rejected this type of noise and thus protected the practitioner from this problem. This very unique bandpass filter, indeed, did a fairly nice job of eliminating noise and many other manufacturers began including this filter selection in their instruments. In fact, it became the biofeedback industry standard. Unfortunately, along with the noise and artifact, this narrow filter also eliminated approximately 80% of the EMG energy spectrum mentioned earlier. The filter which you select to use on your instrument will determine the nature of your investigation of the neuromuscular system. Let us consider the implications of selecting one of two types of filters: The narrow (100-200 Hz) filter and the wide (25-1000 Hz) filter.

The energy of the muscle emanates from two types of muscle tissue: fast and slow twitch fibers. The 100-200 Hz bandpass is monitors from the fast twitch muscle fibers. These fibers control phasic or fast ballistic movements, such as those used by a sprinter. One might consider these the acute fight or flight muscles. The wide filter monitors from the entire spectrum of muscle energy, including that found below 80 Hz. The slower frequency is considered to originate from the slow twitch muscle fibers. These fibers are responsible for (tonic) postural support as well as the muscular endurance needed by the marathon runner. Therefore, the narrow filter deletes the slow twitch fibers from its analysis.

However, these slower frequencies might be our best neuromuscular indicator of prolonged physical or emotional distress. It has been observed that a fast twitch muscle fiber will fatigue secondary to being overworked physically or emotionally. As fatigue occurs, the fast twitch fibers attenuate or stop firing, leaving only the slow twitch fibers to do the work (DeLuca, 1979). Van Boxtel (1984) has proposed and studied an anatomical shift in muscle tissue fiber types in a chronic headache population. He has observed a shift from mixed fast and slow fibers to that of predominately slow twitch fibers in the cephalic musculature of these patients. It appears that as the function of the muscle changes, so does its morphology or structure.

He argues that it is this phenomenon, a shift to slow twitch fiber type coupled with the use of the 100-200 Hz filters, which partially account for the mixed or poor results in surface EMG studies on headache and other pain related disorders.

Below, two single case studies are presented which examine the issue of the sensitivity of filter setting in the diagnostic assessments of chronic pain patients. These patients were studied with both the narrow and wide filter bandpass settings.

In the study, the EMG signal was monitored for noise and artifact and the data was collected and analyzed using the CRAM SCAN (version 1.0) software. This program uses the narrow filtered EMG data base presented by Cram and Engstrom (1986) and the wide filter data base presented by Cram (1986) to determine the abnormality of the source EMG. The determinations of Mild (*), Moderate (**), and Radical (***) elevations are made statistically, using the mean plus the one standard deviation, two standard deviations or three standard deviations respectively.

The results are presented in tabular form. Table 1 presents the results of the two studies conducted on JP, with Table 2 presenting the results of PM. Notice that in both cases the clinical elevations (as indicated by the *'s) are primarily found in the wide filter studies.

**TABLE 1.** NARROW AND WIDE FILTER RECORDINGS FOR PATIENT JP

NARROW FILTER (100-200 HZ) RECORDING

| SITE | SIT | | STAND | |
|---|---|---|---|---|
| | LEFT | RIGHT | LEFT | RIGHT |
| FRONTALIS | 1.1 | 3.1 | 2.3 | 1.4 |
| TEMPORALIS | 1.9 | 1.3 | 1.9 | 2.4 |
| MASSETER | 0.8 | 1.1 | 1.0 | 0.8 |
| S.C.M. | 1.0 | 0.7 | 0.7 | 1.0 |
| C.P.S. | 1.2 | 1.2 | 1.3 | 1.4 |
| TRAPEZIUS | 0.6 | 0.7 | 3.9 | 2.6 |
| T1 P.S. | 1.3 | 0.9 | 2.7 | 2.0 |
| T6 P.S. | 2.2 | 3.1 | 1.9 | 3.4 |
| T10 P.S. | 2.2 | 7.1* | 2.0 | 1.9 |
| L3 P.S. | 2.7 | 3.9 | 4.4 | 0.9 |

WIDE FILTER (25-1000 HZ)

| SITE | SIT | | STAND | |
|---|---|---|---|---|
| | LEFT | RIGHT | LEFT | RIGHT |
| C.P.S. | 2.9 | 2.8 | 2.5 | 2.3 |
| TRAPEZIUS | 4.1 | 13.9* | 7.6 | 4.7 |
| T1 P.S. | 3.8 | 4.3 | 5.9 | 7.5 |
| T6 P.S. | 5.5 | 4.4 | 4.2 | 4.7 |
| T10 P.S. | 9.8 | 10.4** | 2.8 | 4.9 |
| L3 P.S. | 8.6** | 10.3** | 4.7 | 3.0 |

(* = mild, ** = moderate elevations)

**TABLE 2.** NARROW AND WIDE FILTER RECORDINGS FOR PATIENT PM

NARROW FILTER (100-200 HZ) RECORDINGS

| SITE | SIT | | STAND | |
|---|---|---|---|---|
| | LEFT | RIGHT | LEFT | RIGHT |
| FRONTALIS | 1.5 | 1.5 | 0.9 | 1.1 |
| TEMPORALIS | 1.5 | 1.3 | 1.2 | 0.7 |
| MASSETER | 1.6 | 0.9 | 1.4 | 0.8 |
| S.C.M. | 0.6 | 0.6 | 0.8 | 0.6 |
| C.P.S. | 2.6 | 3.0 | 2.5 | 2.7 |
| TRAPEZIUS | 0.9 | 1.4 | 1.8 | 1.3 |
| T1 P.S. | 2.9 | 2.2 | 2.1 | 1.8 |
| T6 P.S. | 3.6 | 5.8* | 2.4 | 2.8 |
| T10 P.S. | 1.9 | 2.7 | 3.0 | 2.6 |
| L3 P.S. | 1.1 | 2.O | 3.3 | 1.5 |

(* = mild elevations)

WIDE FILTER (25-1000 HZ) RECORDINGS

| SITE | SIT | | STAND | |
|---|---|---|---|---|
| | LEFT | RIGHT | LEFT | RIGHT |
| C.P.S. | 2.3 | 2.2 | 3.3 | 2.1 |
| TRAPEZIUS | 14.7* | 13.5* | 18.2 | 9.7 |
| T1 P.S. | 12.2** | 7.6 | 4.8 | 5.1 |
| T6 P.S. | 5.9 | 6.6 | 4.3 | 7.3 |
| T10 P.S. | 5.1 | 10.2** | 3.5 | 4.0 |
| L3 P.S. | 8.9** | 7.0* | 5.8 | 2.6 |

(* = mild, ** = moderate elevations)

It is clear from the study of these two patients that the wide filter recordings are far more clinically sensitive compared to the narrow filter recordings. Were one to draw clinical conclusions based solely upon the narrow filter data, one would be forced to state that only an isolated and mild dysfunction was noted. The examiner might even consider the slight possibility of malingering. These narrow filter studies are not impressive, particularly given the nature of these patients' complaints. The wide filter recordings, on the other hand, follow the pattern of complaints quite nicely.

In essence, as clinicians we are limited in what we can see by the instruments we use. Or perhaps it is more accurate to state: We are limited by how we use our instruments. Filter selection is a very important parameter. The wide bandpass filter selection will allow for the assessment of chronically fatigued muscles through the monitoring of the slow twitch muscle fibers (i.e., slower frequency). The sole reliance on a narrow, high energy (100-200Hz) bandpass filter certainly needs to be reappraised. Because of the possible attenuation, if not frank disappearance of fast twitch muscle fibers in the chronically tense or overused muscle, narrow filter recordings may be inadequate. The use of the "insensitive" narrow filter selection may lead to an increased probability of false negative findings in your EMG study.

## THE PROBLEMS OF EXTERNAL NOISE

Opening up the filter band width to include the slower frequencies of the slow twitch fibers carries with it the price of possible contamination of the source EMG signal with 60 Hz noise and heart rate artifact. Remember, the elimination of this type of noise was the original purpose of the high energy narrow (100-200 Hz) bandpass filter. Unfortunately, such contamination would artificially increase the microvolts levels displayed by your instrument, thus increasing the probability of false positive findings.

If you use a wide filter setting, testing the veracity of your EMG signal is very important. This is best done by learning to listen to your "Raw EMG" audio. This raw EMG audio function is probably the most important of all the possible audio selections and is found on many EMG instruments. If your biofeedback manufacturer has neglected to include this Raw EMG audio feature in their instruments, modifications to install this audio feature to your EMG may be easily and inexpensively done on certain instruments. Contact your manufacturer to find out more information. If the raw EMG audio is available to you, you can easily learn to hear the differences between "noise" and genuine EMG.

In a relaxed and quiet muscle the raw EMG should sound completely silent. In an active muscle or one that is being intentionally contracted, the raw EMG should sound something like "white noise" (the sound of the ocean). Since recruitment of motor unit activity in a muscle normally occurs in a random fashion, this creates the "white noise" effect. With very strong contractions some popping or crackling sounds may be heard.

It is probably best to understand the "sound" of electrical noise by first testing the noise of your EMG instrument, as well as the electrical noise of the room in which it is being used. This is particularly important if you use other electrical devices in your room (computers, cassette tapes, lamps, etc.) or if the room next door utilizes electrical equipment (copiers, Ultra Sound, etc.). Then the potential problem of 60 Hz noise increases. To test the "electrical noise" of your biofeedback environment place a known resistance across the input of the amplifier or preamplifier. Some instruments come with a "dummy electrode" with a fixed (10 Kohm) resistance. It simply snaps onto the three leads of the EMG with the ground lead on the middle snap connector. Once in place, listen to the "Raw EMG".

If you hear a "humming" sound, instead of silence, you have an external 60 Hz noise problem. This could easily enter and contaminate your EMG signal. If your EMG has more than one filter selection, test out the dummy electrode for all filters. Usually 60 Hz and other noise are most easily heard when the wide filter is selected. Setting the filter switch to the narrow filter position should greatly reduce the external noise entering the amplifier. A slightly "higher pitch" hum may occur when the output from florescent lighting is contaminating your EMG signal. To test for this, one can turn off the florescent lights and listen for the disappearance of this sound. One last external source of contamination of your EMG signal is "radio frequency". Since the leads between your surface electrodes and the preamplifier may act as an antenna, radio frequencies may enter and be amplified along with your EMG signal. This type of contamination is the most easily detected in that it sounds like the radio. If you are listening for these external sources of noise when the electrodes are connected over a muscle, you would listen for these distinct sounds superimposed upon the silence or muscle sounds.

When noise is heard on the raw EMG audio, some form of action is required. There are three major strategies to eliminate the "humming" sound. One might consider moving the EMG instrument to a different location in the room or to try out an entirely different room. A second strategy is to shield the leads between the surface electrodes and the amplifier. This may require a new electrode lead. Many manufactures will supply these more costly leads when noise problems emerge. A very costly, but sometimes necessary level of shielding would be to install a copper foil (like wallpaper) for the entire room. Lastly, noise problem may develop when there is a poor electrode contact. Check your electrodes to see if they are firmly attached. Poor skin preparation may be the culprit. Take off the electrodes, abrade the skin again and re-apply the electrodes to assure a sound electrical contact with the skin. If one cannot eliminate the sounds of external noise from the raw EMG audio, it is wise not to proceed with your diagnostic or biofeedback session.

If your EMG instrument doesn't have a Raw EMG audio feature, perhaps it contains an "oscilloscopic" display feature to shed some light veracity of your signal. Please note that the "traditional" computer generated "time series displays" of EMG levels seen on biofeedback instruments will not provide this type of information. These displays are usually so highly conditioned and integrated as to "hide or smooth out" any potential noise problems. This is why some manufactures will make the utterly absurd statement that their instrument has filters or other features which eliminates any and all 60 Hz noise. Just remember, for the world of computer assisted myography: Garbage in--garbage out.

If you are unable to verify your source EMG with a raw EMG audio you may want to play it safe and stay with a narrow filter recording and upgrade your skin preparation techniques. Test your instrument and environment for internal and external noise following the recommended procedures and guidelines suggested by the manufacturer. Sometimes, they will recommend a cutoff value for an acceptable internal noise of the instrument. If the dummy load test of your instrument yields levels of energy within the guidelines given by the manufacture, you can reasonably assume that your environment is relatively free of 60 Hz noise when the impedance at the electrode site is kept quite low. Therefore, make certain that you conduct a vigorous skin preparation procedure and dip your scanning electrodes into a saline pad or electrode paste prior to placing them onto the skin for muscle scanning recording. These two precautions will reduce the factors which commonly contribute to the presence of 60 Hz noise in your EMG signal.

## THE PROBLEM OF BIOLOGICAL ARTIFACT

There are three major types of biological artifact which may contaminate the EMG signal. The biggest offender comes from the heart. The two other, respiration and crosstalk, play a lesser role.

<u>HEARTRATE ARTIFACT</u>. Heartrate artifact becomes more of a problem as the electrodes are placed further apart. With EMG scanning technology, it is typically less of a problem since the electrodes are placed at a closely spaced fixed distance apart. During scanning, heart rate artifact is typically not seen in the cephalic or cervical regions, nor

is it seen in the extremities. However, it may be seen on the torso, and is seen most clearly with a wider bandpass filter setting. In fact, it can sometimes be eliminated by selecting the narrow (100-200 Hz) filter. Interestingly enough, heartrate artifact may be heard as a rhythmic "thud" on the raw EMG audio.

Though it depends on the type of signal processing, the heartrate artifact is typically small in magnitude compared to the energy seen in that of an active muscle. When a muscle is in a mild to moderate level of contraction, the heartrate artifact would more than likely be totally camouflaged by the EMG. As the muscle becomes more and more quiet, the pluses of the heartrate artifact would then emerge. For example, using an RMS procedure with scanning electrodes and a wide filter, the magnitude of heartrate artifact would be seen to "pulse" around 1-3 microvolts in the T1 paraspinal area, with the left aspect being slightly higher than the right. This mild difference in symmetry is due to the heart being located on the left side of the body. The magnitude of the pulsing heartrate artifact would increase in the dorsal and lumbar areas to around 2-6 microvolts. Since the artifact is comprised primarily of the "t" segment of the heart's potential, it is fairly sharp and pointed in its definition, carrying very little area under the "envelope" which is used in the determination of the integrated, digitalized values. While it may appear to be a pervasive artifact seen during a wide filter scanning procedure in a quiet muscle, it probably contributes less than 10-20% to the integrated EMG values.

However, as the electrode spacings get further and further apart the problem of the heartrate artifact gets larger, and may rise to predominate over the EMG signal. Visual inspection of the integrated EMG display (when a reasonable time constant is used) will allow the investigator to determine the EMG levels. Here, the practitioner reads the EMG levels from the screen at the "troughs" of the tracing. This would be read as 10 microvolts in Figure 3. Also note that the resting levels of narrow filter recording is high enough to hide the heart rate artifact.

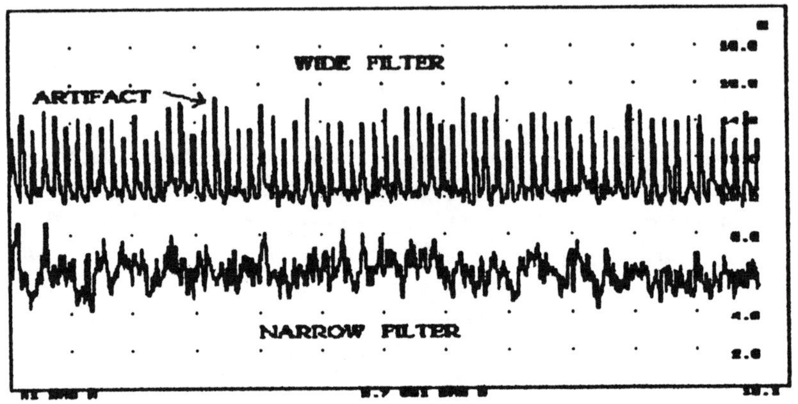


**FIGURE 3.** HEART RATE ARTIFACT AS SEEN ON BOTH THE NARROW AND WIDE FILTER SETTINGS

RESPIRATION ARTIFACT. Respiration is another artifact of which to be aware. It is seen as a slow rhythmic oscillation anywhere, from 5 to 25 times per minute. With the muscle scanning procedure it is seen most commonly in the dorsal (T6-T10) area. However, it may also be seen in the fibers of upper trapezius, particularly in individuals with respiratory problems such as COPD.

Again, the investigator must estimate the relative contribution of this artifact by estimating the true EMG levels from the troughs of the EMG tracing. (See FIGURE 4) Compared to the heartrate artifact, the integrated EMG values will be more elevated due to the larger area under the curve contributing to the integrated value.

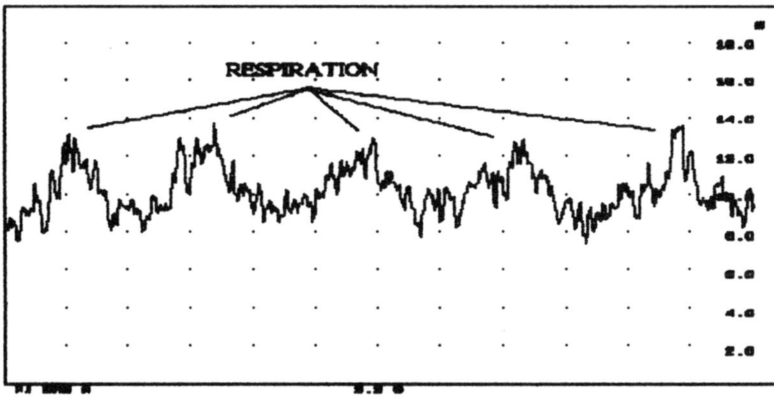


**FIGURE 4.** RESPIRATION ARTIFACT

CROSSTALK ARTIFACT. The last biological artifact is that of "cross talk" from muscles at a distance from the recording electrodes. For example, recordings in the lumbar area may be contaminated by EMG potentials arising from the large Gluteus Maximus muscles. Since the scanning electrodes are closely spaced, this type of cross talk problem is less than would be seen with more widely spaced electrodes. However, the problem does exist and may make it difficult to impossible to get a true reading from a quiet muscle which neighbors an exceptionally active muscle.

## THE EMG SENSORS: ELECTRODES

The electrodes are the sensors for the EMG. Without an adequate understanding of use and care of these sensors, the information which reaches the amplifiers of the EMG may be questionable. In fact, it is estimated that 80% of the errors in surface EMG recordings are due solely to poor electrode contact.

## ELECTRODE ASSEMBLY

There are actually many types of electrode assemblies or housings for the EMG sensors. Two common types of electrodes supplied with EMG instruments are: CUP ELECTRODES and DIRECT CONTACT ELECTRODES. With cup style electrodes, the actual electrode is recessed in the top of a small dome such that it "floats" above the skin, atop a bed of electrode cream. Direct contact electrodes rest directly on the surface of the skin.

**FIGURE 5.** TYPES OF EMG ELECTRODES

The electrodes used in muscle scanning procedures are of the direct contact type. Some manufacturers use a stainless steel snap lead embedded in an electrode housing as the recording surface. The other major model entails post style electrodes located on top of a plastic cylinder whose metal stem plugs directly into the preamplifier.

The direct contact type of electrode is actually an "old style" electrode. Before cup style electrodes were invented, all electrodes were placed directly onto the skin. In fact, I once observed one old time electromyographer use a real silver nickel as his electrode. The reason that cup style electrodes were developed, and temporarily came to be favored over direct contact electrodes during the 1960's, has to do with artifacts which are generated whenever an electrode moves about on the surface of the skin. This was a significant problem in cardiology. The "movement artifact" creates an electrical potential, artificially elevating the EMG recordings indicated by your instrument. This movement artifact may easily occur with post style electrode recordings used in scanning, and one needs to be able to recognize it whenever these are used. This artifact is heard as a "scraping sound" on the raw EMG audio and may be seen as a transient elevation in the EMG signal on visual analogue displays. While conducting a scanning procedure, it is very important to hold the electrodes motionless during the recording.

Many of the current electrodes used for surface EMG recordings use a type of plastic impregnated with silver and then coated with a silver chloride solution. These are commonly called silver-silver-chloride electrodes. The thin layer of silver-chloride helps stabilize and potentiate the biological signal from the skin into the preamplifier. Over time, with the use of the electrode, this thin layer begins to wear off. Initially, it is "blackish" in color. As it is used, it takes on a lighter coloration. As the silver-silver-chloride layer wears off the electrode becomes blackish again, and occasionally one can see slight scratch marks on its surface. When the electrode has returned to its darkened coloration, it is necessary to replace the electrode. The estimated lifetime of these electrodes is 100 applications or scans. The safest strategy for electrode replacement is to estimate how many weeks or months it will take to conduct 100 - 150 applications and to dispose of the used electrodes at that point.

Direct contact and cup electrodes are also used to monitor EMG activity on an "on line" basis, such as during stress profiling, dynamic movement, relaxation or postural training. When the electrodes are attached to the skin it is important to provide adequate "strain relief" to the cables, which attach to the electrodes, by clipping or taping the electrode cable to the shirt or skin of the patient. Without such relief, the cables themselves may move about and create extraneous potentials which would be amplified as artifacts. More commonly, the strain placed on the electrode by the weight of the cable could pull the electrode off of the skin, causing difficulties with your differential amplifier and yielding artificially higher, perhaps dramatically increased readings.

## SKIN PREPARATION AND IMPEDANCE

Skin preparation is a very important part of EMG recordings. An appropriate skin preparation reduces the resistance at the electrode site, allowing more of the biological activity to reach the amplifier and thus giving one a better EMG signal.

The "rules" regarding skin preparation for surface EMG recordings have been to abrade the skin (remove the top horny layer) and to use electrode paste. This paste establishes a strong electrical contact and a cushion between the surface of the skin and the electrode. The general rule of thumb was to make certain that the resistance at the electrode site was 10 times lower than the input impedance of the preamplifier of the instrument. Before 1980, almost all of the preamplifiers had an input impedance of 100 K Ohms. This meant that an aggressive abrading of the skin was needed to reduce the resistance at the electrode site to 10 K Ohms or less. In some research labs, needles were used to puncture the skin to ensure low impedance.

With the advent of new technology and better preamplifiers, the input impedance of current EMG instruments has risen to the 100 Meg and Gig Ohm range. This is a hundred times greater than the older preamplifiers and greatly reduces the need for "vigorous" skin preparation. This is not to say that skin preparation is not necessary. In fact, anyone who proclaims that skin preparation is a thing of the past and no longer necessary, should be considered suspect. In theory and practice, impedance at the skin can easily climb into the infinite range. This is most commonly due to very dry horny skin, excessive body hair (and oils) and/or the presence of makeup. No matter how high the input impedance of an EMG amplifier, it will never be able to handle the case of infinite impedance. The bottom line is that adequate attention needs to be given to skin preparation to ensure that the infinite resistance case does not happen to you. This is most easily done by using a skin preparation "ritual" as part of your EMG setup.

Originally, many clinicians and researchers had concerns about the utilization of the scanning procedure because of an inadequate description of its impedance and electrical characteristics. The characteristics of the interface between the scanning electrode and the skin's surface needed to be better described. Because of this, we conducted a series of studies (Cram and Rommen, 1989; Cram, 1990) to better describe and quantify this relationship.

In the first study, the post style electrodes were plugged into an IG-3 (GSR) preamplifier made by J&J. This output was sent directly to the J&J Thermal/GSR T-68 instrument to quantify the impedance at the electrode site. Four skin preparation conditions were studied in a 2 X 2 design. The two factors in the study were: ABRADE (vs. no abrade) and PASTE (vs. no paste). The abrading condition entailed rubbing the site to be scanned with a B-D alcohol wipe. Four to six rubbing motions with light to moderate pressure were used. The skin was not necessarily reddened. The paste condition entailed dipping the scanning electrodes into Beckman electrode paste, lightly coating them. The four conditions of the study were: No abrade/No paste; No abrade/Paste; Abrade/No paste; and Abrade/Paste.

The procedure was rather simple. Several muscle sites (using the scanning procedure locations) were studied to assure replication and generalization of the findings. The Abrade (and No Abrade) condition was randomly assigned to either the right or left aspect of the muscle group. The No Paste condition was always conducted prior to the Paste condition, so as not to contaminate the skin surface with "electrons" for the No Paste condition. A separate set of scanning electrodes were used for the Paste and No Paste conditions so as not to contaminate the electrode surface with "electrons" for the No Paste condition. Following the "touchdown" of the scanning electrodes to the skin, data was collected every 2 seconds for one minute. Six muscle sites were studied: Frontalis, Temporalis, SCM, CPS, Trapezius and T1 paraspinals.

The findings at each site were the same. These are represented in the graph below, for the Temporalis site. By looking at the graph, one can readily see that the Abrade conditions are associated with a lower impedance, compared to the No Abrade conditions. Within each Abrade condition, the Paste condition showed less impedance, compared to the No Paste condition.

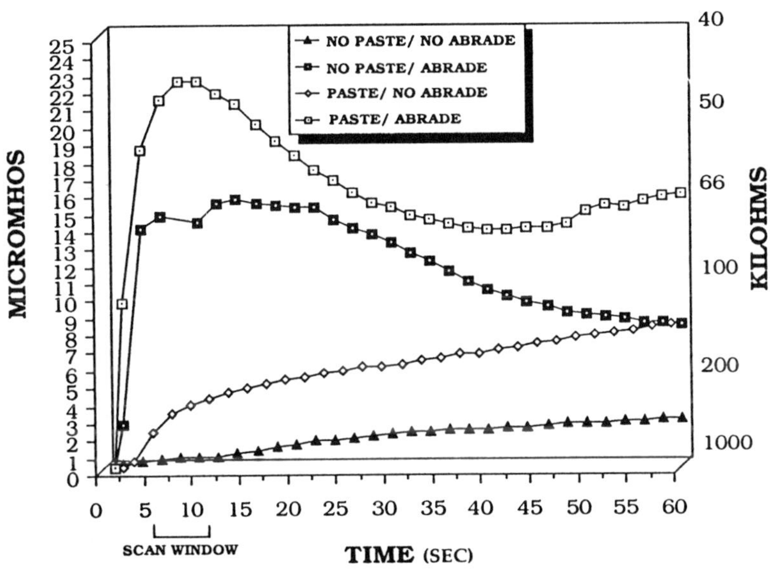


**FIGURE 6.** IMPEDANCE CHARACTERISTIC OF FOUR SKIN PREPARATION TECHNIQUES

When the No Abrade/No Paste condition is examined, one can see that the resistance essentially starts out at infinity. By approximately 30 seconds resistance value of 1000 K Ohms had been achieved. This decrease in resistance is thought to be associated with sweat forming under the electrodes in an effort to make the surface slippery, so as to shed the foreign objects. The level of resistance obtained using this type of procedure, however, is considered poor and inadequate for EMG muscle scanning recordings in particular and EMG in general.

The three other conditions show an adequate impedance of less than 400 KOhms within 5 seconds. This is at least 250 times lower than the input impedance of current EMG amplifier technology, thus passing the criterion test mentioned earlier. All three of these skin prep techniques appeared to be more than adequate for muscle scanning purposes.

## SKIN PREPARATION AND EMG RECORDINGS

In a companion study, we examined the four skin preparation techniques mentioned above on the EMG recordings themselves. Again, we looked at Abrade (vs. No Abrade) and Paste (vs. No Paste) as factors. The four conditions tested were No Abrade/No Paste; No Abrade/Paste; Abrade/No Paste; and Abrade/Paste. The same procedure was used, but this time the post style electrodes were plugged into a J&J IP-5 EMG preamplifier and EMG data were collected every two seconds for one minute. In addition, the scanning electrodes were placed between two attached cup style electrodes. This allowed for a comparison of EMG collected from scanning electrodes to the "gold standard" of EMG collected from attached surface electrodes. The same six muscles were studied to assure generalization of the findings.

The findings from this study were identical for all four sites and are represented in the recordings from the Trapezius muscle presented in the next figure. Here, the EMG activity for the No Abrade condition is plotted across time. In general, it may be seen that the No Abrade/No Paste condition is different from the other three conditions. The other three conditions are all very similar.

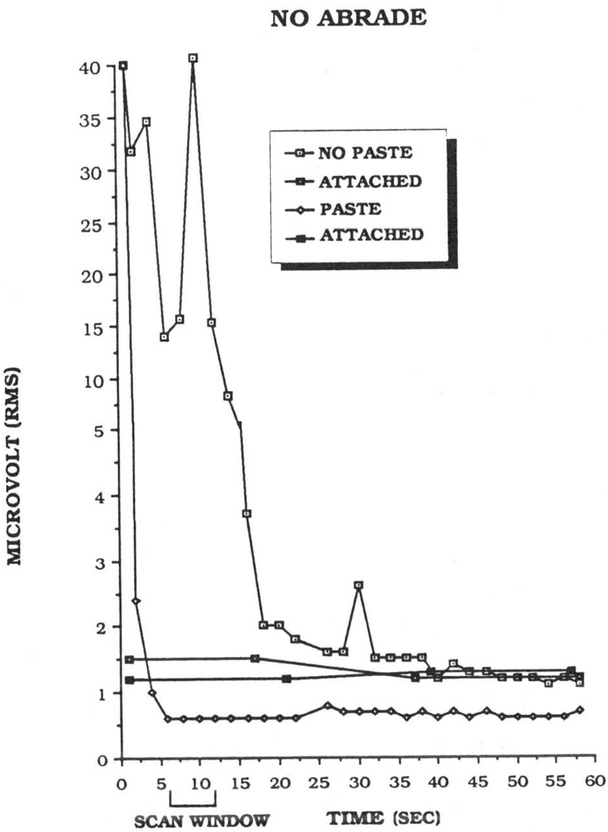


**FIGURE 7.** EMG CHARACTERISTICS OF FOUR SKIN PREPARATION TECHNIQUES

In the No Abrade/No Paste condition, the EMG activity very slowly drifts down from a beginning value of some 30 to 40 microvolts, to a final resting value of 1.1 microvolts at 60 seconds. The readings from the EMG amplifiers for the scanning electrodes, with the exception of the final two data points at 50 and 60 seconds, is greater than the placed electrodes. Since the placed electrodes are set at a further distance apart, the EMG activity from the placed electrodes should always be greater than that observed by the scanning electrodes. When this occurs, the EMG activity from the scanning electrodes is to be considered valid. The recordings seen for the No Abrade/No Paste condition would not be considered a valid indication of EMG activity form the muscles below the sensors. The initial values would be considered "instrument noise", since the skin impedance was so high so as to defeat the input impedance of the EMG amplifier. The slow gradual downward drift of the recording is best explained by the gradual reduction of the impedance at the electrode site, which had been observed in the previous study. Eventually, the impedance drops below that needed for an "input impedance match", and valid EMG value emerge. Here, the final two points of the scanning electrodes fall below those of the attached electrodes. The danger inherent in poor skin preparation conditions is that you may interpret the values displayed on your EMG machine as true EMG values, when they actually represent instrument noise.

The pattern of EMG activity from the other three skin preparation conditions is very similar. There is a very rapid reduction in the electrical activity. Within two to four seconds the activity is stable and the true EMG values may be read. The initial high readings are associated with the "electrical explosion" which happens when the scanning electrodes meet the skin. As the skin electrode interface stabilizes, the EMG values from the scanning electrodes rapidly falls well below those seen by the placed electrodes. This indicates that the recordings are a valid estimate of the EMG activity from the muscles.

One other interesting and important observation can be made from the data. If one follows the curve for the Abrade/Paste condition, one can easily see a "bump" in the curve which occurs at 30 seconds. This usually represents a postural artifact where the subject momentarily makes a postural adjustment. It may also represent (spontaneous) discharge of EMG activity. In conducting the scanning procedure, it has been recommended that the practitioner use a short integration period. We have chosen a two second time period to be ideal for our purposes. In addition, we strongly recommend that the practitioner take a series of 2-5 two second samples, and then record the best estimate of the tonic levels of EMG activity. This recommendation has grown out of the observation that there are artifacts which may occur during this procedure and these should not be integrated into an average EMG value. If one uses an oscilloscopic or computer assisted EMG display, it is not difficult to select a sample of EMG which represents the general tonic levels, and is free of artifacts. One study (Thompson, et. al., 1988) conducted at the Mayo clinic has demonstrated that attached and scanning electrodes yield very significant correlations when the EMG sample is selected after approximately 10 seconds.

It appeared that the three conditions of No Abrade/No Paste, Abrade/No Paste, and Abrade/Paste, yield similar results. The No Abrade/No Paste condition should be avoided. Initially, I gave my preference for using the No Paste/Abrade condition in which the patient was scanned while the alcohol abrade was still wet. This was primarily for hygienic measures. It seemed "less messy" than the Paste condition. Unfortunately, the wet condition is not without its problems.

What I discovered was that on occasions I would get bad EMG values. Using the wet alcohol abrade method, on some individuals I would observe values that were excessively low (.00 or .02). It was as if I had a "short circuit" between my electrodes. If you run a bead of electrode paste on a towel and touch your scanning electrodes on the bead such that it bridges the electrodes, the values on the EMG amplifier should read around 00 (or at your instrument's internal noise level). In fact, when one uses the electrode paste skin prep condition, one must be careful not to use too much paste so as to create this type of bridging on the skin of the patient. As it turns out, the moisture of the alcohol abrade technique would mix with the body salts of some individuals to create the functional equivalence of an electron bridge between the two active electrodes, thus leading to excessively low values. On the J&J amplifiers this would happen approximately 1 out of 50 individuals. On other amplifiers I have studied, the base rate of bad samples may increase upwards towards 1 out of 15.

Because of the above experience, I have gone back to the original procedure. I use the alcohol abrade first and let it dry completely. I then dip my scanning electrodes into a bead of electrode paste, lightly coating them. I then apply this to the skin of the patient, making sure that I don't accidentally form an electrode cream bridge. If a bridging accidentally happens, I can easily see it. I then merely clean the area with an alcohol wipe again, let it dry, and scan again. Finally, I take my EMG readings when the signal is stable. This procedure is preferred in that it seems to give the most reliable EMG readings. In addition, the electrode paste left behind from the scanning of a site may be used to replicate the placement of the electrode when one scans the patient in an additional posture.

## ELECTRODE SITE SELECTION

After you have ascertained that your signal is clean and clear of noise and artifacts, where you place your electrodes is the next most important order of business. This will determine which muscle group(s) you see with your EMG amplifiers. The recommended locations for electrode placement will be described in a later section of this book. These may be complimented with the electrode placement system suggested by John Basmajian (1989). At this point, I want to review some of the basic principles which may guide your selection.

## GENERAL GUIDELINES FOR ELECTRODE SITE SELECTION

There are four general rules which guide electrode placement. They are based upon an interaction of the characteristics of the electrophysiology of muscle activity, and the characteristics of differential amplification and common mode rejection used in EMG instrumentation. The four rules are listed below, along with commentary.

**1. When the two active electrodes are placed such that they run parallel to the muscle fibers of the muscle of interest, that muscle will contribute most heavily to the EMG signal.**

This is due to the characteristics of differential amplification and common mode rejection. When muscles give off energy, this energy emanates out along the line of the muscle fibers. There is a "directionality" to this energy, just as there is a directionality to the force of the muscle. If the electrodes are placed parallel to the muscle fibers, the energy associated with a muscle contraction will reach ELECTRODE 1 at a time different than it will reach ELECTRODE 2. See FIGURE 8.

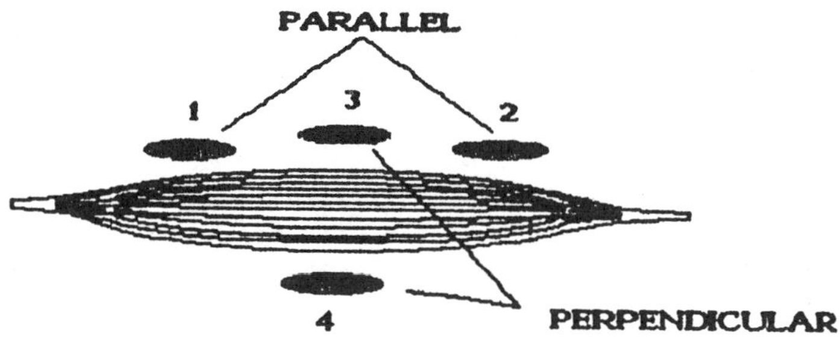


**FIGURE 8.** RELATIONSHIP OF MUSCLE FIBERS TO ELECTRODE PLACEMENT

Because common mode rejection (a characteristic of differential amplification) deletes from further amplification any signal which is common to both active electrodes, and since the energy from the muscle reaches the two active electrodes at different times, this energy should be favored and passed on for further amplification. In other words, parallel placement enhances the probability of monitoring a specific muscle group. Placing the scanning electrodes such that they run parallel to the masseter muscle fibers, for example, increases the probability of recording primarily from this muscle group.

**2. The closer the two active electrodes are in proximity to one another, the more specific and more shallow the source of the EMG signal.**

When you use surface EMG electrodes, you record from populations or pools of motor units. Therefore, the further apart the electrodes, the larger the potential population of motor units which contribute to the signal. When the electrodes are close together, as they are with the scanning electrodes (2.5 cm.), the primary source of motor unit activity will be found directly under the electrodes. While a large amount of motor unit activity at some distance may contribute to the amplified signal, the further away these signals are, the more likely they are to reach the two active scanning electrodes at the same time, thus being canceled out by the common mode rejection. As the electrode spacing becomes farther apart, the more susceptible it is to volume conduction (i.e.; motor unit activity from other muscle groups). One may want to intentionally place the electrodes far apart in order to record from a group or region of muscles. In fact, one common "trick" used in the biofeedback arena is to place one active electrode on the left wrist and the other on the right wrist. This is called the "wrist to wrist" placement. This is the antithesis of narrow electrode spacing. With this exceptionally wide spacing, the EMG amplifiers will receive volume conducted EMG activity from the contributing motor units of the arms, neck, shoulders, chest and back. This is a wonderful placement for conducting general relaxation training for the arms and torso. Narrow placements are recommended for recording from specific muscle groups and wider placements are recommended for recording from larger, more general areas.

**3. When the two active electrodes are placed perpendicular to the muscle fibers of interest, then all of the muscles in the general area of the electrodes will contribute to the EMG signal.**

In Figure 8, the active ELECTRODES 3 & 4 run perpendicular to the muscle fibers. As can be seen, the energy from the discharging muscle fibers reaches both electrodes at the same point in time and thus contributes little to the amplified EMG signal. It is considered that with this electrode array you will primarily be recording from a general region rather than a specific muscle group. The Frontalis and bilateral Trapezius placements used in the muscle scanning procedure are good examples of this. The Frontalis placement runs perpendicular to the muscle fibers. Thus, the energy from fibers between the electrodes would cancel itself out, while fibers to the left or right of the electrode array would tend to contribute more to the amplified signal. In particular, if emotional displays involved the Corrugator muscle group, these would contribute to the amplified signal as well.

**4. The ground reference electrode should be placed on the body.**

Since the ground electrode is used as the basis for common mode rejection in a differential amplification system, it is one of the most important electrodes in the configuration. At one point in time, it was believed that the ground electrode should be placed either on an electrically inactive site such as a bony prominence or half way between the two active electrodes. This would clean up the differential amplification process. With the instrumentation currently available, this is not such an important factor. However, it is important that the ground electrode be securely placed on a well abraded site on the patient being studied! In the muscle scanning procedure, one common array entails the two active electrodes being fixed on a plastic shell. The ground electrode is attached to a separate site which is connected to the preamplifiers via a 36 inch wire or lead to permit freedom of electrode placement. This is the preferred method. In some muscle scanning systems, a tripod arrangement is used with the two active electrodes and the ground electrode being physically located in a triangular arrangement on the same frame. While this might afford a "tad-bit" of convenience, I have difficulty believing that the ground electrode is always adequate. The problem has to do with the shape of the human body. It has curves. The tripod arrangement assumes contact with a flat plane. In the cervical region, a site commonly scanned, such a flat plane does not exist. This increases the probability of faulty ground or active electrode connection, thus leading to faulty EMG values appearing on the digitalized output.

In conclusion, the way in which you affix your electrodes over the muscles you wish to study will determine the source of muscle activity contributing to your signal. You may use this to your advantage during a muscle scanning procedure to "isolate" particular muscle groups. With attached electrodes, both specific or very general regions may be studied.

## SINGLE VS. MULTIPLE CHANNEL MONITORING AND SCANNING

Muscle scanning may be conducted using one or two scanning electrode arrays. When two scanning electrode arrays are used simultaneously, both the right and left aspect or both the agonist and antagonist muscle groups may be studied simultaneously. While there is no data on this phenomenon, simultaneous collection of data may provide more accurate and meaningful information, compared to collecting the same data in a serial fashion using a single channel of scanning. Dual channel scanning is the preferred method for scanning the paraspinal muscles of the back.

With attached electrode arrays, dual channel placements provide much more information than does a single channel of EMG. As time proceeds, four channel EMG arrays will become more common. This would allow you to monitor, for example, the right and left aspects of both the upper and lower back. Since the back is a well orchestrated system, the diagnostic and treatment implications of multichannel monitoring are clearly evident. The work by Donaldson & Donaldson in this book is a very good example.

## SECTION 3. AN OVERVIEW OF DIAGNOSTIC CONSIDERATIONS

A fundamental premise of any healing art is that diagnosis should lead to treatment. Differential diagnosis should lead to differential treatment. One of the problems which has plagued surface EMG (biofeedback oriented) work, is the very limited range of diagnostic techniques and companion therapeutic procedures.

Historically, the diagnostic techniques for surface EMG have been dominated by activation theory (Duffy, 1962) and its companion diagnostic procedure: stress profiling. Initially surface EMG research appeared fixated on the study of the frontalis (frontal) muscle group. However, this very narrow focus provided only the very simplest view of possible diagnostic and treatment considerations.

The neuromuscular system is much richer and should not only be limited to emotional displays on the face. To adequately work with this system, one should at least appreciate and understand some of its complexities. As noted earlier, the sensory system for the muscles (muscle spindles, golgi tendon organs, ruffini terminals, and free nerve endings) informs the CNS of the instantaneous status of each muscle, while strongly modulating the output of the final common pathway of the motor unit, and by pushing and pulling on the two motor systems (alpha and gamma). The primary function of the skeletal muscles is to provide postural support. Once posture is assumed, movement (a secondary function) can take place. Posture and movement patterns may be altered by a shortened resting length of the muscle, by muscle splinting or by a learned protective guarding pattern to prevent pain. All of these elements should at least be considered in diagnostic evaluation of this system.

Since most of the problems of a neuromuscular system commonly present as a pain related problem, an appreciation of the nature of pain is also needed. Loeser (1978), for example, has proposed a four-tiered system for the perception of pain. Table 3 elaborates on his four levels by adding surface EMG diagnostic and treatment considerations. As can be seen, EMG diagnostic and treatment procedures would vary depending upon the level of pain one is assessing and/or treating.

**TABLE 3.** LEVELS OF PAIN

| LEVEL OF PAIN | DESCRIPTION | EMG DX | EMG TX |
|---|---|---|---|
| ENVIRONMENTAL (CORTICAL) | OPERANT CONTROLLED, TIME LOSS COMP., AVOIDS CONSEQUENCE, PAIN BEHAVIORS | PROTECTIVE GUARDING PATTERNS | DYNAMIC POSTURALLY BASED |
| PSYCHIC (LIMBIC) | ANXIETY, DEPRESSION | EMOTIONAL DISPLAYS, ACTIVATION | RELAXATION BASED, CONTROL |
| CENTRAL (SPINAL) | PHANTOM LIMB | NONE | NONE |
| NOCIOCEPTION (SENSORY) | SOFT TISSUE INJ., ISCHEMIA, EDEMA, METABOLITES (ACID), MECHANICAL | ACTIVATION, SPLINTING PATTERNS | SPECIFIC RELAXATION, POSTURALLY BASED |

The purpose of this section is to review and consider a set of diagnostic concepts to assist the practitioner in weighing the possible contributions to the neuromuscular problems. The diagnostic frameworks presented in Table 4 are considered relevant to all pain related disorders. As can be seen, the outcome of the diagnostic procedures lead to the consideration of procedures for differential treatment.

**TABLE 4.** DIAGNOSTIC FRAMEWORKS

| DX LEVEL | DX PROCEDURE | TX OPTIONS |
|---|---|---|
| NO DIAGNOSIS | NONE | TREAT EVERYBODY THE SAME |
| EMOTIONAL | | |
| GENERAL | STRESS PROFILE/ TYPICALLY FRONTAL SITE | GENERAL RELAXATION/ RECLINER CHAIR THERAPY |
| SPECIFIC | SCAN/STRESS PROFILE/ SITE SPECIFIC RECORDING | SITE SPECIFIC RELAXATION/ TRANSACTIONAL PSYCHOPHYS. |
| ORGANICITY | | |
| METABOLIC | SCAN/DYNAMIC MOVEMENT/ SYMMETRIES ASSESSED | MYOTHERAPY/ SITE SPECIFIC RELAXATION |
| RESTING LENGTH | SCAN/DYNAMIC MOVEMENT/ SYMMETRIES ASSESSED | STRETCH TO RELAX/ SITE SPECIFIC FEEDBACK |
| POSTURAL | SCAN/DYNAMIC MOVEMENT/ GRAVITATIONAL | POSTURAL CHANGE/ BIOMECHANICS |

## THE NO DIAGNOSIS MODEL

Currently, there are no data on the prevalence of the types of surface EMG diagnostic or treatment procedures used in the field. However, a cursory view of the diagnostic aspects of the field indicates both dynamic and static procedures, along with rehabilitation and stress profiling models. Treatment procedures seem to include the range of general relaxation, rehabilitation, postural and transactional psychophysiology models.

If we assume that there is more than one type of treatment procedure for surface EMG work, we may begin to consider the conditions upon which a "No Diagnosis" model would be observed. This diagnostic framework would be invoked when a practitioner or researcher always uses the same muscle site (i.e., Frontal), the same procedure (i.e., guided imagery) and/or the same posture (i.e., recliner chair) with all patients who present for evaluation, treatment, or study. In other words, diagnostic and/or treatment procedures depend more on the repertoire of skills of the practitioner/scientist than upon the presenting complaint of the patient.

It is this limited repertoire which has weakened the foundations of surface EMG studies and therapies. As Turk, et. al. (1979), point out in a review of biofeedback and the treatment of pain, the field has generally suffered from the over utilization of both the "patient and treatment uniformity myths" (Kiesler, 1966). It is this lack of differential diagnostic thinking which has led earlier reviewers (i.e., Turner and Chapman, 1982) to conclude that there is no difference in clinical efficacy of EMG assisted or other relaxation therapies in the treatment of pain. More recent reviews, (i.e., Dolce and Raczynski, 1985), however, have cast away this simplistic thinking and have clearly acknowledged and elaborated upon the multiple etiologies of myogenic (back) pain, while considering multifaceted diagnostic and treatment procedures.

## THE EMOTIONAL MODEL

The emotional model is the oldest and best known model for the surface EMG and biofeedback researcher and clinician. There are two aspects which may be described. The first has grown out of general activation theory (Cannon, 1932; Malmo, 1959; Duffy, 1962), while the second stems from the response specificity theory (Sternbach, 1966; Lacey and Lacey, 1958). Both of these theoretical frameworks grew out of observations similar to those of Jacobson (1938) and Sargent (1946). They noted that ineffective coping with environmental and emotional stress led to increased muscle tension and therefore painful muscles. Increased anxiety and a concomitant increase in general muscular tension led to a state of muscular fatigue characterized by initial muscle ischemia, followed by hyperaemia (Bancroft and Millen, 1939).

The first aspect is called the general emotional framework. It has used psychophysiological stress profiling of frontalis (Malmo and Shagass, 1949) as its primary diagnostic technique. The popularity of this diagnostic procedure, in all probability, grew out of anxiety related research such as that of Lader and Mathews (1968). Here, the researcher/clinician attempts to identify and then treat emotional displays of anxiety. EMG recordings were therefore made from the muscles of the upper face, well known for its richness of negative emotional displays of anger, fear, and sadness (Ekman, 1972). Typically, the impact of a variety of real (i.e., cold pressor test), imagined (i.e., imagined phobic stimulus) or performance oriented (i.e., serial sevens) stressors were studied. If dysfunctional emotional displays were observed, they were to be treated with general muscle relaxation techniques modeled after the work of Jacobson (1938) or Wolpe (1958). EMG assisted relaxation training for the frontal region was directed at creating a state of "cultivated low arousal" (Budzynski and Stoyva, 1974). In the reciprocal inhibition fashion postulated by Wolpe, the anxiety related disorder was treated by replacing the hyperarousal with a newly acquired general state of low arousal.

The second aspect is called the specific emotional model. Its development stemmed from the observation that the "key muscle" hypothesis of activation did not adequately describe the recordings observed on all patients or subjects (Alexander and Smith, 1979). Lacey and Lacey (1958), for example, presented data on individual patients who responded in only one physiological system to a variety of stressors over many years of study. The headache literature provides another example. Reviews of the area (i.e., Haynes, 1980) are characterized by mixed results. Some studies verify the myogenic theory of headache, while others do not. The specificity model essentially states that the EMG activation pattern may be found in specific muscle groups, rather than the more general Frontal recordings. Dysfunctional activation patterns, then, may be specific to one muscle group rather than indicative of a general arousal problem. The Specificity Paradigm utilizes the psychophysiological stress profiling procedure described above, but is directed at specific muscle groups or sites.

Consider the EMG recordings from the right and left aspects of Trapezius from a single patient shown in Figure 9. Here, the patient was asked to rest, engage in a serial mental arithmetic task out loud (mild stress) and then relax to the best of her abilities. The recordings clearly show a "specific" EMG activation pattern. The left aspect responds to the mild stressor, while the right aspect does not. In this particular case, the patient had injured her left shoulder in a fall some six months prior. She reports a very strong relationship between stress and the intensity of the pain in her left shoulder. This type of response pattern might fall under the "weakened organ" or the "Diathesis Stress" models (Turk and Flor, 1984), where a structurally weakened organ becomes the primary pathway for physiological responses to stress.

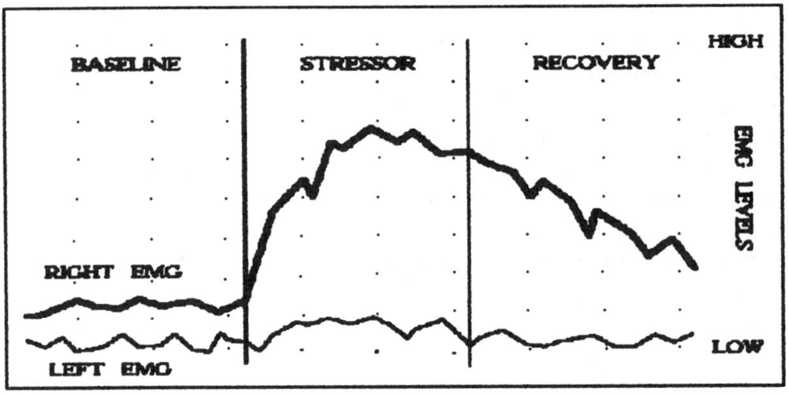


**FIGURE 9.** PSYCHOPHYSIOLOGICAL STRESS PROFILE ON SINGLE PATIENT FROM RIGHT AND LEFT TRAPEZIUS

The treatment offered this particular individual should logically follow from the diagnostic findings. Given the lack of generalization of EMG assisted relaxation training from the Frontal site to muscles of the torso or limbs (Alexander and Smith, 1979; Cram and Freeman, 1986), the Frontal site would be a poor selection. More specific training of the left Trapezius would be indicated. This patient was initially offered specific relaxation training for her left upper back, while monitoring her right upper back. In addition, stress/pain management and psychotherapy were conducted while the left aspect of trapezius was monitored using the "Transactional Psychophysiology" model proposed by Lynch (1985). During a traditional psychotherapeutic session, each time the patient displays an "emotional" response with her left Trapezius, the therapist might pose the question "I wonder what this means?" to allow for accessing of the emotional event. Or the therapist might interrupt the psychotherapy to practice a brief relaxation in order to break up the dysfunctional activation pattern.

Specific emotional activation patterns have been presented in a very well controlled group study by Flor, et. al. (1985). In this study Frontal EMG, along with the right and left aspects of the L3 paraspinals, were monitored in a well matched set of patients/subjects. While sitting upright in a chair, Chronic back pain patients, General pain patients, and Healthy controls were asked to engage in the four events shown in Figure 10. Since Frontal EMG did not differentiate between the groups, it is not shown. The L3 recordings indicated no significant differences between groups during the reciting of the alphabet forwards or while conducting mental arithmetic. Only during the two personally relevant stressors (discuss your pain; discuss a recent stressful event), did the chronic back pain group demonstrated a specific activation pattern. While these dysfunctional activation patterns were observed on both sides of the back, the left aspect was far more robust.

The meaning of a "more robust" left sided activation needs further exploration. Cram and Engstrom (1986) have noted, for example, that the left low back clearly separates chronic back pain patients from normal subjects, but only in the standing posture. Two theoretical frameworks may be used to explain these left sided findings. The first is an emotional one. It is possible to argue that the emotions (or pain) are represented primarily in the right hemisphere of the brain. It is known, for example, that negative emotions (i.e., anger) are found in the right hemisphere (Davidson, 1984).

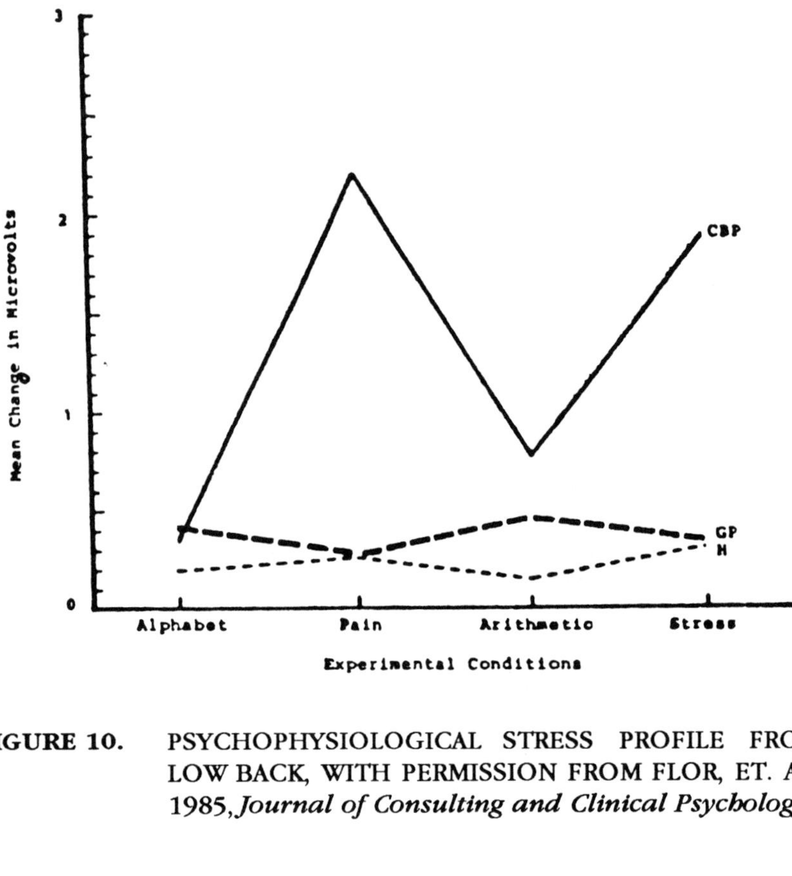


**FIGURE 10.** PSYCHOPHYSIOLOGICAL STRESS PROFILE FROM LOW BACK, WITH PERMISSION FROM FLOR, ET. AL. 1985, *Journal of Consulting and Clinical Psychology*

In addition, it has been observed that patients with "hysterical" symptoms tend to display these symptoms primarily on the left side of the body. The second, more parsimonious framework is a biomechanical one. It essentially assumes that the majority of the population is right handed, and thus "over-utilize" the left low back muscles as they move off their center of gravity. This overuse pattern sets the stage for a neuromuscular or mechanical disturbances on the left aspect. The diathesis stress model proposed by Flor, et. al. (1985) takes this one step further, suggesting that environmental or psychological stress would tend to find its expression in these weakened muscular structures. The treatment planning for such a left sided diagnostic finding would again utilize specific relaxation training, coupled perhaps with stress management from a transactional psychophysiological perspective.

**THE ORGANICITY MODEL**

The organicity model has grown out of interest in the muscle as an organ in and of itself. When one attempts to understand the phenomena of muscle tension pain, several parameters should be considered: the nature of the pain receptor, the circulation and metabolism of the muscle, nodules or other structural changes in the muscle, the resting length of the muscle and splinting guarding patterns.

For muscle related pain to exist, there needs to be an anatomical prerequisite. The mechanical and chemical aspects of the pain receptor were described in an earlier section. Swelling is one source of mechanical nociception. This swelling may be the result of trauma or overuse. It is important to remember that pain associated with a swelling or edema is associated with a "temporal dislocation" of pain. This would mean that overuse of the muscle would lead to a disturbance in the circulation metabolism. However, it may take hours after the over-utilization before the actual experience of the pain emerges. This temporal shift in pain has created some problems in the study of headache and other related pain (Phillips, 1977; Bischoff and Traue, 1983).

To complicate things further, prolonged diminished oxygen supply can cause necrosis of muscle tissue creating a nodule or "Trigger Point" (Travell, 1976). It has been clearly documented that these trigger points can refer pain to locations away from the source of stimulation. Travell and Rinzler (1952), for example, have demonstrated how pain in the head or face may be associated with trigger points in the neck or shoulders. This "spatial dislocation" of pain has created problems in establishing a one to one relationship between pain and EMG activity, especially in pain related research and diagnostic procedures. Phillips (1977) for example observed only a 40% concordance between the strongest pain, and the highest level of muscle activity in headache patients.

Lastly, one needs to consider the phenomena of splinting and protective guarding (Price, 1948). When a joint or area is initially injured, the muscles of the surrounding area tighten down to provide a "muscular splint". This "reflexive spasm" has a strong biological advantage, preventing further injury to the area or joint. If the splinting persists for a prolonged period of time, however, it may merge into a self-sustaining "pain-spasm-pain" cycle (Bonica, 1957). Or it may merge into a protective guarding pattern (Price, 1948) in which the patient learns to favor the injured area, using the muscles to move away from the source of pain. In the splinting or pain-spasm-pain cycle one would expect to see specific EMG activation patterns at the reported site of pain. In a protective guarding pattern, however, one would expect to see the specific EMG activation pattern in a homologous muscle group away from the pain. Diagnostically, one would want to consider asymmetrical activation patterns as a sign of splinting or guarding.

Given these phenomena, it may seem naive to assume that the EMG activation pattern will be found at the site of reported pain. Perhaps it was the above "dislocation" which led Goldstein (1972) to suggest that one "sample from the muscular system" rather than just study one site or muscle group. Hudzinski and Lawrence (1988), for example, in the study of headache patients reported that "electrode placement" was a major factor in separating headache patients from non-patients, and headache periods from non-headache periods.

The muscle scanning technique allows one to rapidly sample the "static" activation patterns of the neuromuscular system for: Sites of activation; Asymmetrical pattern of splinting or guarding; and Postural disturbances. This is considered only one level evaluation for the identification of possible "organicity" factors. Careful history, palpation (for trigger points) and/or dynamic movement oriented EMG procedures provide additional diagnostic information. The total constellation of diagnostic information may suggest that myotherapy (Prudden, 1980), or spray and stretch (Travell, 1983) may be indicated. Specific or even general relaxation training could easily accompany these more structural type of therapies.

The surface EMG assessment under static conditions also plays a strong role in identifying a muscle with a shortened resting length. As was mentioned at the beginning of this chapter, a muscle may feel hard to the touch due to muscle contraction or contracture. If the muscle is hard to the touch and is electrically active, relaxation oriented biofeedback may be indicated to treat the muscle spasm. If the muscle is hard to the touch and electrically silent, a shortened resting length (contracture) is indicated and stretching, not relaxation, is indicated. With shortened resting length one might consider using EMG biofeedback during a stretching program (Cram, 1986). If one encourages relaxation into the stretch, one would expect to see reductions of EMG activity during the stretch. A single subject example of this is presented in Figure 11. The treatment session is directed at the cervical paraspinal muscles, and begins with a postural change strategy. Here, the patient is asked to notice how she holds her head above her shoulders, while seeing if she can lower her levels of EMG activity by changing the posture of her head and shoulder. This "postural" training is somewhat akin to the Alexander technique. As can be easily seen, alteration in posture leads to what would appear to be a clinically relevant reduction in EMG activity of the Cervical Paraspinal muscles. Next, the patient is asked to go out onto a series of gravity guided stretches for a period of 30 - 60 seconds. These stretches (forward flexion and lateral bending) are very similar to those suggested by McKenzie (1985), but without the manual assist. Instead, the patient is asked to "breathe into the stretch; relax into the stretch". At the end of the stretches for this patient, the resting baseline levels are reduced by an additional 50 percent. This form of EMG feedback might be considered as the only way in which the gamma motor system may be trained using feedback.

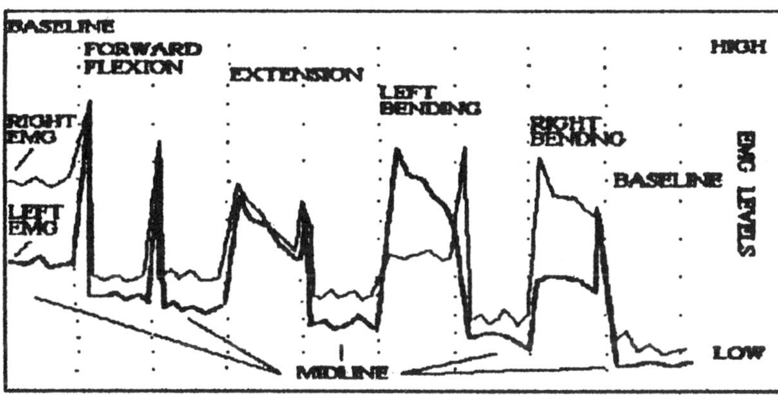


FIGURE 11. STRETCH TO RELAX PROTOCOL ON A SINGLE PATIENT (WITH PERMISSION FROM CRAM, 1986, CLINICAL RESOURCES)

## THE POSTURAL MODEL

The postural aspects of the neuromuscular system should not be ignored. This may involve both emotional and biomechanical aspects. Patton (1976), an early investigator of the gamma motor system, has stated; "Mammals react to stressful events with three mechanisms-- Glandular secretions; Movements; and Postural Adjustments (p. 69)". Reflection on how the cat arches its back as part of its fixed action pattern for territorial defense is an obvious example. One might also consider the anxious patient who has elevated shoulders or the depressed patient with slumped shoulders. Whatmore (1959, 1974) has investigated and proposed a theory of "dysponesis" to help conceptualize the role of the neuromuscular system in pathological affective states.

From a biomechanical viewpoint, the intrinsic muscles of the back are considered to have the primary function of stabilizing the spine (Basmajian, 1976b). In general, the muscles of the back should come into play only when the spine moves off its center of gravity (Calliet, 1977). Otherwise, it is the vector of force at the ankle and the activation of the soleus muscle group which keeps us erect. The spine should be adequately supported with ligamentous support (Carmichael and Burkart, 1979).

In the classic paper by Floyd and Silver (1955), the back muscles are described as being quiet during "static", quiet standing. This has been supported by Basmajian (1978b). These type of findings, however, are probably found only in "well postured" individuals. It was observed by Floyd and Silver that even minor postural adjustments of the head, arms or torso were enough to recruit paraspinal activity. In a recent study by Cram and Engstrom (1986) a significant increase in tonic paraspinal activity was noted between quiet, unsupported sitting and a quiet standing posture in a large sample of normal subjects. Perhaps it is not as easy to "go out onto ligament support" or to find our biomechanical center of gravity as we would like to believe. When we are in pain, the matter only worsens. Wolf, et. al. (1982), for example, have noted that chronic back pain patients often assume abnormal static postures. They typically stand with the trunk bent forward, the shoulders rounded and protracted and the pelvis rotated. Hoyt, et. al. (1981) noted higher levels of EMG activity in chronic pain patients compared to normals during 10 minutes of quiet standing. Cram and Engstrom (1986) have also observed that L3 paraspinal EMG activity separates normals from chronic pain patients, but only in the standing posture. It would appear that having the muscular system bear weight increases the likelihood of observing abnormal patterns of "static" EMG activity in the back.

While the "static" diagnostic information concerning posture may be obtained from a muscle scanning procedure, a "dynamic" evaluation of the back musculature yields a different set of diagnostic information. This procedure incorporates biomechanical postural principles. Price (1948) has commented upon the observation that abnormal "patterns" of EMG activity contribute to back pain. These were described as abnormally high or low EMG levels, as well as left-right asymmetries observed during normal posturing and movement. Wolf and Basmajian

(1978) report abnormal patterns of muscle activity in low back patients compared to normal controls during the structured movement patterns of forward flexion and rotation of the trunk. The "flexion-relaxation" phenomenon essentially assesses the patient's ability to go out onto ligament support as the patient goes beyond the initial 45 degrees of forward flexion. This is shown in Figure 12. An inability to demonstrate decreased EMG activity towards the end of this range of motion is considered abnormal.

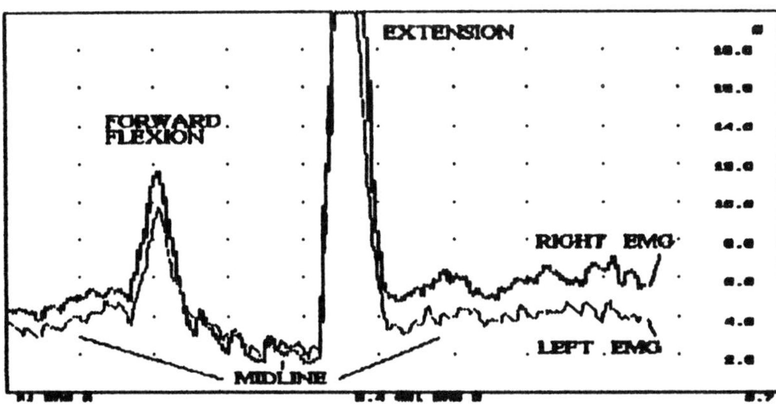


**FIGURE 12.** FLEXION RELAXATION OF THE LOW BACK MUSCULATURE

During the rotational patterns, the hips are stabilized, and synergistic muscle patterns are assessed. Separation of function is considered normal, while co-contractions of the right and left aspects during rotation are considered abnormal.

Once posture is considered diagnostically relevant, the surface EMG procedures described above may be used to clearly identify abnormalities for either the dynamic or static aspects. Once identified, treatment plans may begin to consider the "postural training" protocols introduced by Jones and Wolf (1980) and Wolf, et. al. (1982). In this treatment protocol, the patient is taught to display "normalized" patterns of EMG activity during both static and dynamic movements observed during every day life. Strengthening, stretching, and postural changes are usually needed to assist the patient in using their back in a biomechanically efficient and safe way.

## SUMMARY AND CONCLUSION

This section has attempted to present a diagnostic framework for surface EMG recordings. It has placed an emphasis upon the role of differential diagnostic considerations for emotional, organ related and postural aspects. These are not meant to be viewed in an orthogonal fashion and may indeed overlap. The diagnostic techniques reviewed included static muscle scanning, dynamic movement evaluations and psychophysiological profiling. Each of these techniques samples from and develops diagnostic information about a different level of the neuromuscular system. Negative findings from any one technique does not necessarily rule out the possibility of findings from a second assessment technique.

The diagnosis of pain related problems is complex and necessitates a comprehensive assessment approach. As Dolce and Raczynski (1986) point out in the conclusion of their review of the area, "...a complete and in-depth electromyographic assessment may be useful when dealing with myogenic back disorders to differentiate accurately the extent and type of muscular pathology" (p. 518). In addition, one should consider other maintaining or causative aspects of pain. Discogenic factors may play the primary role in the report of pain, with muscle activation patterns occurring as a secondary response. After an adequate passage of time to allow for healing (estimate 6 months), pain related disorders move from the acute to the chronic phase. It is after this point that one wants to consider how environmental/operant factors may have come to maintain pain behaviors in the absence of nocioception (Fordyce, 1977).

In addition, I have attempted to link each level of diagnostic findings with a companion therapeutic technique. This is based on the belief that a differential diagnosis should lead to differential treatment of the problem. The techniques reviewed have included traditional general relaxation training, specific relaxation training, transactional psychophysiology, myotherapy, stretch to relax, and postural change. The effects of EMG biofeedback treatments have been observed to be very specific (Cram and Freeman, 1986). This specificity includes not only the site of EMG feedback, but also the posture in which it was trained. There is very little "nonspecific" generalization of EMG reduction from one muscular region to the next. There is very little spontaneous transfer to EMG reductions from one posture to the next (i.e., reclining to standing). Because of this, one needs to carefully match the site(s) and posture(s) of treatment to the nature of the dysfunctions.

## SECTION 4. THE CLINICAL PROCEDURE OF MUSCLE SCANNING

EMG Muscle Scanning is a test for static patterns of splinting and guarding commonly associated with soft tissue injury and pain. It allows for a quantified assessment of muscle spasm, and differentiates this from muscles which feel hard because of contracture. It systematically allows one to quickly and easily sample from the neuromuscular system, aiding in the identification of chronically tense muscle groups. Once identified, these sites would become the subject for further study.

In the sections that follow, an in depth view of its instrumentation, the clinical procedure, the sites and the applied clinical research that has emerged over the course of the last eight years will be provided. I will review the normative data, reliability, sensitivity and specificity associated with the technique. Lastly, I will look at a variety of patient protocols.

### INSTRUMENTATION AND ELECTRODE CONSIDERATIONS

A detailed description of the EMG amplifies, filters and electrodes was presented earlier. Briefly, the apparatus used in muscle scanning involves surface electrodes at a fixed distance and differential amplification of Muscle Action Potentials (MAPs). Since the energy of muscles is given off along the lines of the muscle fibers, when the "active electrodes" are oriented so that they are parallel to the lines of the muscle fibers, this maximizes the amplification of energy from those specific muscle fibers. In conducting a muscle scan for the normative data sites, it is VERY IMPORTANT that you orient the electrodes along the lines of the muscle fibers of interest.

Some EMG amplifiers allow you to choose from different filter band widths. One common amplifier allows a choice between a narrow high energy bandpass, and an even wider full spectrum bandpass. The high energy narrow (100-200HZ) filter preferentially amplifies the fast twitch muscle fibers. These are the fibers most commonly associated with "immediate" action or muscle activation. When a muscle is activated as the result of an acute injury, the fast twitch fibers would be activated along with slower twitch fibers and elevations of EMG activity would be seen on the narrow filter. However, as the muscle activation pattern is sustained over prolonged periods of time, the fast twitch fibers may fatigue and drop out of the recruitment pattern. A frequency shift in the energy spectrum of muscles is known to occur during states of muscular fatigue. The wide (25-1000 HZ) full spectrum filter amplifies MAPs from both the slow and fast twitch fibers. If activation patterns are seen only in the wide filter and not the narrow filter, muscle fatigue is strongly suspected. Muscle scanning which allows for spectrum analysis will be the next leap in muscle scanning technology.

## THE GENERAL PRINCIPLES OF SCANNING

While the general principles of muscle scanning are quite simple, these do not translate into a "fool proof" procedure. One must adhere to the principles outlined below or the EMG study will be significantly flawed, providing a false view of the neuromuscular system.

First of all, it is important to keep the resistance (impedance) at the site as low as possible. One must abrade the site to be scanned, so that it is free of oils and dead or horny layer of skin. Coat the electrodes with a light coating of electrode paste or cream prior to placing them onto the site. This will add electrons to the electrode skin interface, thus lowering the resistance, while potentiating the biological signal.

Secondly, hold the sensors in place with a light pressure. Do not press too firmly, else you might push the patient off their center of gravity, thus creating an artificial recruitment pattern. Some practitioners have asked whether excess pressure might place the muscle on stretch. I have attempted to demonstrate this on many occasions, in many ways, and have not been impressed with this as a source of artifact.

Lastly, since scanning electrodes are not held in place with an adhesive tape, the electrodes need to be held motionless by hand for a period of 6-20 seconds. It will take the signal some 4-8 seconds to stabilize from the electrical explosion which occurs when the scanning electrodes initially make contact with the skin. In addition, it has been my experience that going much beyond 30 seconds does nothing but add movement artifact into the signal and it seems to become more variable. Intentional or accidental movement or sliding of the scanning electrode must be avoided at all costs, in that it will generate an offset potential which will artificially elevate the levels of energy seen on your instrument.

## THE CLINICAL PROCEDURE

The clinical procedure described below has evolved over eight years of refinement. I suspect that it will continue to evolve as more and more experience is gained from its applications. Many different strategies have been tried over the years. The procedure described below is based both upon applied clinical research and clinical use. It represents the current "state of the art" and is incredibly simple. This is not to say, however, that it is "fool proof." You must have some knowledge of muscle anatomy and you must be able to determine if the EMG signal is free of artifacts and noise.

### STEP BY STEP DESCRIPTION

| | |
|---|---|
| **STEP 1.** | Identify the muscle group of interest. Think about the direction of the muscle fibers. Visually or tactually, locate the belly of the muscle. |
| **STEP 2.** | Abrade and clean the scan site with a "rough" alcohol swab. No cotton balls please. Let it air dry. It is usually wise to clean all of the sites you plan to scan at this time. |
| **STEP 3.** | Ask the patient to sit or stand comfortably. Do not otherwise verbally or physically prompt their posture. Make visual observations of their "natural" posture. |
| **STEP 4.** | Run a bead of electrode paste or cream on a tissue. Dip the scanning electrodes in the bead and lightly coat them with this paste. Avoid ending up with globs of paste on the electrodes. |
| **STEP 5.** | For recording from specific muscle groups, place the scanning electrodes such that they run parallel to the muscle fibers of the muscle of interest. If you are going to compare the data collected on a patient to that of a normative sample, be certain to use the electrode placement guide used in the normative study. Once on the skin, hold the electrodes motionless. |

**STEP 6.** Observe the EMG recording. Allow it to settle from its initial contact artifact. This usually takes 4-8 seconds. The signal is considered stable and good when it is +/- 10% in its fluctuations and free of artifacts (swallowing, movement) and noise (60 Hz). The level of heartrate artifact, if present, should be noted.

**STEP 7.** When you are satisfied with the veracity of your EMG signal, record that EMG value as the best representative sample of EMG activity for that site. The total scanning procedure for a given site commonly takes about 10 seconds.

(Please note, with computerized systems which use the CRAM SCAN software, the mere press of the data record button places the data into memory for a later compilation of a computer generated report. For stand alone instruments, I recommend the use of a voice activated dictaphone to verbally record the sites and their values for transcription at a later time.)

**STEP 8.** If a site shows values in the "severe" range, see if you can verify that this is due to EMG rather than noise of some type. Listen to the raw EMG if it is available. Consider rescanning it to see if it remains stable in its elevation.

**STEP 9.** Study the right and left aspects of each muscle site. If you have a dual channel scanning capacity such that you could monitor both the right and left side simultaneously, use it. If postural muscles are involved, conduct the scan in both the sitting and standing postures.

**STEP 10.** If multiple filters are available, conduct the study first with the narrow high energy filter (100-200 HZ). Then open up the filter to the wide full spectrum filter (25-1000HZ), primarily to let in the slower twitch fiber frequencies. Consider the differences between the findings of the two filters for muscle fatigue.

## POTENTIAL PROBLEMS AND PITFALLS

There are five problems which commonly enter into the scanning procedure: Exceptionally low values, exceptionally high values, variable EMG values, obesity, and physical appliances.

Exceptionally low readings are in the 0.0 to 0.3 range on most amplifiers. These values are down in the range of the noise of the instrument itself, and have very little to do with MAPs.
These readings occur when there is a "bridging" of paste between the two active electrodes. If the procedure is done while the skin is still moist from the alcohol abrade, the natural salts of an individual patient may join the moisture of the alcohol to form an electron bridge. Exceptionally low values should always be suspect. Clean the site again, let it dry, and scan it a second time.

Exceptionally high EMG values are above 25-40 microvolts on most amplifiers. EMG values seldom go above these amplitudes, except in active muscles or muscles in "spasm". High readings may also be due to 60 cycle noise entering your amplifiers. This is commonly associated with poor contact of the electrode with the skin. If the skin has not been adequately abraded (i.e., natural oils or oils from makeup have not been removed), 60 cycle can easily invade your signal. One common problem is associated with scanning at the hairline. Here, the practitioner has adequately abraded the skin in the cervical region, but not into the hairline. Or adequate electrode paste is used for the skin, but is inadequate to penetrate through the hair. As it turns out, if there is a "radical" difference in the level of impedance between the two active electrodes, the differential amplifier is defeated and becomes very vulnerable to 60 cycle noise. I encourage practitioners not to scan up into the hairline. Again, if you can listen to the raw EMG for the muscle action potentials, versus the hum of the 60 cycle noise, you can increase your confidence in the validity of a high EMG reading. If the raw EMG is not available to you, check out your skin preparation and adequacy of electrode paste and rescan the site again.

The third problem has to do with excessive variability of the EMG signal. You place the scanning electrodes down onto the skin and the EMG values just don't stabilize enough. It is difficult to tell which sample is the best representation of the true EMG value. Ninety percent of the time this complaint is associated with a person who is new to scanning, staying on the site for too long. Typically, the signal drops rapidly and then stabilizes nicely for some 10-20 seconds. After which, it becomes variable. This type of variability, I believe, has to do with the subtle effects of the light pressure from holding the scanner in place by hand on the postural regulation of the individual. In essence, the patient begins to "sway", as he or she subtly presses back against the pressure of the scanning electrode. In general, one should stay on the site for as short of time as possible. As soon as the signal has stabilized, record that value.

There is one more case of variability to consider. It has to do with respiration artifact. Some individual (especially smokers, and patients with COPD) tend to use their ancillary muscles instead of the diaphragm for respiration. Variability in the muscles of the upper trapezius and the dorsal region in general, seem to be most affected by this type of artifact. When it is observed, one can easily note the correlation between the variability of the EMG and the respiration cycle of the individual. If this is the case, the correct EMG value to record would be the one associated with the "pause" between expiration and inspiration.

The level of obesity of an individual is of great interest to the surface electromyographer. Since adipose tissue is known to be a good insulator it will attenuate or reduce the amount of EMG signal reaching the skin's surface. The EMG levels one would see on an obese individual would be much lower due to this. One should exercise extreme caution in the interpretation of scan values taken from an obese individual. If one is comparing an individual patient's data to that of a normative population, the problem is one of a high probability for a false negative finding. Future scanning systems will want to utilize a systematic estimate of obesity (perhaps an impedance testing system) as a major moderator variable in scan data interpretations.

Physical appliances such as dentures, glasses, shoes, lifts, etc, may introduce an alteration in muscle activity. Recordings from the temporalis muscle group provides a good example of this. We have noticed that patients who wear dentures have a higher probability to brace with these muscles. Where this is clearly seen the patient is having problems with loose fitting dentures and is using their muscles to hold them into place. We have also noted problems with frontalis and cervical muscles in patients who are having difficulties with glasses (usually bifocals). Shoes and/or lifts may also introduce another level of variance in the scan data collected. It is up to the practitioner to make inquiry and astute observations to enter factors such as these into their interpretive frameworks.

## DIAGNOSTIC SCANNING PROTOCOLS

The initial development and research on muscle scanning began in 1981 at the Swedish Hospital Pain Center in Seattle. With the opportunity to do "touchdown" recordings using the new post style scanning electrodes , we began looking at patterns of muscle activity in a variety of pain related conditions. In the process of this initial development, three dimensions of clinically relevant information clearly emerged: SITE, SYMMETRY, and POSTURE.

The first was that of "site" or muscle group. The muscle scanning protocol easily allowed us to locate where the "habitual bracing or tensions patterns" were located. The second dimension was "symmetry". As we began to systematically sample from both the right and left aspects of each muscle group we began to notice how radical imbalances or asymmetries seemed to play a role in some of our patient's pain complaint. This was most dramatically evident in the pattern of neuromuscular activity in some of our neck and back pain patients. The third dimension was that of "posture". Because of the work of Steven Wolf on postural abnormalities in back pain, we began to conduct the muscle scanning in two postures (sitting and standing). Much to our surprise and pleasure, this simple postural variable provided a wealth of information about protective guarding and splinting patterns in neck and back pain. In some of our patients, for example, these bracing patterns were only evident while in a standing posture.

## THE ORIGINAL STANDARD PROTOCOL

Over the course of the first six months we came to establish a standardized protocol for conducting the muscle scanning procedure and began collecting normative and clinical data in a systematic fashion. The initial protocol entailed the scanning of the right and left aspect of 11 muscle groups, first in the sitting and then in the standing postures. In order to compare the EMG values you collect on an individual patient to that of the normative data base presented in this chapter, it is essential that you follow the procedure and protocol described below. The filter selection, skin preparation, instructions to the patients, posture and electrode placement which you use, need to be exactly the same as those used in the collection of the original normative data. The step by step procedure outlined in the previous section should be followed closely. Below is a refined description of the postures used in the study and an extensive description of the original muscle sites.

---

SITTING POSTURE: The therapist asks the patient to sit sideways on a chair such that there is no support given to the back. It is best if the sitting surface is level to the ground, and lightly padded. If its height sitting surface can be altered to accommodate the leg length of the patient, all the better. If the height is fixed, you will need to use devices such as pillows or boxes so that you seek a 90 degree angle at the knee and the patient's feet resting comfortably on a surface. Patients are asked to place their hands in their lap, and to "sit in a comfortable fashion". Avoid words such as "sit up straight". They are requested not to alter their posture so as to make the scanning easier. In other words, they are not to help you. If the patient looks up so that you can scan their frontalis muscle, or leans their head forward so that you can scan their cervical paraspinals, the therapist gently redirect the patient to "sit comfortably". No further instructions concerning their posture is given. The object is for the patient to take on as natural a sitting posture as possible.

---

STANDING POSTURE: The therapist asks the patient to "stand comfortably" in an unsupported fashion with his hands at his sides. Again the patient is requested not to alter their posture to help make the scan easier. If this occurs, the therapist redirects the patient to stand comfortably. No further instruction as to posture is given. The object is for the patient to take on the natural/habitual posture for standing quietly.

---

I have observed a growing interest in the attempt to scan patients in postures other than the "neutral" sitting and standing described above. Postures such as forward flexion or lateral bending have been proposed and used clinically. Unfortunately, I have some difficulty with the use of these postures during a scanning procedure. My first concern is one of time. In order to scan several paraspinal muscles sites during forward flexion, for example, the patient would need to be off their center of gravity during the procedure, perhaps for a few minutes. The interaction between gravity and the neuromuscular system may change over time. For example, the patient might relax or brace into the stretch associated with forward flexion. My second concern has to do with the reliability of the procedure and its findings. The patient might execute a slightly different range of motion on a subsequent testings of the same movements. Differences in the range of motion would activate different levels of EMG recruitment. Until a reliable method of monitoring or controlling the range of motion is developed, I see the procedures associated with other postures to be highly suspect. My impression is that EMG recordings obtained during a range of motion should be conducted using attached electrodes with the pattern of recruitment being monitored throughout the actual range of motion. This concept is covered in a later section.

**THE STANDARD 11 SITES**

This section is intended to help clarify the placement of the scan electrodes for each of the standard 11 sites. In addition, an interpretive framework for the recordings of each site. As the placement for each site is presented, consider the anatomical markers described to aid in the placement. Also examine the graphic depiction to the sites displayed in Figure 13.

**1. LOW FRONTALIS.**

PLACEMENT: The scan electrodes are placed parallel to the eyebrow, and approximately 1/2 inch above it. The space between the two electrodes is centered over the iris of the eye. This configuration of the active electrodes runs perpendicular to the muscle fibers of the Frontalis muscle. It records from a small region on the forehead, including: Frontalis, Corrugator, Orbicularis Oculi, and Temporalis.

INTERPRETATION: The primary function of this muscle or region is the non-verbal display of emotions. The emotions of anger, fear, sadness and surprise activate this region. It may also be associated with "intense concentration" or "squinting of the eyes". Elevations noted here may represent an "emotional" contribution to the presenting problem. It has been suggested that elevations on the right aspect may be seen in patients with cardiac proneness.

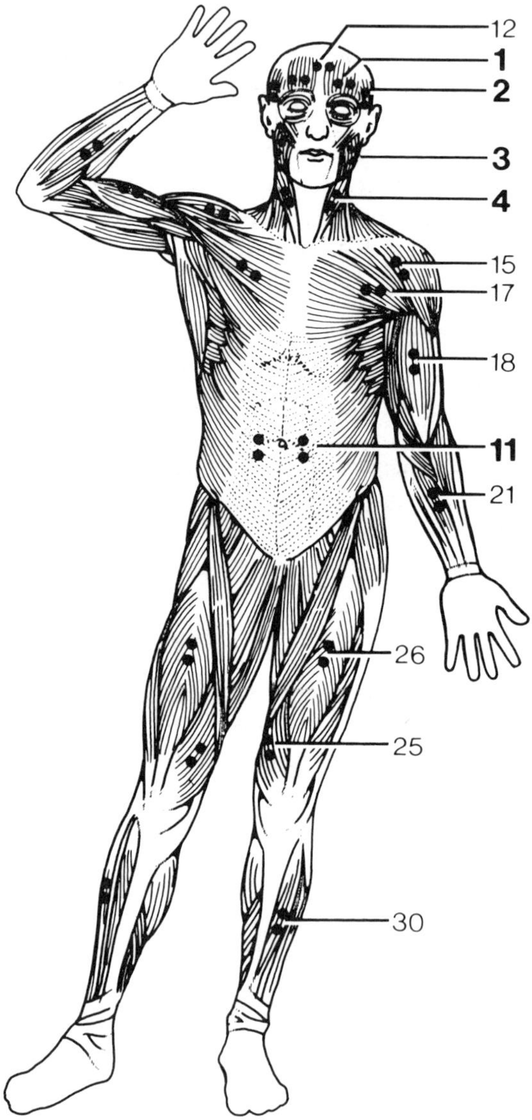


**SCAN SITE CODE**

1. Low Frontalis
2. Temporalis
3. Masseter
4. SCM
5. Cervical P.S.
6. Upper Trapezius
7. $T_1$ Paraspinals
8. $T_6$ Paraspinals
9. $T_{10}$ Paraspinals
10. $L_3$ Paraspinals
11. Abdominals
12. High Frontalis
13. Lat. Neck
14. Capitus
15. Ant. Deltoid
16. Post. Deltoid
17. Pectoral Major
18. Biceps
19. Triceps
20. Wrist Extensors
21. Wrist Flexors
22. Lower Trapezius
23. Latissimus Dorsi
24. Gluteus Maximum
25. Vastus Medialus
26. Quadriceps (Fem.)
27. Med. Hamstrings
28. Gastrocnemius
29. Soleus
30. Ant. Tibialis

**FIGURE 13a** SCAN SITE PLACEMENT (FRONT)

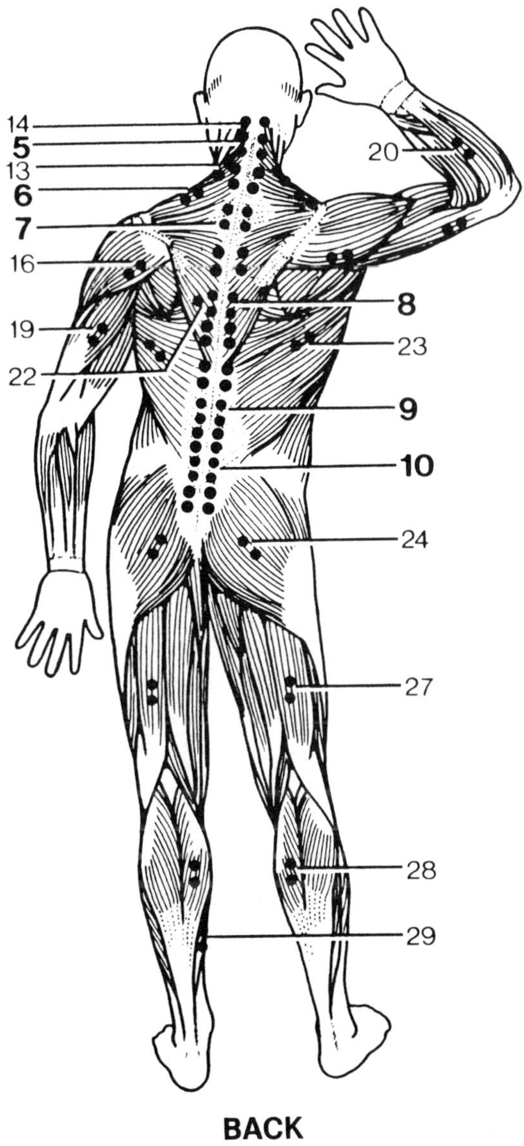


**FIGURE 13b** SCAN SITE PLACEMENT (BACK)

**2. ANTERIOR TEMPORALIS.**

PLACEMENT: The scan electrodes are placed in a vertical direction. The "lowest" of the active electrodes is placed across from the notch of the eye socket approximately 1/2 inch laterally. The electrodes run parallel to the muscle fibers of the Anterior Temporalis.

INTERPRETATION: The Temporalis is a major muscle of mastication and stabilization of the mandible. It pulls the jaw bone up (elevates it). Patients who tend to "clench" their teeth during times of stress may show elevations here. Asymmetrical patterns (R vs L) may suggest TMJ involvement in headache. In addition, we have noticed that patients who wear dentures may show excessive activation in this muscle group. They appear to use the Temporalis muscle to stabilize a loose fitting denture. Noted elevations in this muscle group should be followed by an inquiry into the use of dentures.

**3. MASSETER.**

PLACEMENT: The Masseter muscle may be easily palpated by placing your fingers on the patient's jaw and asking him to clench his teeth. The goal of this placement is to place the scanning electrodes over the belly of the masseter muscle with the electrodes running parallel to the muscle fibers. To do this, draw an imaginary line from the "corner of the jaw" to the cheek bone and place the scanning electrodes along this line over the belly of the muscle.

One common problem in scanning this site in men is the presence of beards. Scanning of this site on men with beards is impossible and should not be attempted.

INTERPRETATION: The Masseter muscle is the major muscle of mastication. The vectors of force form concentric pressure on the teeth and allow chewing. When one is not chewing, swallowing, or talking, this muscle should be quiet. Elevations and/or asymmetries in this muscle are commonly associated with dental problems (i.e., TMJ, MPD, BRUXING) and may suggest an etiology for headache.

**4. STERNOCLEIDOMASTOID (SCM).**

PLACEMENT: This muscle runs from the back corner of the skull (Mastoid Process) to the collar bone (Clavicle). If you have the patient rotate their head to the "right", and elevate it slightly, the belly of the SCM on the "left" may be easily seen. The scanning electrodes are placed parallel to the muscle fibers of the SCM and half way between the Mastoid Process and the clavicle. The head is in the midline position during the scanning.

INTERPRETATION: The major function of this muscle is the rotation of the head. The SCM on the side opposite the direction of the head turn is active. The SCM also plays a minor role in respiration. Rhythmic oscillations which follow the patient's respiration pattern may suggest excessive use of the ancillary muscles of respiration. Asymmetrical patterns are clearly seen in Torticollis, and may be seen in patients with acceleration injuries to the neck. In general, the activity in this muscle is typically quit low.

**5. CERVICAL PARASPINALS (CPS/C4).**

PLACEMENT: The scan electrodes are placed parallel to and approximately 1/4 to 1/2 inch out from the spine over the belly of the muscle. If you place the electrodes too far laterally, slightly off the belly of the muscle, the EMG values will be significantly lower. The upper scanning electrode is below the hairline with the general placement being approximately in the vertical center of the neck. This differs from the CAPITIS or C2 placement, where the upper scanning electrode may reach into the hairline. In conducting the scan of this site, it is important to notice whether or not the patient attempts to help you by leaning his head slightly forward. This is a common error that the patient makes since we have all been trained by the barber or beautician to help in this way. If this occurs, simply instruct the patient that it is not necessary to help out in this way, to return to a comfortable sitting or standing posture, facing forward, while you monitor from the site.

INTERPRETATION: There are many layers of muscles in this region and the scanning electrodes collect from the upper fibers of the Trapezius, the Splenius Capitis, the Splenius Cervicis and the Semispinalis Cervicis. These muscles are primarily involved in the support and rotation of the head. With unilateral activations of these muscles, rotation or lateral flexion of the head may occur. With bilateral activation, extension of the head may occur. In patients with "poor posture", where the head is held "forward" of its center of gravity, these muscles are required to provide "chronic" muscular support for the 15 pound weight of the head. In such cases, fatigue of these muscles may occur and high energy narrow filter recordings may not adequately represent the nature of muscular dysfunction. In another common variation, the cervical muscles may appear to have very low EMG values, especially in comparison to trapezius, T1 or T6.

When this occurs, the patients' posture may be such that they have "unloaded" the cervical muscles, transferring the work to these lower muscles. This pattern of activity is considered to be an inappropriate muscle synergy. In other words, even though low EMG values are present, activations in other related muscle groups make these values suspect. It is always good to consider pattern analysis of one site to other sites, as one develops their diagnostic impressions. Lastly, the cervical muscles have been indicated in headaches located in the occipital and frontal regions. In fact, I have noted a concordance in activation between the frontalis muscle and the cervical paraspinal muscles. Such a concordance may indicate an emotional overlay functioning in the patient's case.

**6. UPPER TRAPEZIUS.**

PLACEMENT: The scanning electrodes are placed such that they run parallel to the muscle fibers of upper Trapezius. Their direction follows the "ridge" of the shoulder. They are placed approximately one inch from this ridge, posteriorly towards the back. They are located approximately halfway between the spine and the lateral edge of the shoulder, over the belly of the muscle. The belly of the muscle may easily be determined by palpation.

INTERPRETATION: The primary function of the Trapezius at this site has to do with stabilization or movement of the shoulder. In addition, it is used in lateral bending or rotation of the head. Lastly, the shoulder region is considered a strong emotional display system. Patients may elevate their shoulders when they feel threatened or are anxious and "up tight". When EMG activation or asymmetries are noted on Trapezius, one or both shoulders are probably elevated.

This muscle is strongly affected by posture. Its "static" level of discharge may be radically different during standing compared to sitting. The position of the arms in relationship to the torso may be a major factor. In addition, we have noted that there appears to be a left sided dominance in this region, with the left aspect being generally more active than the right. Some have argued that this is associated with "handedness". Others have suggested that this is associated with heart rate artifact. Since this left sided dominance is more predictably seen during the standing state, compared to the sitting state, it is difficult to imagine how heart rate artifact might change as a function of posture.

Lastly, the activity of trapezius should always be compared to the activity of other muscles, particularly those of the lower and mid back. Compensatory shifts may be seen in some cases, where a left sided asymmetry at the L3 level is associated with a right sided asymmetry at Trapezius. In addition, it may be overly active due to too much inhibition of the dorsal (T8/T10) muscle groups. Elevations in Trapezius may be associated with temporal headache, neck or upper back pain.

**7. T1/T2 PARASPINALS.**

PLACEMENT: The scanning electrodes are placed such that they run parallel to the spine and approximately 3 cm. or 1/2 inch out from the vertebral ridge. The "landmark" for this placement is the C7/T1 prominence. In most adults, the vertebral process of the last cervical vertebrae clearly stands out. Follow the spinal processes in the cervical region down into the upper back. When you notice the first prominent spinal process, the chances are good that this is the C7/T1 prominence. The upper most scanning electrode is placed across from this prominence.

INTERPRETATION: Again, there are multiple muscles at this recording site. Contributions from Trapezius (upper/middle fibers), Longissimus, Semispinalis Splenius and Multifidus muscle groups may contribute to the surface EMG recordings. The primary use of these muscles is the stabilization and support of the head and neck. This appears to be a major stress point for postural support.

Elevations at this site commonly covary with the trapezius muscle group. Activations here, with the cervical muscles being very low may suggest that they are doing some of the work of the cervical muscles. Elevations and asymmetries at this site may be noted in patients with complaint of headache, neck or upper back pain.

**8. T6 PARASPINALS.**

PLACEMENT: The scanning electrodes are placed such that they run parallel to the spine and are approximately 3 cm. out from the vertebral ridge. The vertical level of placement is approximately between the shoulder blades (or scapula). If the spinal processes are clearly seen or palpated on your patient, one could easily count down six vertebrae from T1 to find the location. If your bottom scanning electrode is below the bottom edge of the scapula you are too low. The bottom edge of the scapula may be easily palpated.

INTERPRETATION: Again, there are many layers of muscles at this site. Contributions from Trapezius (inferior fibers), Rhomboids, Longissimus, Semispinalis, and Multifidus may determine the surface EMG readings. The function of these muscles are two fold: 1) To stabilize the thoracic spine, and 2) To stabilize or move the scapula. In the free sitting posture (no back support) these muscles may be very active. This is one of the few sites where higher levels of activation are expected during free sitting compared to free standing. Pain in the mid back or neck is commonly associated with elevations or asymmetries at this site.

**9. T10 PARASPINALS.**

PLACEMENT: The scanning electrodes are placed such that they run parallel to the spine. Again, they are approximately 3 cm. out from the vertebral ridge. The vertical placement along the spine may be counted down from T1, if the spinal processes are prominent. One can palpate for the ribs, and place the scanning electrodes slightly above where the rib cage disappears. Another method is to take approximately one half the distance between the L3 placement and the bottom of the scapula.

INTERPRETATION: Again, many layers of muscles. The primary contributors appear to be Longissimus and Multifidus. There is a possibility of contributions from latissimus dorsi and posterior serratus. The primary function of these muscles is postural support, and stabilization of the spine. Since this is the region of the back where lateral bending is possible, this site may reflect this action. They may play a role in rotation of the back. It is of interest to note that protective guarding and splinting patterns or asymmetries are commonly observed in this region for back pain patients. This phenomena appears to be augmented in patients whom have had surgery (particularly fusions) at lower levels. Posture strongly effects this site, and elevations or asymmetries may not be observed until the patient has assumed the standing posture.

Lastly, as you will read later on in this chapter, this region seems to correlate most highly with the total amount of muscular energy observed in a scan. If a "key muscle" were to be selected to be indicative of general muscle tension, this would be the one to choose.

**10. L3 PARASPINALS.**

PLACEMENT: The scanning electrodes are place such that they run parallel to the spine and approximately 3 cm. out from the vertebral ridge. The iliac crest (top of the pelvic girdle), is used as the landmark to guide the placement. Palpate for the Iliac Crest. Once this has been found, draw an imaginary line, horizontally across the back towards the spine. Place the active scanning electrodes, such that the lower electrode is at this level.

INTERPRETATION: Again, many layers of muscles. The muscles thought to contribute most heavily to the surface EMG recording are the Iliocostalis Lumborum and Multifidus. The major function of these muscles is postural support, the stabilization of the trunk during standing. In addition, they are the extensor of the lower back and also come into play with an eccentric contraction during forward bending (flexion) and an activating concentric contraction during the return to the upright position. Protective guarding/splinting is commonly seen at this site in low back pain patients. The site is strongly effected by posture, and the surface EMG recordings may radically shift as one goes from the sitting to standing postures. Interestingly enough, during quiet standing only minimal muscle activity in the low back region is needed as long as you are over your center of gravity. Go slightly off your center of gravity (i.e., lean forward a little), and these muscles come into play.

It is important to note whether or not surgery has been conducted in this region. If so, the surgeon may have removed or denervated some of the muscles or in some way altered the neuromuscular structure. Needle, not surface, EMGs may be done to assess for denervation. While the effects of denervation on surface EMG potentials has not been adequately studied, one should exercise caution in their interpretation of scan data in a region where surgery has been conducted.

**11. ABDOMINALS.**

PLACEMENT: The scanning electrodes are placed approximately 3 cm. to the left or right of the umbilicus (belly button). The space between the scanning electrodes is placed opposite the umbilicus. The direction of the electrodes is vertical, and they run parallel to the muscle fibers of the Rectus Abdominous.

INTERPRETATION: These muscles play a role in the stabilization, elevation, and rotation of the pelvis. They are commonly underdeveloped due to lack of use. They may also be covered with a thick layer of fat. If this is true, the recordings would be lower due to the attenuation of the signals by the adipose tissue. One of the common therapy programs for low back pain (William Flexion), focuses on the strengthening of these muscles so as to pull the front of the pelvis up and take some of the excessive lordotic curve out of the low back. Asymmetries may be noted in low back pain patients, but this is very uncommon. Elevations may be noted in patients with pain of abdominal origin.

## SUPPLEMENTAL SCAN SITES

Over the seven to eight years since the introduction of the standardized scanning protocol, the limitations of the set of standard sites became readily apparent, as professionals outside the biofeedback arena began asking me if the scanning protocol had been extended to various other muscle sites. In the last two years, I have been developing clinical information concerning supplemental sites. Below are some additional sites requested by the various disciplines.

## SUPPLEMENTAL SCAN SITES FOR DENTISTRY

### 31. DIGASTRIC.

PLACEMENT. The scanning electrodes are placed under the chin, such that they run in an anterior/posterior direction, approximately 1/2 inch to the left or right of the middle of the chin, following the muscle fibers of digastric. It is important that the patient not raise their head so as to help you with this placement. They are asked to sit comfortably with no specific instructions regarding the position of the mandible. Men with beards may not be scanned at this site.

INTERPRETATION: This muscle is part of a group of muscles called the suprahyoids. Their primary function is to elevate the hyoid bone during swallowing when the mandible is stable, or to open the jaw when the hyoid is stable. It is an important muscle to assess in patients with TMJ oriented problems.

### 32. POSTERIOR TEMPORALIS.

PLACEMENT: The scanning electrodes are placed slightly above towards the posterior flap of the ear. The electrodes are oriented downward at approximately a 45 degree angle relative to the ground. They are clearly above the mastoid process,and below the hairline.

INTERPRETATION: This muscle, like the anterior temporalis, plays a role in elevating and stabilizing the mandible. But, because of its more posterior location, the vector of force also includes the retraction of the jaw. Again, this is an important muscle group to consider in the TMJ oriented patient.

## SUPPLEMENTAL SCAN SITES FOR CHIROPRACTIC

PARASPINAL MUSCLES (C2,C6,T2,T4,T8,L1,L5).

PLACEMENT (General): The scanning electrodes are placed such that they run parallel to the spine, approximately 2-3 cm. out from the vertebral ridge over the belly of the paraspinal muscle. These muscles may be visually seen in some individuals, but may need to be palpated in others. The vertical placement for each site is as follows:

**33. C6 PLACEMENT:** For the cervical region, it is probably best to start at the bottom and move upwards. The electrode placements, then, go from easy to more difficult. The C6 electrodes are placed such that the lowest scanning electrode is directly across from the C7/T1 prominence.

**5. C4 PLACEMENT:** This is the site referred to in the original scan site series as CPS. Refer to that section if need be. The most accurate method for vertical placement is to count up from the C7 prominence three vertebra and place the electrodes such that the space between the scanning electrodes is directly opposite this spinous process. An alternate placement rule would be to place the bottom electrode for this site at the spot where the upper electrode had been placed for the C6 recording. If you have used the suggested electrode paste method, this should be highlighted by the paste residue left behind from the previous C6 scan. This procedure will also minimize problems with bridging of electrode paste between the two active scanning electrodes.

**34. C2 PLACEMENT:** This is a very important, yet very difficult site to scan. The potential problems lie with the hair line. The most accurate method for vertical placement is to count up 5 vertebra from the C7 prominence and place the middle of the scanning electrode opposite this spinous process. An alternate rule would be to place the lower electrode for the C2 scan at the spot where the upper scanning electrode had been for the C4 scan. The residue left behind should help guide this placement, while avoiding problems with the bridging of electrode paste.

The problems with the hairline need to be carefully examined. Make certain that the abrade for the site is rough and adequate, particularly around the hairline in particular. If the patient's hairline is too low, such that the upper scanning electrode clearly reaches into a thick patch of hair, you will be pushing scanning technology to its limits. The problem is this: If the scanning electrode which reaches into the hairline doesn't make a good, clean contact with the skin (scalp), then this poor contact will artificially drive up the EMG readings due to interfering with the differential amplifier. Thus, you are then faced with the dilemma of deciding if a high reading on your instrument is real EMG or just instrument noise.

As an alternate means of monitoring this area, one might want to consider orienting the electrodes such the upper electrode goes off at a lateral angle towards the mastoid process, rather than up into the hairline. In fact, it might be argued that this arrangement might better reflect the activity of the group of muscles which cross the atlas/axis joint. If you were to utilize this concept, you would probably want to give it a different label, such as "axis", to differentiate it from the C6 configuration.

**7. T2 PLACEMENT:** To locate this site, simply count down two vertebra from the C7/T1 prominence, and place the space between the scanning electrodes directly across from the vertebral process. As an alternate method, place the upper scanning electrode just slightly below the C7/T1 prominence.

**35. T4 PLACEMENT:** To locate this site, simply count down four vertebra from the C7/T1 prominence and place the space between the active electrodes directly across from the vertebral prominence.

**36. T8 PLACEMENT:** Again it is typically very easy to count down from the C7/T1 prominence placing the space between the active electrodes directly across from the vertebral prominence. The location of this site is approximately where a woman's bra strap would run. You will typically need to have the woman remove the brasserie prior to the examination.

**37. L1 PLACEMENT:** If you have been using the counting method so far, it would essentially be the case were you would count down 13 vertebra from the C7/T1 prominence. If you are familiar with the use of the iliac crest to determine the L3 vertebra (see the L3 placement in the original scan site descriptions), you would essentially go up two vertebra from that point.

**38. L5 PLACEMENT:** Again, you may count down 17 from the C7/T1 prominence or 2 down from the L3 vertebra determined using the iliac crest method described above. This site is typically below the belt line.

INTERPRETATION: In general, when monitoring from the paraspinal muscles there are many layers of muscles to consider: Longissimus, Spinalis, Semispinalis, and Multifidus may contribute to the signal. All of these muscle play a substantial role in stabilization of the spine. In addition, when both the right and left aspects are activated, extension of the spine occurs. Unilateral contractions of one side will lead to either rotation or lateral bending of the spine.

It is believed by some chiropractic that an asymmetry in EMG at one level would indicate a rotated or subluxated vertebra at that level. This may or may not be true. A recent review of 25 chiropractic patient records for surface EMG (neutral posture) and radiographic findings (using the Gonsted method) indicated that the relationship varied from level to level. The strongest relationships seem to be found at sites of the greatest spinal instability: the Axis and L5 levels. At the Axis, 21 out of 25 patients showed abnormal radiographic findings. Of the 23 whom were scanned at this site, nine (40%) showed significant EMG asymmetries. Of those with EMG asymmetries, six (75%) had concomitant radiographic evidence. At L5, 18 out of the 25 patients showed radiographic evidence. Of the 22 whom were scanned at this site, only 6 (28%) had clinically significant asymmetries. Of those with asymmetries, five (83%) had concomitant X-ray findings. At a site like the T1 paraspinals, only four abnormal X-ray findings were identified for the 25 patients. Of the 14 patients scanned at that site, only four (29%) had clinically significant EMG asymmetries. Of those four, only one (25%) had a concomitant X-Ray finding. At a similar site, T7, there were four patients with abnormal X-rays, and of the 14 patients receiving an EMG scan at this site, only two (15%) were asymmetrical. Both of these two patients (100%) also had radiographic findings.

From this brief study, it may be concluded that things are more complicated that just saying that an EMG asymmetry at a given level indicates a rotated or subluxated vertebral body. As was seen, X-ray evidence of rotation, is not always associated with asymmetrical EMG activity. In addition, when clinically significant asymmetries of EMG activity are noted, they are not always associated with radiographic evidence of rotation.

In addition to rotation, one would want to consider other diagnostic concepts such as scoliosis, pain displays of splinting/guarding, functional problems, and emotional displays.

## SUPPLEMENTAL SCAN SITES FOR PHYSICAL MEDICINE

The muscle scanning procedure may be applied to virtually any muscle group which is shallow, and has a strong enough definition to be palpated. Below are some common sites to be considered for both the upper and lower extremities.

### 15. ANTERIOR DELTOID

PLACEMENT: The scanning electrodes are placed such that they run parallel to the muscle fibers in the anterior deltoid group. The scanning electrodes are placed on the anterior medial surface of the upper arm, just below the shoulder joint. The upper scanning electrode is placed approximately 3 cm. below the clavicle bone, with the lower electrode going laterally at approximately a 25 degree angle from vertical so as to follow the muscle fibers.

INTERPRETATION: This muscle group plays a role in the flexion, horizontal adduction and medial rotation of the humerus or upper arm.

### 16. POSTERIOR DELTOID

PLACEMENT: The scanning electrodes are placed so that they run parallel to the muscle fibers. When the patients arms are hung to the side, the scanning electrodes are placed on the posterior lateral surface of the upper arm, just lateral to the shoulder joint. The upper scanning electrode is placed 3-5 cm. below the spine of the scapula, with the lower electrode going laterally at approximately a 25 degree angle from vertical to follow the orientation of the muscle fibers.

INTERPRETATION: The posterior deltoid group provides extension, horizontal abduction and lateral rotation of the humerus or upper arm.

### 17. PECTORALIS MAJOR

PLACEMENT: The scanning electrodes are placed such that they run parallel to the middle fibers of this rather large and flat muscle group. The electrodes are placed below the clavicle and above the breast and nipple, with the scanning electrodes running close to parallel to the ground.

INTERPRETATION: The primary function of this muscle group is the adduction and medial rotation of the humerus or upper arm. Activation patterns have also been noted in post MI patients who continue to have a fear of a future MI.

**18. BICEPS**

PLACEMENT: With the arms hung to the sides, the scanning electrodes are placed on the anterior surface of the humerus in a vertical plane, such that the scanning electrodes run parallel to the muscle fibers. They are placed over the muscle belly, approximately 2/3 the distance between the shoulder and the elbow.

INTERPRETATION: The primary function of this muscle it the flexion of the elbow and supination of the forearm.

**19. TRICEPS**

PLACEMENT: With the arms hung to the side in the standing posture, the scanning electrodes are placed on the posterior lateral surface of the upper arm. The scanning electrodes are placed over the belly of the muscle, approximately half the distance between the shoulder and the elbow. The electrodes are oriented in a vertical plane so as to follow the fibers of the muscle.

INTERPRETATION: The primary function of this muscle is to extend the elbow.

**20. WRIST EXTENSORS**

PLACEMENT: The scanning electrodes are placed on the middle of the dorsal forearm about 6 cm. down from the elbow. Asking the patient to engage in a wrist extension will help clarify the muscle belly. The scanning electrodes are placed over the muscle belly in a vertical plane so that they follow the muscle fibers.

This site may be scanned in either standing or sitting postures. In the standing posture, have the arms hung naturally at the sides. During sitting, rest the arms on the patient's lap or some other surface.

INTERPRETATION: This muscle group primarily causes an extension of the wrist and fingers. It may be indicated in cases of carpal tunnel or arm pain in general. Elevations at this site in patients with pain conditions away from the arm may indicate an overall arousal problem which involves the neuromuscular system.

**21. WRIST FLEXORS**

PLACEMENT: The electrodes are placed on the anterior surface of the forearm approximately 6 cm. from the elbow. Again, its muscle belly is easily identified by having the patient flex the wrist. The electrodes are placed over the muscle belly in a vertical direction (while standing), such that the scanning electrodes follow the muscle fibers.

INTERPRETATION: The primary function of this muscle group is to flex the wrist and fingers.

**24. GLUTEUS MAXIMUS**

PLACEMENT: The scanning electrodes are placed in the center of the posterior surface of the buttock. The scanning electrodes are oriented laterally at approximately a 25 degree angle off of vertical, such that the electrodes follow the direction of the muscle fibers.

INTERPRETATION: This very powerful muscle is used primarily in rising from sitting, going up stairs, running, etc. It's primary function is the forceful extension of the hip and lateral rotation of the extended hip.

**25. VASTUS MEDIALIS AND LATERALIS**

PLACEMENT: The scanning electrodes are placed on the anterior, medial or lateral surface of the lower third of the thigh, approximately 6 cm. above the knee cap. The scanning electrodes are oriented laterally at approximately a 20 degree angle from vertical so as to parallel the muscle fibers.

INTERPRETATION: These large strong muscles act to extend the knee. It is interesting to note that significant reductions in EMG recruitment are noted in this, and the other Quadriceps muscle groups, in patients presenting for evaluation of total knee replacement. Asymmetries between the lateral and medial aspects have been noted in patients with patella femoral pain.

**26. QUADRICEPS FEMORIS**

PLACEMENT: The scanning electrodes are placed on the medial anterior surface of the thigh approximately half the distance between the hip and the knee. The scanning electrodes are place in a vertical plane from the ground such that the electrodes follow the muscle fibers.

INTERPRETATION: The primary function of this very strong muscle is the extension of the knee and an assistive role in the flexion of the hip. It helps to bring the leg forward during walking.

**27. MEDIAL HAMSTRING**

PLACEMENT: The scanning electrodes are placed on the medial posterior aspect of the thigh, about half the distance between the hip and the knee. The electrodes are placed in a vertical plane relative to the ground such that they run parallel to the muscle fibers.

INTERPRETATION: These muscles primarily extend the hip and flex or rotate the knees.

**28. GASTROCNEMIUS (GASTROC)**

PLACEMENT: The scanning electrodes are placed medially, on the upper half of the posterior aspect of the calf. There are two rather large muscle bellies. Palpate the muscle to identify the muscle belly. Place the scanning electrodes in a vertical plane relative to the ground such that they run parallel to the muscle fibers of interest.

INTERPRETATION: These muscles assist in the flexion of the knee and play a major role in the plantarflexion of the foot (pointing of the toe).

**29. SOLEUS**

PLACEMENT: The scanning electrodes are placed on the lateral, posterior side of the lower leg. Since this muscle lies underneath the gastrocnemius, one must attempt to monitor it from the lateral or outside portion of the lower third of the calf, right around where the gastrocnemius begins to narrow in its width. The electrodes are oriented in a near vertical plane relative to the ground, with the electrodes running parallel to the muscle fibers.

INTERPRETATION: This muscle is a very strong plantarflexor. It is actually the only muscle we need to use to stand in an erect posture (all other muscles should remain quiet while we hang on our ligaments). One would expect activity during standing.

**30. ANTERIOR TIBIALIS**

PLACEMENT: The electrodes are placed about a third of the distance from the knee to the ankle, lateral to the tibia on the anterior surface of the lower leg. The electrodes are oriented in a vertical plane relative to the ground, such that they follow the muscle fibers.

INTERPRETATION: The primary action of this muscle is the dorsiflexion of the ankle and inversion of the foot. This is the muscle which stops us from dragging our toe on the ground as we swing through the foot as part of normal gait. Paralysis may cause "foot drop".

## NORMATIVE DATA BASE

When data is collected on an individual patient, the question may be raised as to how to interpret this. There are three major themes or strategies which abound: Clinical experience; Idiographic approaches; and Nomothetic nets.

The first interpretive framework, clinical experience, essentially states that only by scanning literally dozens of patients and coupling this new source of information with a wealth of clinical experience, can one come to any sound conclusions. A psychologist might scan only the face and neck muscles, adding this to a diagnostic interview. A Physical therapist might palpate the muscles and place the scanning electrodes based upon their knowledge of the muscular system. Because of anatomical variations among individuals, some therapists might even suggest that the use of a "standardize" electrode placement system be discouraged. In addition, they might be able to feel things with their hands which they can't see or verify with the EMG instrument. Therefore, EMG data might be considered suspect.

While this type of model is an extremely useful one, I consider it a "preliminary" level of inquiry. When one is first learning to use a new technology or even to scan a new muscle site, one must use their existing knowledge base to guide their (scanning) activities. With experience, where to place the electrodes soon becomes far more than a random event. Rules begin to develop. As this occurs, one should be able to articulate these to others. Then, the strict clinical experience model begins to lend itself to one of the other two models.

Instances of incongruencies of the traditional data source (i.e., palpation) to the new data source should challenge inquiry. Rather than disbelieving the EMG scan data when the EMG readings are low and the muscle feels hard to the touch, consider the filter width of the instrument you are using. As was discussed earlier, low readings may indicate muscle contracture rather than contraction. Also, consider moving out of the static mode, attach electrodes to the muscle of interest and begin testing it in a dynamic fashion through its range of motion. This will be discussed later. As rules which may be shared with others develop, a paradigm shift occurs towards a different model.

The idiographic model essentially states that one should use the patient as his or her own basis of comparison. The pattern of EMG activity observed on a patient should be compared to other observations from that patient, rather than with an external reference group. In such a model, the average and standard deviation might be calculated for all of the EMG sites on a given patient. Then any particular scan site is compared to the average. If one of the sites is elevated, relative to the statistics on all sites, it would be interpreted as such. This method has its merits, and one might consider using it when one studies patients with obesity or when one introduces a new site or two into their all ready existing scan protocol. What it lacks is the elegance of comparing the scan findings on a patient to that of a peer reference group.

The last model, nomothetic nets, essentially states that there are general rules or constructs which may be used to describe or even quantify populations. Given a person's sex and age it is possible, for example, to know where their height (in inches) would place them within the population. In order to understand the EMG data collected on any given patient, it is extremely useful to compare the patient's data to that collected on a reference population.

The approach to muscle scanning which I commonly utilize in clinical practice is that of the nomothetic net. Initially, I utilized the clinical model, and drew the conclusion after many scans that EMG values above three (for my equipment) were abnormal. Unfortunately, once I developed a normative data based and inspected the data, the rule of 3.0 just did not do justice to the different muscles and the postures in which they could be studied. In this chapter, I present three reference groups: 1) A "normal" reference group; 2) A chronic pain population as its reference group; and 3) some adjusted values based on clinical experience and representing our best educated guess for that given muscle site.

## NORMATIVE DATA BASE: THE J&J NARROW FILTER

The normative data presented in this section is a combination of population data collected on 104 non-pain individuals ("normals") at the University of California, Irvine; and some adjusted values for the supplemental sites. The adjusted values represent my best educated guess at what the values for a the new scan site "should" look like, coupled with a small sample (usually an N of 10 or so) to verify that the guess was in the right range. The adjusted values are clearly labeled with an "*" following their site title and values.

It is very important to note that the values presented below are accurate for only the J&J instruments, and for only the narrow filter. These data may be transformed mathematically to fit the filter definitions of other EMG amplifiers.

The normative table (Table 5) presents the mean level of EMG activity for each muscle site and posture for the group. The cutoff points for the levels of elevations are based on the mean plus first standard deviation (MILD), the second standard (MODERATE), and the third standard deviation (SEVERE). As can be seen, the level of EMG activity to reach the "mild elevation" status varies as a function of muscle site and posture. From a normative point of view, side or symmetry is not an issue and therefore is not presented. In other words, there were no significant differences between the right and left aspect of muscle activity for the normative, non-pain group.

**TABLE 5.** NORMATIVE DATA BASE FOR THE J&J NARROW FILTER

| MUSCLE SITE | POSTURE | MEAN | MILD | MODERATE | SEVERE |
|---|---|---|---|---|---|
| FRONTALIS | SIT | 1.9 | 3.9 | 5.9 | 7.9 |
| | STAND | 2.1 | 4.1 | 6.1 | 8.1 |
| ANT. TEMPORALIS | SIT | 2.4 | 4.5 | 6.6 | 8.7 |
| | STAND | 2.3 | 4.4 | 6.5 | 8.6 |
| POST. TEMPORALIS * | SIT | 1.9* | 3.1* | 4.3* | 5.5* |
| | STAND | 1.8* | 3.7* | 5.6* | 7.5* |
| MASSETER | SIT | 1.7 | 3.0 | 4.3 | 5.6 |
| | STAND | 1.6 | 3.1 | 4.6 | 6.1 |
| DIGASTRIC * | SIT | 1.5* | 3.2* | 4.9* | 6.6* |
| | STAND | 1.6* | 3.4* | 5.2* | 7.0* |
| SCM | SIT | 1.3 | 2.8 | 4.3 | 5.8 |
| | STAND | 1.4 | 3.0 | 4.6 | 6.2 |
| C2 PARASPINALS * | SIT | 1.9* | 3.8* | 5.7* | 7.6* |
| | STAND | 1.8* | 3.6* | 5.4* | 7.2* |
| CERVICAL P.S. (C4) | SIT | 1.9 | 3.1 | 4.3 | 5.5 |
| | STAND | 1.8 | 3.7 | 5.6 | 7.5 |
| C6 PARASPINALS * | SIT | 2.0* | 4.0* | 6.0* | 8.0* |
| | STAND | 2.5* | 5.0* | 7.5* | 10.0* |
| TRAPEZIUS | SIT | 2.2 | 4.8 | 7.4 | 10.0 |
| | STAND | 3.1 | 5.9 | 8.7 | 11.5 |
| T1 PARASPINALS | SIT | 2.2 | 4.1 | 6.0 | 7.9 |
| | STAND | 2.9 | 5.7 | 8.5 | 11.3 |
| T2 PARASPINALS * | SIT | 2.2* | 4.1* | 6.0* | 7.9* |
| | STAND | 2.9* | 5.7* | 8.5* | 11.3* |
| T4 PARASPINALS * | SIT | 2.3* | 4.6 | 6.9* | 9.2* |
| | STAND | 2.3* | 4.6* | 6.9* | 9.2* |
| T6 PARASPINALS | SIT | 2.5 | 5.2 | 7.9 | 10.6 |
| | STAND | 2.3 | 5.2 | 7.5 | 9.8 |
| T8 PARASPINALS * | SIT | 2.4* | 4.8* | 7.2* | 9.6* |
| | STAND | 2.7* | 5.4* | 8.1* | 10.8* |
| T10 PARASPINALS | SIT | 2.2 | 5.0 | 7.8 | 12.3 |
| | STAND | 3.0 | 6.1 | 9.2 | 12.3 |
| T12 PARASPINALS * | SIT | 2.0* | 4.0* | 6.0* | 8.0* |
| | STAND | 3.1* | 6.2* | 9.3* | 12.4* |
| L1 PARASPINALS * | SIT | 2.0* | 4.0 | 6.0* | 8.0* |
| | STAND | 3.1* | 6.2* | 9.3* | 12.4* |
| L3 PARASPINALS | SIT | 1.9 | 4.5 | 7.1 | 9.7 |
| | STAND | 3.3 | 6.7 | 10.1 | 13.5 |
| L5 PARASPINALS * | SIT | 2.0* | 4.0* | 6.0* | 8.0* |
| | STAND | 3.1* | 6.2* | 9.3* | 12.4* |
| ABDOMINALS * | SIT | 1.0* | 2.5* | 4.0* | 5.5* |
| | STAND | 1.1* | 2.1* | 3.1* | 4.1* |

* Represents adjusted filter values

## NORMATIVE DATA BASE: WIDE FILTER

The normative data presented in this section were collected using the J&J (Model M-53) EMG amplifiers, set at the wide filter (25-1000 Hz) setting. As has been noted previously, the wide filter collects from more of the EMG spectrum and is thought to collect from more of the deeper layers of muscles along with the more superficial layers. In virtually all cases, the wide filter recordings for surface EMG are significantly higher than their narrow filter equivalents.

The population on which these wide filter data were collected is a chronic pain population. Compared to the narrow filtered data, when these wide filter norms are utilized for normative data comparisons purposes they should be considered the more conservative test. This is because it is known that the chronic pain population shows significantly higher EMG levels of activation (means) and greater levels of dispersion (standard deviations) compared to normals (see Cram and Engstrom, 1986). Because of this difference in the data base, the rules concerning the interpretation of abnormality (mean plus standard deviations) would necessitate that an individual patient's EMG value exceed the even higher criterion of his peer "pain population" group.

The population of this data base consisted of 62 chronic pain patients evaluated on the first day of admission to the inpatient pain therapy program at The Swedish Hospital Pain Center in Seattle. Their mean age was 43.3 years, with a range of 21 to 79, and a Standard Deviation of 12.8 years. Fifty five percent were males and 45% were females. Ninety three percent were right handed. The primary presenting symptoms of the population were as follows: 2% Headache, 18% Neck/Shoulder Pain, 58% low back pain, and 22% Pain in other locations. Thirty percent had had surgery on the Low Back one time, and 9% had had surgery on the Low Back two or more times.

**TABLE 6.** NORMATIVE DATA FOR THE J&J WIDE FILTER

| MUSCLE SITE | POSTURE | MEAN | MILD | MODERATE | SEVERE |
|---|---|---|---|---|---|
| FRONTALIS | SIT | 5.0 | 8.1 | 11.2 | 14.3 |
| | STAND | 4.9 | 7.9 | 10.9 | 13.9 |
| ANT. TEMPORALIS | SIT | 6.5 | 12.8 | 18.6 | 24.4 |
| | STAND | 6.5 | 11.3 | 16.1 | 20.9 |
| POST. TEMPORALIS * | SIT | 4.0* | 6.1* | 8.2* | 10.3* |
| | STAND | 4.1* | 5.9* | 10.7* | 9.5* |
| MASSETER | SIT | 3.1 | 5.2 | 7.3 | 9.4 |
| | STAND | 3.0 | 4.8 | 6.6 | 8.4 |
| DIGASTRIC * | SIT | 2.9* | 5.0* | 7.1* | 9.2* |
| | STAND | 3.0* | 4.8* | 6.6* | 8.4* |
| SCM | SIT | 2.7 | 4.7 | 6.7 | 8.7 |
| | STAND | 2.8 | 4.4 | 6.0 | 7.6 |
| C2 PARASPINALS * | SIT | 5.3* | 8.9* | 12.5* | 16.1* |
| | STAND | 5.5* | 8.6* | 11.7* | 14.8* |
| CERVICAL P.S. (C4) | SIT | 4.3 | 6.9 | 9.5 | 12.1 |
| | STAND | 4.5 | 6.7 | 8.9 | 11.1 |
| C6 PARASPINALS * | SIT | 4.5* | 7.5* | 10.5* | 13.5* |
| | STAND | 5.8* | 9.6* | 13.4* | 17.2* |
| TRAPEZIUS | SIT | 5.1 | 10.8 | 16.5 | 22.3 |
| | STAND | 13.3 | 25.6 | 38.1 | 50.6 |
| T1 PARASPINALS | SIT | 4.6 | 7.6 | 10.6 | 13.6 |
| | STAND | 6.8 | 10.0 | 13.2 | 16.4 |
| T2 PARASPINALS * | SIT | 4.6* | 7.6* | 10.6* | 13.6* |
| | STAND | 6.8* | 10.0* | 13.2* | 16.4* |
| T4 PARASPINALS * | SIT | 5.3* | 9.8* | 14.3* | 18.8* |
| | STAND | 6.0* | 10.3* | 14.6* | 18.9* |
| T6 PARASPINALS | SIT | 6.0 | 10.5 | 15.0 | 19.5 |
| | STAND | 5.5 | 9.8 | 14.1 | 18.4 |
| T8 PARASPINALS * | SIT | 4.6* | 8.4* | 12.2* | 16.0* |
| | STAND | 5.6* | 8.5* | 11.4* | 14.3 * |
| T10 PARASPINALS | SIT | 4.5 | 8.2 | 11.9 | 15.6 |
| | STAND | 5.6 | 8.5 | 11.4 | 14.3 |
| T12 PARASPINALS * | SIT | 3.8* | 7.1* | 10.4* | 13.7* |
| | STAND | 5.0* | 8.6* | 12.2* | 15.8* |
| L1 PARASPINALS * | SIT | 3.8* | 7.1 | 10.4* | 13.7* |
| | STAND | 5.0* | 8.6* | 12.2* | 15.8* |
| L3 PARASPINALS | SIT | 2.9 | 5.4 | 7.9 | 10.4 |
| | STAND | 4.8 | 8.4 | 12.0 | 15.6 |
| L5 PARASPINALS * | SIT | 2.7* | 4.9* | 7.1* | 9.3* |
| | STAND | 4.1* | 7.8* | 11.5* | 15.2* |
| ABDOMINALS* | SIT | 2.0* | 3.2* | 4.4* | 5.6* |
| | STAND | 2.7* | 4.3* | 5.9* | 7.5* |

* Represents adjusted filter values

## TEST-RETEST RELIABILITY

In order to establish the utility or validity of a measurement technique the criterion of reliability must first be met (Nunnally, 1967). The reliability of surface EMG procedures have been assessed for various EMG assessment techniques. The reliability of stress profiling of selected muscle sites, for example, has been reported by (Blanchard, et. al. 1980). The reliability of recordings from the erector spinal muscles (Wolf, et. al., 1978; Ahern, et al 1986) and the cervical/thoracic muscles (Spector, 1979) during dynamic movement have also been investigated. Both of these studies conclude that surface EMG may be reliably measured with attached electrodes during dynamic movement.

We conducted a study (Cram, et. al., 1990) on a large clinical population to assess the test-retest reliability of the "static" muscle scanning procedure described above, using the two neutral postures. In addition, we assessed for the effects of gender and age. The study consisted of 104 patients presenting for an initial evaluation in an outpatient pain treatment center in Spokane, Washington. There were 50 males (48%) and 52 females (52%), with a mean age of 37.6 years (SD = 12.69). The symptoms reported by the patient population varied. Essentially, all of the patients were experiencing pain of a chronic rather than acute nature, with a minimum of six months of duration. Sixty percent of the sample reported a single primary symptom: 37% neck pain; 1% head pain; and 22 % low back pain alone. The remainder of the sample (40%) were poly-symptomatic: 10% neck and face pain; 5% neck and upper back pain; 17% neck and low back pain; 3 % upper back and low back pain; and 5% neck, upper back and lower back.

Each patient was seen on three occasions during the same day of evaluation. Each scanning procedure took approximately one-half hour, and was separated by a minimum of one and a maximum of two hours. In between the muscle scanning procedures, the patient was evaluated by other pain clinic staff.

The results of the study indicate that the test-retest reliability coefficients were moderate to high at all 20 muscle sites for both of the "static" postures of sitting and standing. The range of the Pearson correlations was 0.41 to 0.94, with a median correlation of 0.64. This indicated that 50% of the repeated samples shared 64% or greater of the variance. A viewing of Tables 7 & 8 suggests that the rank ordering of the correlations differs with the two postures. In both postures, the facial sites show the highest correlations. In the sitting posture, the upper back tended to have higher correlations compared to those of the lower back. In the standing posture, the lower back tended to have higher correlations compared to the upper back. These postural observation tend to suggest a greater reliability in the upper back while sitting, with a greater reliability in the lower back while standing. The upper back observations might best be explained due to the natural support given the arms by the lap during sitting, thus reducing the need for stabilization of the shoulder by the muscles of the upper back. The increased reliability in the low back region during standing is most likely attributable to the synergistic patterns of muscular support given by the hip stabilizers.

The test-retest study occurred across a single day. One might have expected lower EMG values across the course of the day as a function of habituation and familiarization of the patient with the procedure. The findings of this study do not support this expectation and are similar to those found by Ahern (1986) using attached electrodes while monitoring the erector spinae.

The effects of age and gender were also studied and found to have no impact or reliability. This is not surprising given previous research by Wolf (1978) which found no differences in EMG levels for different gender and age in static postures.

In general, the findings of this study lend further evidence to the robust nature of the muscle scanning procedure. With adequate attention given to skin preparation, EMG sensors held in place by hand with a light pressure provide reliable results.

## PART-WHOLE RELATIONSHIP OF THE SCAN DATA

Because the EMG scanning procedure is intended to give the therapist a more holistic view of the status of the neuromuscular system, the Biofeedback Institute of Seattle in conjunction with Swedish Hospital Pain Center conducted a series of analyses on how each neuromuscular region of the body related to the system as a whole.

**TABLE 7.** CORRELATION BETWEEN THE FIRST TIME PERIOD TO SECOND TIME PERIOD

| Scan Site | Sit L | Sit R | Stand L | Stand R |
|---|---|---|---|---|
| Frontalis | .66 | .70 | .74 | .74 |
| Temporalis | .77 | .75 | .85 | .60 |
| Masseter | .74 | .75 | .80 | .82 |
| SCM | .58 | .76 | .75 | .81 |
| Cervical P.S. | .61 | .61 | .44 | .55 |
| Trapezius | .63 | .67 | .53 | .43 |
| Upper Thoracic (T1) | .53 | .42 | .49 | .45 |
| Mid-Thoracic (T6) | .72 | .51 | .58 | .55 |
| Lower Thoracic (T10) | .51 | .57 | .54 | .64 |
| Lumbar | .57 | .58 | .70 | .71 |

**TABLE 8.** CORRELATION BETWEEN THE FIRST PERIOD TO THE THIRD PERIOD

| Scan Site | Sit L | Sit R | Stand L | Stand R |
|---|---|---|---|---|
| Frontalis | .72 | .61 | .73 | .76 |
| Temporalis | .77 | .56 | .94 | .77 |
| Masseter | .76 | .80 | .76 | .76 |
| SCM | .54 | .70 | .67 | .68 |
| Cervical P.S. | .57 | .76 | .54 | .56 |
| Trapezius | .56 | .59 | .57 | .41 |
| Upper Thoracic (T1) | .50 | .48 | .56 | .61 |
| Mid-Thoracic (T6) | .57 | .65 | .63 | .70 |
| Lower Thoracic (T10) | .48 | .47 | .53 | .47 |
| Lumbar | .61 | .64 | .63 | .71 |

The procedure entailed running a series of correlations on approximately 100 patient samples. Because of this relatively small number of samples, initially the individual scan sites were clustered into regions and then individual site correlations were considered. To do the analysis on activation patterns, we summed the scan data from the right and left aspect of each muscle for each region and correlated these to the grand total of all of the scan data points. To conduct the analysis on asymmetry patterns, the absolute difference between the right and left aspect of each muscle group was summed for each region. These were then correlated to the sum of the absolute differences for all sites. The regions were defined by the following muscle groups:

| | |
|---|---|
| FACE: | Frontalis, Temporalis, and Masseter |
| NECK: | SCM and CPS |
| UPPER BACK: | Trapezius, T1 and T6 Paraspinals |
| LOWER BACK: | T10 and L3 Paraspinals |
| ABDOMINALS: | Abdominals |
| TOTAL EMG: | All scan sites. |

The initial analysis was conducted on the standing posture only and the results are presented in the next two tables:

**TABLE 9.** CORRELATIONS BETWEEN EACH REGION AND THE TOTAL EMG: ACTIVATION PATTERNS (STANDING POSTURE)

| REGION | TOTAL EMG | (RANK) | TOTAL ASYMMETRY | (RANK) |
|---|---|---|---|---|
| FACE | +.46 | 3 | +.46 | 3 |
| NECK | +.45 | 4 | +.45 | 4 |
| UBACK | +.84 | 1 | +.59 | 1 |
| LBACK | +.66 | 2 | +.54 | 2 |
| ABDOM | +.18 | 5 | -.05 | 5 |

Values are significant at .20

**TABLE 10.** CORRELATIONS BETWEEN EACH REGION AND THE TOTAL EMG: ASYMMETRICAL PATTERNS (STANDING)

| REGION | TOTAL EMG | (RANK) | TOTAL ASYM-METRY | (RANK) |
|---|---|---|---|---|
| FACE | +.37 | 4 | +.59 | 3 |
| NECK | +.41 | 3 | +.42 | 4 |
| UBACK | +.60 | 1 | +.71 | 1 |
| LBACK | +.48 | 2 | +.67 | 2 |
| ABDOM | +.24 | 5 | +.15 | 5 |

Values reach significance at .20

From these preliminary analyses, one can begin to develop a framework of how each neuromuscular region influences the system as a whole. The above data strongly suggest that the neuromuscular activity in the upper back tends to correlate most strongly with the total EMG and the total amount of asymmetry in the system. It is not possible, however, to determine if excessive tension or asymmetry in the upper back causes more EMG activity in the system as a whole or if it merely reflects the amount of tension or asymmetry in the system as a whole.

An additional analysis (Cahn, et. al., 1990) was conducted for each site and posture to see if we could better identify a "key muscle" (Alexander and Smith, 1979) for activation patterns. The results are presented in the next table:

**TABLE 11.** CORRELATIONS BETWEEN INDIVIDUAL MUSCLE GROUPS AND TOTAL EMG ACROSS POSTURE ("THE KEY MUSCLE HYPOTHESIS")

| MUSCLE GROUP | SIT | RANK | STAND | RANK |
|---|---|---|---|---|
| FRONTALIS | .49* | 9 | .25* | 9 |
| TEMPORALIS | .45* | 11 | .35* | 7 |
| MASSETER | .46* | 10 | .05 | 11 |
| SCM | .57* | 5 | .20 | 10 |
| CERVICAL | .50* | 7 | .40* | 6 |
| TRAPEZIUS | .49* | 8 | .52* | 5 |
| T1 | .63* | 3 | .59* | 3 |
| T6 | .54* | 6 | .59* | 2 |
| T10 | .72* | 1 | .73* | 1 |
| L3 | .65* | 2 | .54* | 4 |
| ABDOMINALS | .57* | 4 | .34* | 8 |

* p< .01 2Q

In this finer analysis, it was the T10 paraspinals group which correlates most heavily with the TOTAL EMG for both the sitting and standing postures. In the past, the Frontalis muscle group would have been considered the best indicator of general neuromuscular arousal (the "key muscle hypothesis"). This data tends to strongly suggest that the key muscle belongs in the back, not the face. And, it indicated that the site where the spine is finally able to move in all planes is the most sensitive barometer of general tension.

The next analysis looked at the relationship of each region to every other region, both for activation and asymmetry. The data were clustered and analyzed using the same method described above. The results of this analysis are presented in the table below.

## TABLE 12. CORRELATIONS BETWEEN ACTIVATION AND ASYMMETRY FOR EACH REGION IN THE STANDING POSTURE

| | ACTIVATION | | | | | : ASYMMETRY | | | |
|---|---|---|---|---|---|---|---|---|---|
| REGION | FACE | NECK | UBACK | LBACK | ABDOM | FACE | NECK | UBACK | LBACK |
| A | | | | | | | | | |
| C | | | | | | | | | |
| T NECK | +.24 | | | | | | | | |
| I | | | | | | | | | |
| V UBACK | +.13 | +.29 | | | | | | | |
| A | | | | | | | | | |
| T LBACK | +.14 | +.26 | +.30 | | | | | | |
| I | | | | | | | | | |
| O ABDOM | -.10 | -.12 | +.15 | +.13 | | | | | |
| N | | | | | | | | | |
| : | | | | | | | | | |
| A FACE | +.75 | +.18 | +.14 | +.16 | +.10 | | | | |
| S | | | | | | | | | |
| Y NECK | +.30 | +.62 | +.33 | +.17 | -.04 | +.18 | | | |
| M | | | | | | | | | |
| M UBACK | +.07 | +.24 | +.71 | +.24 | +.01 | +.09 | +.25 | | |
| E | | | | | | | | | |
| T LBACK | +.17 | +.38 | +.20 | +.68 | -.11 | +.22 | +.20 | +.20 | |
| R | | | | | | | | | |
| Y ABDOM | -.03 | -.00 | +.19 | +.15 | +.66 | +.03 | +.11 | +.11 | +.02 |

SIGNIFICANCE AT .01 LEVEL OCCURS AT AN R OF 20

From this analysis very strong, robust correlations may be noted between the level of activation and the level of asymmetry for each region. It appears that the more energy in the neuromuscular region, the greater the probability of asymmetry. Of course, the opposite could also be argued as well. What these findings do suggest, is that one could approach reducing asymmetries by lowering the general levels of activation. Or one could approach lowering the general levels of activation by working on better symmetry.

**TABLE 13.** CORRELATIONS BETWEEN EACH REGION AND THE TOTAL EMG FOR THE SITTING POSTURE

| REGION | FACE | NECK | UBACK | LBACK |
|---|---|---|---|---|
| NECK | +.25* | | | |
| UBACK | +.10 | +.40* | | |
| LBACK | +.25* | +.33* | +.56* | |
| ABDOM | +.03 | +.39* | +.33* | +.29* |

* $p < .01$ 2Q

When one compares the activation patterns alone for the sitting and standing postures, very strong relationships may be noted between the lower and upper back, the neck and the upper back, and the neck and the lower back. These relationships are there to remind us of how well orchestrated the muscles of the back are.

## PATIENT PROTOCOLS

In order to better appreciate the idiographic or individual nature of the scan data base, a series of patient protocols are presented. Two of the more common disorders, low back pain and headache are presented. Study each protocol, compare it to the normative data base and begin to develop a sense for the clinical concepts of SITE of activation, degree of ASYMMETRY and the impact of POSTURE.

## PATIENT PROTOCOL: LOW BACK PAIN

Below is an example of a scan protocol collected using the J&J narrow filter on a low back pain patient. His pain was localized to the lumbar region and he has had no surgeries. His diagnosis was that of Lumbar Strain. He has had a painful back for some seven months and was currently unable to work at his usual occupation. He appeared to be moderately well adjusted from a psychological point of view.

**TABLE 14.** LOW BACK PAIN PATIENT PROTOCOL

| MUSCLE SITE | SIT | | STAND | |
|---|---|---|---|---|
| | LEFT | RIGHT | LEFT | STAND |
| FRONTALIS | 3.5 | 3.8 | 1.2 | 1.3 |
| TEMPORALIS | 3.5 | 3.5 | 4.6 | 3.6 |
| MASSETER | 1.3 | 1.5 | 1.5 | 1.6 |
| SCM | 1.2 | 1.5 | 0.8 | 0.9 |
| CERVICAL P.S. | 1.6 | 1.7 | 1.9 | 1.9 |
| TRAPEZIUS | 2.2 | 2.3 | 3.0 | 2.5 |
| T1 PARASPINALS | 1.5 | 1.5 | 1.3 | 1.1 |
| T6 PARASPINALS | 3.7 | 3.5 | 4.7 | 4.4 |
| T10 PARASPINALS | 9.6* | 19.2* | 44.8* | 59.2* |
| L3 PARASPINALS | 2.2 | 11.2* | 19.2* | 30.4* |
| ABDOMINALS | 1.3 | 1.5 | 1.6 | 1.8 |

INTERPRETATION: MODERATE TO RADICAL ASYMMETRICAL ACTIVATIONS ARE NOTED IN THE T10 AND L3 PARASPINAL MUSCLES. A PROTECTIVE GUARDING PATTERN IS NOTED WITH THE RIGHT ASPECT BEING GREATER THAN THE LEFT. A STRONG POSTURAL COMPONENT IS PRESENT.

The starred (*) items represent levels of activation which fall outside of the "normal range". As can easily be seen, these are localized to the T10 and L3 sites. There is an asymmetrical pattern of muscle activity noted at the T10 and L3 levels. An interpretation of asymmetry can be made with confidence when one side is one standard deviation greater than the other side. Also note the impact of posture on the levels of splinting and guarding patterns at the T10 and L3 levels. As the patient bares weight, the patterns of dysfunction become more dramatic. In this patient's case, one would want to comment upon a postural disturbance.

Given the above interpretation, further diagnostic workup would be done with attached electrodes placed at either or both the T10 or L3 paraspinal muscles, with the right and left aspect being monitored independently (dual channel). A dynamic postural evaluation would be conducted to confirm the findings from the scan and to further explain the nature of the disturbance. Given that the additional workup supports the findings of the scan, treatment would be directed at either the T10 or L3 sites using a posturally based model.

## PATIENT PROTOCOL: LOW BACK PAIN WITH NORMAL SCAN

Not every patient evaluated using this procedure will show abnormal findings in a static posture. The table below presents a protocol collected using the J&J narrow filter on a 32 year old female with a two year history of low back pain and disability. She had been through multiple "conservative" therapies without successful control of either her pain or her pain behaviors.

**TABLE 15.** LOW BACK PAIN PATIENT WITH NORMAL SCAN

| MUSCLE SITE | SIT | | STAND | |
|---|---|---|---|---|
| | LEFT | RIGHT | LEFT | RIGHT |
| FRONTALIS | 2.5 | 3.2 | 1.1 | 1.1 |
| TEMPORALIS | 1.5 | 1.5 | 1.6 | 1.6 |
| MASSETER | 1.0 | 1.5 | 0.9 | 1.1 |
| SCM | 1.1 | 1.5 | 1.8 | 1.9 |
| CERVICAL P.S. | 2.6 | 1.9 | 2.9 | 2.6 |
| TRAPEZIUS | 1.4 | 1.3 | 2.0 | 2.5 |
| T1 PARASPINALS | 1.9 | 1.8 | 1.6 | 1.5 |
| T6 PARASPINALS | 3.4 | 3.2 | 2.7 | 2.4 |
| T10 PARASPINALS | 3.6 | 3.2 | 3.8 | 3.2 |
| L3 PARASPINALS | 2.2 | 1.2 | 3.2 | 3.4 |
| ABDOMINALS | 1.0 | 1.0 | 0.6 | 0.8 |

IMPRESSIONS: EVALUATION CONDUCTED DURING STATIC POSTURES IS WITHIN NORMAL LIMITS. FURTHER ASSESSMENT USING DYNAMIC PROCEDURES SHOULD BE CONSIDERED TO RULE OUT OTHER POSSIBLE NEUROMUSCULAR CONTRIBUTIONS TO PAIN.

The findings from this study are all within normal limits. There are no sites with excessive activation, there is very little in the way of asymmetries and there are no postural abnormalities. Even when this patient was evaluated using the dynamic assessment to be described later, the findings of the evaluation were well within normal limits. Given these types of findings, there is very little that the therapist can offer the patient in the way of biofeedback training. Essentially, there is nothing to work with. As it turns out in this patient's case, there were significant family dynamics operating in maintaining both the patient's level of pain and disability. Pain therapy was directed towards this arena, rather than expended on biofeedback therapy.

## PATIENT PROTOCOL: HEADACHE

The following three protocols provide examples of utilizing an abbreviated scan to assess the neuromuscular contributions to headache. All three of the headaches are described by the patient as being dull in nature and located across the frontal region. In reviewing the scan data base on these patients, the first headache appears to have its origin in the frontal muscles, the second headache appears to have its origin in the cervical or trapezius muscles, and the third headache demonstrates little, if no, neuromuscular contribution to the headache. The findings from the second headache represents a case of "referred pain", pain that originates in one region and is experienced in another region.

**TABLE 16.** HEADACHE OF FRONTAL ORIGINS

| MUSCLE SITE | SIT | | STAND | |
|---|---|---|---|---|
| | LEFT | RIGHT | LEFT | RIGHT |
| FRONTALIS | 7.3* | 9.2* | 8.2* | 9.4* |
| TEMPORALIS | 6.0* | 6.4* | 6.1* | 6.8* |
| MASSETER | 1.8 | 1.6 | 1.6 | 1.8 |
| SCM | 1.0 | 1.1 | 1.1 | 1.1 |
| CERVICAL P.S. | 1.9 | 1.6 | 2.2 | 2.3 |
| TRAPEZIUS | 0.8 | 0.9 | 0.7 | 0.9 |
| T1 PARASPINALS | 1.9 | 1.8 | 1.9 | 1.9 |

IMPRESSIONS: PATIENT DEMONSTRATES MODERATE TO RADICAL ACTIVATION OF FRONTALIS MUSCULATURE. MUSCLE CONTRACTION HEADACHE SUGGESTED.

**TABLE 17.** HEADACHE OF CERVICAL OR TRAPEZIUS ORIGINS

| MUSCLE SITE | SIT | | STAND | |
|---|---|---|---|---|
| | LEFT | RIGHT | LEFT | RIGHT |
| FRONTALIS | 3.1 | 3.0 | 3.2 | 3.0 |
| TEMPORALIS | 2.8 | 3.2 | 3.0 | 3.3 |
| MASSETER | 0.9 | 0.8 | 0.9 | 0.9 |
| SCM | 0.6 | 0.5 | 0.6 | 0.6 |
| CERVICAL P.S. | 7.7 | 2.7 | 9.6 | 3.3 |
| TRAPEZIUS | 10.7* | 3.0 | 14.0* | 8.0 |
| T1 PARASPINALS | 4.9 | 3.8 | 5.6 | 4.2 |
| T6 PARASPINALS | 5.5 | 4.7 | 5.2 | 4.3 |

IMPRESSIONS: RADICAL ASYMMETRICAL PATTERNS OF ACTIVATION IN THE CERVICAL PARASPINALS AND TRAPEZIUS MUSCLE GROUPS ARE NOTED. LEFT ASPECT GREATER THAN RIGHT. REFERRED PAIN FOR FRONTAL HEADACHE SUGGESTED. TRIGGER POINT LOCATED IN LEFT TRAPEZIUS.

A "referred pain" indicates that the headache pain located in the frontal region can be stimulated by firmly pressing on a "nodule" in the left trapezius region. This patient tends to fall more clearly into the "specificity model" of diagnostic frameworks.

Medical treatment of a trigger point might involve dry needling or injection of the trigger point with an anesthetic. Massage or physical therapy treatment of a trigger point might involve manual/mechanical stimulation of the trigger point. A biofeedback therapist might approach the treatment of the trigger point by recommending mobilization exercises for the shoulders, and using EMG biofeedback to teach the patient to release the excessive muscle activity while the muscle is gently placed on stretch.

**PATIENT PROTOCOL: HEADACHE WITH NORMAL SCAN**

It should be noted that the evaluation of the headache patient is not complete, having conducted only an EMG scanning procedure. In some instances, the scan data will be within normal limits for both the narrow and wide filters, and the dynamic assessment for the head and neck regions or stress profiling of frontalis may be very abnormal.

This is entirely possible given the very different nature of these assessment procedures and the pathophysiology which they assess. Please refer to the other assessment sections in this manual.

**TABLE 18.** HEADACHE: NORMAL SCAN

| MUSCLE SITE | SIT | | STAND | |
|---|---|---|---|---|
| | LEFT | RIGHT | LEFT | RIGHT |
| FRONTALIS | 2.3 | 2.2 | 1.8 | 2.0 |
| TEMPORALIS | 2.0 | 2.2 | 2.1 | 2.5 |
| MASSETER | 1.6 | 1.5 | 1.4 | 1.5 |
| SCM | 1.0 | 1.0 | 1.4 | 1.2 |
| CERVICAL P.S. | 1.7 | 1.3 | 1.2 | 1.3 |
| TRAPEZIUS | 0.9 | 0.8 | 0.9 | 0.9 |
| T1 PARASPINALS | 1.2 | 1.1 | 1.2 | 1.2 |

IMPRESSION: STATIC EVALUATION OF NEUROMUSCULAR SYSTEM IS WITHIN NORMAL LIMITS. FURTHER EVALUATION USING EITHER A DYNAMIC ASSESSMENT OF THE NECK AND SHOULDER MUSCLES OR STRESS PROFILING SHOULD BE CONSIDERED TO RULE OUT OTHER POSSIBLE NEUROMUSCULAR CONTRIBUTIONS TO HEADACHE.

## PATIENT PROTOCOL: HEAD, NECK AND BACK PAIN

Most patients are not mono-symptomatic. Instead, they present with a variety of pain complaints. As was noted earlier, the longer pain has remained a problem the more likely it is to migrate from one area to another. Thus, what started out to be a solitary mono-symptomatic problem may indeed blossom into multiple symptoms.

The following patient represents this type of patient. She is 28 year old female who had received a flexion-extension injury from a motor vehicle accident some three years prior. She has received conservative therapy on a yearly basis, improving somewhat during each therapy. However, once the therapy stopped she began to lose some of her gains. This was described as something like taking two steps forwards and one step backwards. Slow gradual improvement. She continued to use the slow, long stretches during long hot showers, as she had learned to do over the course of treatment. Her current physician has been using manipulation therapy and gentle reassurance. She notes reduction in pain immediately following one of his manipulations, with a gradual return of the symptoms over the next two days. He is at a loss as to what else he might do to help her. She is referred for consultation.

The results of her study are presented in two forms: A Scan Summary Report, and two Graphic descriptions. Each of these has been created by a computer based muscle scanning program. This program systematically allows for input of demographic information, the collection of data, interpretation of the patient data by comparison to the normative data presented earlier in this volume, and finally the output of the findings in both analog and digital form.

## TABLE 19. SCAN SUMMARY REPORT

CLINICAL RESOURCES

M U S C L E   S C A N   S U M M A R Y

File: File Date: 04/29/90 Identification Number: 535-74-0333

Name: JUDY Q. PATIENT Age: 28 Sex: FEMALE

Height: 5'7" Weight: 118 Handedness: LEFT Dentures: NO

Surgery Code: 0 Surgery: None

Diagnosis Code: MIXED Diagnosis: NECK, UPPER BACK, HEADACHE

Comment:

The patient has the above symptoms on a daily basis. The severity follows the rank order seen above. The range of pain can go from 2 to 5 on a 5 point scale. It is made worse with activity, and helped by rest. Symptoms have been present since the mva of 5/87. They have improved generally over time with conservative care, but they continue to adversely effect her lifestyle.

---

Using: ADJUSTED NARROW FILTER File:

| Site | Sit L | Sit R | Stand L | Stand R |
|---|---|---|---|---|
| Frontalis | 2.7 | 1.4 | 4.7 * | 5.1 * |
| A Temporalis | 5.3 * | 4.6 * | 5.2 * | 2.8 |
| Masseter | 3.7 * | 3.5 * | 5.1 ** | 2.4 |
| S.C.M. | 4.0 * | 5.0 ** | 6.4 *** | 2.5 |
| C.P.S. | 9.3 *** | 5.7 *** | 5.2 * | 4.3 * |
| Trapezius | 4.6 | 3.1 | 7.4 * | 4.1 |
| T1 P.S. | 7.8 ** | 7.2 ** | 5.9 * | 6.1 * |
| T6 P.S. | 15.2 *** | 17.9 *** | 4.3 | 2.9 |
| T10 P.S. | 19.5 *** | 12.3 *** | 14.1 *** | 5.6 |
| L3 P.S. | 10.9 *** | 5.5 * | 19.1 *** | 9.5 * |

| --ACTIVATION-- | |
|---|---|
| mild | * |
| moderate | ** |
| radical | *** |

Interpretation:

The * sites were observed to have mild to severe elevations indicative of muscle spasm. Abnormal levels of asymmetry were noted at the A Temporalis, Masseter, S.C.M., C.P.S., Trapezius, T6 P.S., T10 P.S., and L3 P.S. sites, suggesting chronic protective guarding patterns associated with pain. Postural disturbances are noted at the Frontalis, S.C.M., Trapezius, T6 P.S., T10 P.S., and L3 P.S. sites, suggesting that the muscle spasm is easily detected both with and without axial loading. The findings of the study clearly indicate that further treatment is necessary. EMG biofeedback using both relaxation and functional/postural training would be suggested within the context of a more comprehensive therapy Rx'd by the attending physician.

JEFFREY R. CRAM, PH.D.

FIGURE 14. GRAPHIC DISPLAY OF SCAN RESULTS

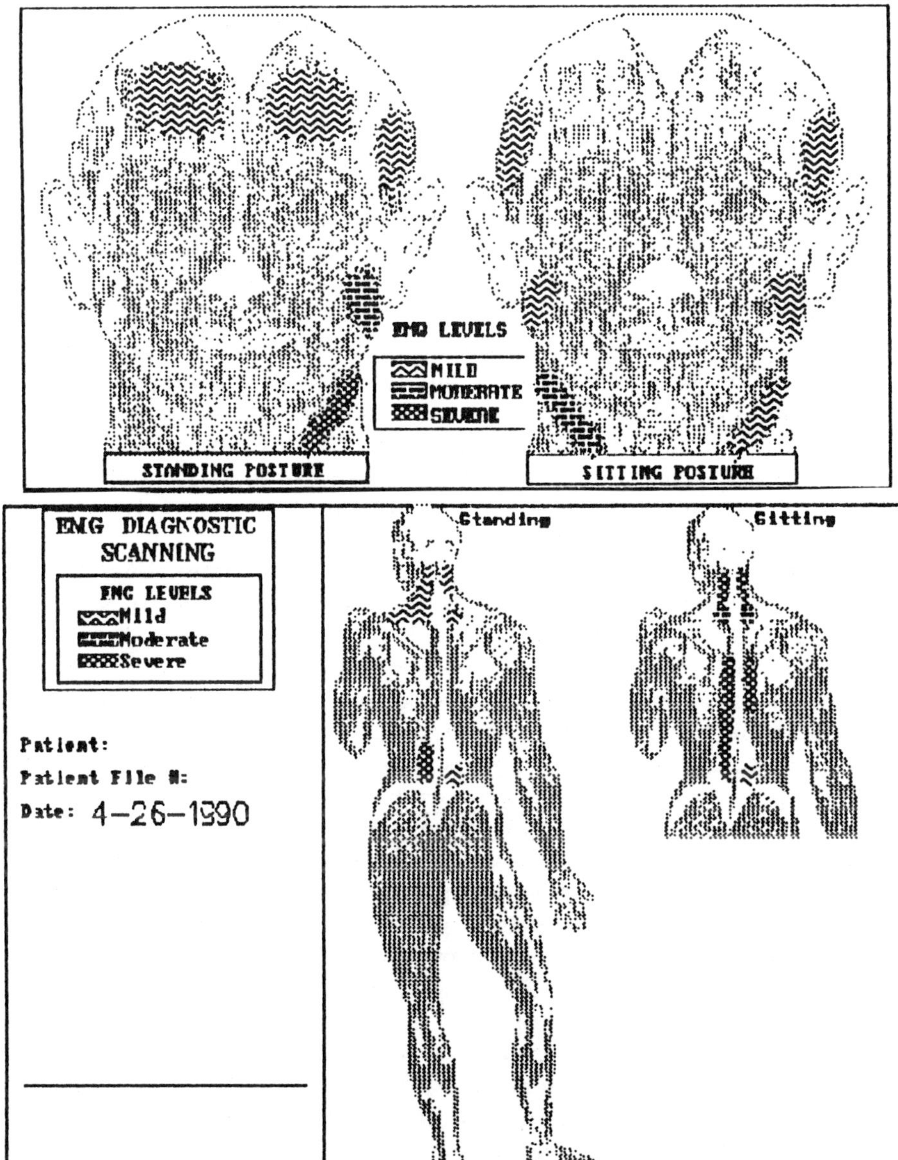

The study was congruent with the patient's self report of discomfort. The "left" sided dominance of the findings is quite striking for both the face and back. As is common with many patients, these findings were the first she had encountered which seemed to verify her symptoms. She felt a sense of relief at "being believed," and took on the treatment recommendations with a firm commitment.

## STUDIES ON THE SPECIFICITY OF MUSCLE SCANNING

The issue of specificity refers to the ability of a test to discriminate between different clinical populations. The first study on EMG muscle scanning (Cram and Steger, 1983) investigated the pattern of EMG scan data for a headache and low back pain population. At two pain clinics in Seattle, a full EMG scanning protocol was conducted on each patient at the time of their initial evaluation. The headache and low back pain patients were categorized and statistically compared. The results of the study indicated that headache patients, compared to low back pain patients, tended to show excessive levels of EMG activation in the frontal and masseter muscle groups. The low back pain patients, on the other hand, tended to show asymmetrical patterns in the erector spinae and cervical muscle groups. This involvement of both the lumbar and cervical groups was interpreted to indicate a strong postural component for the low back pain patients. In addition, it supports the notion of a holistic view of the impact of back pain on the neuromuscular system. Many years ago, Price, et. al., (1948) made the observation that pain may migrate from one area to another.

The second specificity study (Cram and Engstrom, 1986) compared the normal population, referred to in the normative data section of this chapter, to the large chronic pain population at the Swedish Hospital Pain Center. The chronic pain population (N = 200) presented with low back pain as their primary complaint (69%), neck or shoulder as their primary complaint (21%), with the remaining patients (10%), having other primary sites of pain. The mean age of the population was 40.5 years, 51% were female and 49% were males.

Below, Table 19 represents the comparisons of these two groups. As you glance through this table, please note that significant differences emerged at some sites and not others, in some postures and not others, and on some sides but not others. Pay particular attention to the trapezius, T6 and L3 sites.

## TABLE 20. COMPARISON OF STATIC EMG FOR A NORMAL POPULATION (N = 104) AND PAIN POPULATION (N = 204)

| SITE | SIDE | POSTURE | PAIN $\bar{X}$ | PAIN (SD) | NON-PAIN $\bar{X}$ | NON-PAIN (SD) | SIGNIF. |
|---|---|---|---|---|---|---|---|
| FRONTALIS | LEFT | SIT | 2.6 | (2.4) | 2.0 | (2.1) | |
| | RIGHT | | 2.5 | (2.4) | 1.8 | (1.9) | * |
| | L | STAND | 2.4 | (1.7) | 2.0 | (2.0) | |
| | R | | 2.4 | (1.6) | 2.1 | (2.0) | |
| TEMPORALIS | L | SIT | 3.8 | (4.5) | 2.4 | (2.0) | |
| | R | | 3.7 | (3.7) | 2.4 | (2.2) | * |
| | L | STAND | 3.4 | (4.0) | 2.5 | (2.1) | * |
| | R | | 3.3 | (3.4) | 2.1 | (2.0) | * |
| MASSETER | L | SIT | 2.1 | (2.5) | 1.7 | (1.4) | |
| | R | | 2.0 | (1.8) | 1.8 | (1.2) | |
| | L | STAND | 1.8 | (1.6) | 1.4 | (1.8) | |
| | R | | 1.7 | (1.9) | 1.7 | (1.3) | |
| SCM | L | SIT | 1.5 | (1.7) | 1.3 | (1.2) | |
| | R | | 1.3 | (1.3) | 1.3 | (1.9) | |
| | L | STAND | 1.3 | (1.1) | 1.3 | (1.7) | |
| | R | | 1.4 | (1.4) | 1.6 | (1.6) | |
| CPS (C4) | L | SIT | 2.8 | (1.5) | 1.9 | (1.2) | * |
| | R | | 2.6 | (1.5) | 1.9 | (1.1) | * |
| | L | STAND | 2.3 | (1.8) | 1.8 | (2.0) | |
| | R | | 2.1 | (1.1) | 1.8 | (1.9) | |
| TRAPEZIUS | L | SIT | 2.7 | (3.8) | 2.2 | (2.5) | |
| | R | | 3.1 | (5.3) | 2.3 | (2.8) | |
| | L | STAND | 5.3 | (5.5) | 3.1 | (2.7) | * |
| | R | | 4.6 | (5.4) | 3.3 | (2.9) | * |
| T1 | L | SIT | 2.7 | (1.7) | 2.4 | (2.1) | |
| | R | | 2.8 | (1.8) | 2.2 | (1.7) | * |
| | L | STAND | 3.5 | (3.5) | 2.7 | (2.7) | * |
| | R | | 3.3 | (3.1) | 3.1 | (3.0) | |
| T6 | L | SIT | 3.9 | (3.7) | 2.6 | (2.9) | * |
| | R | | 3.4 | (3.7) | 2.5 | (2.6) | * |
| | L | STAND | 3.0 | (3.9) | 2.2 | (2.9) | |
| | R | | 3.1 | (4.0) | 2.3 | (3.0) | |
| T10 | L | SIT | 3.3 | (3.2) | 2.4 | (2.9) | * |
| | R | | 2.6 | (2.7) | 2.1 | (2.8) | |
| | L | STAND | 3.9 | (4.2) | 2.9 | (3.1) | |
| | R | | 3.3 | (3.7) | 3.2 | (3.1) | |
| L3 | L | SIT | 2.4 | (4.3) | 2.1 | (2.7) | |
| | R | | 2.2 | (3.0) | 1.8 | (2.6) | |
| | L | STAND | 4.7 | (5.4) | 3.3 | (3.6) | * |
| | R | | 4.0 | (4.2) | 3.3 | (3.2) | |

* $p < .01$

As can be seen from this table, some muscle sites are more specific than others. In the facial muscles, for example, temporalis seems to be the muscle which separates the two groups the best. Notice the effects of posture on trapezius. It does not separate the groups in the sitting posture, but does in the standing posture. Lastly note the effects of posture and symmetry at the L3 site. Here, a significant separation is seen only in the standing posture and only on the left aspect. The significance of this asymmetry will be discussed later.

## A STUDY ON THE SENSITIVITY OF MUSCLE SCANNING

Another study (Cram and Freeman, 1985) was conducted to assess the sensitivity of the scanning procedure to treatment effects. Changes in the resting or tonic EMG levels detected with the scanning electrodes were evaluated for three different EMG biofeedback treatment groups: 1) Relaxation based frontal EMG feedback training; 2) Posturally based EMG feedback for treatment of trapezius; and 3) Posturally based EMG feedback treatment for L3 paraspinals.

Patients at the Swedish Pain Center were scanned on their first day of admission and then again at the end of a seven week pain therapy program. Treatment was offered sometime during this seven weeks. The major findings of this study are presented in Table 20. The three treatment conditions are presented. A "+" represents a significant reduction of EMG activity from pre to post treatment. A "-" represents no significant change in EMG activity.

**TABLE 21.** RESULTS OF THE OUTCOME STUDY

| MUSCLE SITE | FRONTAL | | (N=12) | | TRAPEZIUS | | (N=25) | | LUMBAR | | (N=24) | |
|---|---|---|---|---|---|---|---|---|---|---|---|---|
| | SIT | | STAND | | SIT | | STAND | | SIT | | STAND | |
| | LT | RT | LT | RT | LT | RT | LT | RT | LT | RT | LT | RT |
| FRONTALIS | + | + | + | + | - | - | - | - | - | - | - | - |
| TEMPORALIS | + | + | - | - | - | - | - | - | - | - | - | - |
| MASSETER | - | - | - | - | - | - | - | - | - | - | - | + |
| SCM | - | - | - | - | - | - | - | - | - | - | - | - |
| CERVICAL | - | + | - | - | - | - | - | - | + | - | - | - |
| TRAPEZIUS | - | - | - | - | - | + | + | - | + | - | + | - |
| T1 P. S. | - | - | - | - | - | - | + | - | + | - | - | - |
| T6 P. S. | - | - | - | + | - | - | - | - | - | - | - | - |
| T10 P.S. | - | - | - | - | - | - | - | - | - | - | + | - |
| L3 P. S. | - | - | - | - | - | - | + | - | - | - | + | + |
| ABDOMINALS | - | - | - | - | - | - | - | - | - | - | - | - |

The findings of this study clearly demonstrated that the EMG scanning protocol was sensitive to treatment effects. The frontally trained group, received EMG feedback from primarily the muscles of the upper face. The significant reductions in EMG activity were noted only in the upper face (frontalis and temporalis). In addition, these reductions appeared to be locked into the sitting posture, the posture utilized during training. From my point of view, this is an elegant description of how exquisitely sensitive and specific the CNS is to the information it receives. Similar results are found for the trapezius trained group. Again, significant reductions of EMG activity were primarily isolated to the area given feedback: the upper back (trapezius and T1 paraspinals). This was also primarily seen in the standing posture, the posture in which they were trained. The outcome for these two groups strongly suggest a high level of specificity, both for the site of treatment as well as the posture in which they were trained. For the group that was trained at L3, significant reductions in EMG were noted in the lower back, upper back, neck and jaw. This outcome suggests that posturally based feedback directed to the "base of the back" appears to telegraph up to higher regions due to biomechanics.

In addition to demonstrating the sensitivity of the scanning protocol, this study clearly states that if one is interested in promoting generalization of treatment effects, one should direct EMG biofeedback to multiple sites and in a variety of postures.

## SECTION 5. INTRODUCTION TO THE SECOND TIER OF EVALUATION AND ANALYSIS

The scanning procedure is considered the first stage of the evaluation. The procedure and its data base provide you with essential information concerning dysponesis, splinting/guarding patterns, asymmetries, postural components and potential emotional overlays. The bottom line of the study, however, suggests muscle groups or regions which require further clinical investigation.

This second stage of investigation is intended to provide you with another level of information concerning the organization (or disorganization) of the neuromuscular system. The major competing "theories" or diagnoses which you will want to rule in or out include: Emotional Reactivity, Postural Disturbance, and Muscle Shortness and Trigger Points.

The emotional reactivity component primarily entails cortical and limbic system input to the neuromuscular system. It may represent the conversion of emotional conflict into dysfunctional neuromuscular patterns. This is one element of Whatmore's (1974) dysponesis. If these patterns may find their way into a weakened or injured muscle group they would tend to fit the "diathesis stress" model presented earlier. Under either model, the emotional display system has run amuck. It may represent an over active or overly dramatic emotional display or it may concern the patient's fear or anxiety about pain. A patient I am currently working with, for example, has extreme difficulty identifying or tolerating anger at a significant other. As he thinks or talks about this individual, dramatic anger displays appear on his shoulders which eventually lead to sore trapezius muscles and headaches.

The postural disturbance component interacts with all other components but is primarily controlled by the cerebellum, basal ganglia, and spinal cord. As we grow, develop and learn new motor acts (such as standing, walking, sitting, skiing, playing tennis, etc.) we essentially learn to integrate our reflexes. We soon develop habitual patterns of integration of these reflexes, and learn to take on certain postures upon which we conduct the finer movements of an action. The manner in which I sit in front of the keyboard as I write this manuscript is done in an habitual fashion, on automatic pilot. Poor or inappropriate posture may lead to excessive strain on the muscles, tendons, ligaments and bones. This, in turn, may lead to a painful muscle or irritate a tendon, ligament or nerve. It is estimated that 80% of back pain is due to poor or inappropriate posture. Therefore, assessing the patterns of neuromuscular activity in a variety of postures can provide a clue as to this component of a pain related complaint.

A second aspect of the potential postural disturbance results from pain itself. When an area is injured, the postural muscle surrounding the injured site clamps down to provide a splint and reduce the range of motion. This is called "splinting" and reduces the probability of further injury to the tissue by reducing movement in the area. All of this happens reflexively, outside of your own awareness and perhaps, control. In concert with this, the neuromuscular system wants us to move away from the pain. We shift our posture, our weight, away from the injured area. This is why we commonly see the side opposite from the injury being more active (higher levels of EMG) during the scanning procedure. The postural evaluation allows us to get a better handle on these phenomena.

The component of resting muscle length interacts primarily with the postural component but may grow out of a very strong emotional component. Muscles have an exquisitely sensitive and well developed sensory system which is highly integrated with reflexive servo-mechanisms. One major sensory field is that of the "stretch receptor". This receptor is activated whenever the muscle is forced to be longer than its resting length or its projected contracted length. When the stretch receptor is activated its sensory information loops through the spinal cord, which sends a command back to the originating muscle to contract. The most common example of this occurs when the physician tests our reflexes by tapping the patella tendon which crosses our knee. In response to the momentary lengthening of our quadriceps (femoris) muscle due to the tap, the quads contract sending our foot in the forward direction. In general, whenever a muscle is stretched it becomes activated producing more EMG activity.

As mentioned earlier in this manuscript, a muscle may take on a shortened resting length due to excessive use or lack of use. A chronic pain patient may develop a shortened muscle due to prolonged "splinting" or "protective guarding". Protective guarding encompasses the cortical and limbic systems, in that there appears to be some higher level "intention" not to move the painful area. Once a muscle has a shortened resting length, it is more likely to be placed on stretch and have a disorganized movement pattern. This is particularly true, and most easily seen, when only one side of a muscle group has a shortened resting length. When assessing for this component, one asks the patient to systematically engage in movement patterns which place the muscle on stretch. If one side is "shortened", the therapist will see an "irritability" or over activation of the EMG on the shortened side. Assessing for this component assists the therapist in determining the type of treatment regimen he or she might want to consider.

The sections which follow suggest systematic ways in which the therapist might want to investigate and rule in or out these various components. As the nature of the neuromuscular dysfunction is clarified, the treatment options should follow.

## SECTION 6. THE CLINICAL PROCEDURE OF PSYCHOPHYSIOLOGICAL STRESS PROFILING

Psychophysiological or stress profiling has been the work horse of diagnostic protocols for many years. Because of this, there has been quite a bit written about this procedure and the reader should familiarize himself with this literature to obtain more depth on the subject. For this manuscript, a brief description of the procedure and its interpretation will be given.

## THE ISSUE OF SITE SELECTION

In the past, diagnostic studies have concentrated primarily on the frontalis muscle site. The diagnostic frameworks outlined in this chapter have strongly stated that site selection should vary depending upon the patients' presenting problem and the findings of the initial muscle scan protocol. When assessing the general model of emotional arousal with stress profiling, the electrode configuration for the selected site is usually bilateral in nature. Here, the active electrodes are placed over the right and left aspect of the belly of the muscle group of interest, with the ground electrode in the middle. With the more specific models of emotional arousal, one would independently monitor the right and left aspect of a muscle group using a dual channel EMG. The electrodes would be placed such that they run parallel to the muscle fibers of the muscle of interest. With computer driven graphics displays, the therapist's ability to discover patterns of emotional reactivity is greatly enhanced.

If your EMG amplifier has a filter selection, consider using the wide filter. I recommend that you move to the narrow filter only if there is too much noise or artifact observed on the wide filter selection.

## THE STRESS PROFILING PROCEDURE

This procedure is usually conducted in either the sitting or reclining postures. Once the patient has accustomed himself to that posture, he is asked to engage in a series of activities.

Below is a sequence of events commonly used in clinical practice:

---

RESTING BASELINE: Here, the therapist asks the patient to "sit quietly" for a few minutes (2-5 minutes is usually ample).

---

MENTAL STRESSOR: The therapist asks the patient to engage in a mildly stressful cognitive event. This is commonly a serial subtraction task, such as counting backwards from 1000 by sevens as quickly and accurately as possible. It is my impression that this task needs to be done out loud with some periodic "evaluative or critical comments" from the therapist. This adds an additional social stressor into the task and provides a more meaningful response pattern to examine.

---

RECOVERY: Next the therapist asks the patient to discontinue the mental arithmetic and to rest for a few minutes. Again, 2-4 minutes is usually adequate to determine the type of rebound pattern. If the patient does not rebound quickly, however, more time may be given.

---

SYMPTOM IMAGE: Next, the therapist asks the patient to engage in imagery associated with his or her symptom. This imagery may reflect the accident associated with the onset of the symptom. This is particularly powerful in personal injury assessments. Or it may entail having the patient imagine what the impact of the pain has been on their lifestyle. Another variation includes imagining their pain at its worst. The length of the imaging task may be important. Allow the patient a few minutes to get into this image and then a few minutes more to sustain it. At the end of this image, ask the patient about his ability to successfully recall and experience this image. The discussion or review of this event could be considered an additional stressful event and the patient's EMG response to the discussion should be observed.

---

RECOVERY/RELAXATION: Lastly the therapist asks the patient to "relax to the best of his abilities". The therapist remains quiet during this time allowing approximately 5 minutes for the relaxation to settle in.

---

As the patient engages in these activities, the therapist records the levels of EMG activity by hand or computer. It is useful to observe the rapidity of the patient's initial response, the peak of the response, the mean levels, as well as the amount of variance which the task throws into the muscle systems of interest. The computer directed clinical applications provide an elegant method for the practitioner to conduct, display, and analyze this type of stress profiling.

## CLINICAL INTERPRETATION OF THE STRESS PROFILE

For this type of evaluation the therapist conducts a "time series analysis". In this analysis, a comparison is made between resting levels and those seen during the stressful or relaxing events. One can inspect the EMG levels they have recorded during the study. If one is using a computer driven graphics program, an examination of the time series graphics may be clearly displayed on the screen for analysis and review. The next figure provides an example of such a time series display.

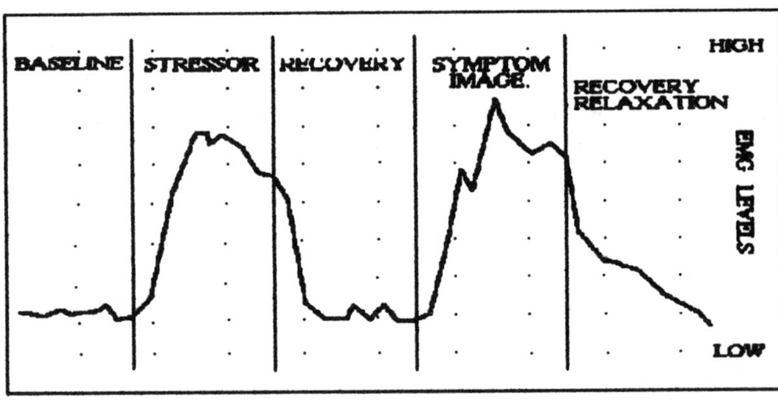


**FIGURE 15.** STRESS PROFILE EXAMPLE

An examination of the initial baselines may lead one to consider a high level of tonic EMG activity. In the more general model, this type of patient would typically be labeled anxious or "up tight". They appear to constantly be in a muscular vigilance, ready to fend off the world or flee from it. In the more specific framework, you might talk about a muscle group being "hyperactive" in its resting tone. The EMG response pattern to the mildly stressful event may occur in three directions. They are presented in Figure 15 and described below:

1) An increase in EMG activity usually reflects that the patient tends to activate the muscle being monitored when under stress. This may highlight the mechanism by which the muscle comes to play a role in their presenting symptom. Specifically, the muscle may have a very reasonable resting level, but it has a very high probability of responding to stressful events by "bracing" in preparation for fight or flight. If this happens frequently enough the muscle will begin to ache.

2) A decrease in EMG activity during the stressful event may reflect the patient's "cognitive ruminative style". If the baseline is elevated and the EMG levels drop as a function of the mental stressor, this may mean that the patient has more difficulty dealing with the uncertainty of what to expect during the rest period. Given that backdrop, the patient may actually find some psychological and physiological comfort in being given the "structure" of the stressful task (such as mental arithmetic). The task, then, becomes a focused, redundant task similar to a mantra. If the "social stressor" aspect of the mental arithmetic stressor is used wisely, a second level of stress may be introduced and examined. Here, the initial drop in EMG activity may reverse during the social criticism.

3) Lastly, there may be no change or very little change in the level of EMG activity as a function of the mental stressor. If the initial baselines are well within normal limits and the EMG response is small, negative findings for that muscle site are suggested.

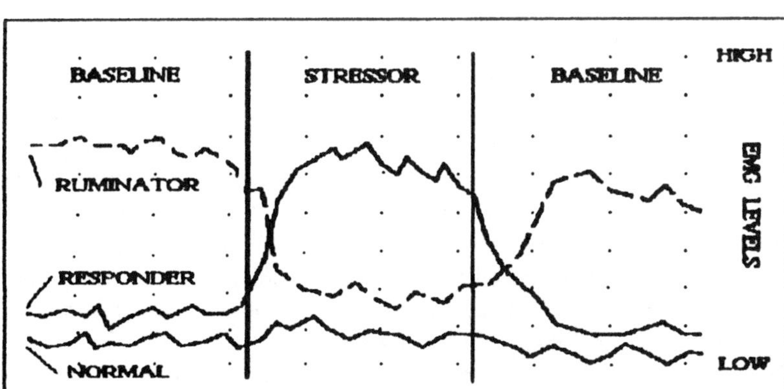


**FIGURE 16.** PATTERNS OF RESPONSE TO MENTAL STRESSORS

Next, the therapist examines the rebound or recovery from the stressful event. A rapid recovery is healthier than a slow or delayed recovery. Again a delayed recovery pattern may suggest the major mechanism for the emergence of the pathology. Poor recovery should not be labeled as the muscle being chronically tense. Instead, it suggests a more reactive process where the patient has difficulty "letting go" of the stress response, thus increasing the probability of a tense muscle. If this happens often enough, the muscle will begin to ache.

The symptom image attempts to examine an emotional overlay as part of the symptom complex. If the patient is "afraid" of his or her symptom it is not uncommon for the muscles in the affected region to become more active whenever the patient "thinks" about the impact of the pain. Thus, we may have the patient's anticipation or fear of the pain playing a role in promoting sustained muscular contractions, and therefore promoting the internal environment responsible for the onset of dull pain. This may become a vicious cycle.

In addition, the emotional display may represent anger at the pain, its impact on the patient's lifestyle and the blocking of a return to that lifestyle. Inquiry and/or differences in the way the patient is asked to imagine this image may help to tease out these two very different sources of dysfunctional arousal. Each, of course, would be treated differently from a psychological point of view.

If the EMG activity goes down during the symptom image phase, the cognitive/ruminative style is suggested. If there is no change noted, negative findings are indicated.

The recovery/relaxation phase allows for the assessment of two components. Again, it allows the therapist to assess the pattern of recovery from the symptom image. A slow recovery is problematic, suggesting the patient has difficulty "letting go" of the fear, anxiety or anger. In addition, it allows the therapist to assess the initial relaxation skills of the patient. Also, how these skills may be manifested following a series of stressful events. A patient who can demonstrate a good ability to quiet and relax the affected region following the symptom image has a better prognosis. The length of training will generally be shorter, and the symptoms will abate more quickly. With a patient who shows an activation of EMG activity during relaxation, careful selection of the relaxation procedure is in order.

The treatment protocols which emerge from this level of diagnosis are usually relaxation based in their orientation. The site or sites of treatment are typically clarified, as are the potential mechanisms of pain and dysfunction. The types of relaxation to be used (progressive relaxation, autogenics, guided imagery) may also be suggested. Lastly, one may want to choose a "Transactional Psychophysiological" approach to treatment (Lynch, 1985) if reactivity tends to center around specific themes.

## SECTION 7. THE CLINICAL PROCEDURE FOR POSTURAL EVALUATION

The postural evaluation concept has been developed primarily through the work of Steven Wolf and John Basmajian. For more in depth information about this procedure, please refer to their publications. The essence of this type of evaluation is the assessment of the patterns of neuromuscular activity during "dynamic movement". As you might recall, the muscle scanning procedure provides a "snapshot" of potential postural contributions. The dynamic evaluation, on the other hand, provides something more like a motion picture and therefore provide a more comprehensive picture of the neuromuscular system in action.

Whenever one uses the dynamic evaluation process, one should be aware that there is a major flaw in the procedure which needs serious consideration. It essentially has to do with the degree or extent of movement. Muscles either move or stabilize bones. When we ask a patient to do an action (such as rotating the head to the right), the level of EMG recruitment will be directly proportional to the extent of that movement. In other words, if a patient turns their head only 5 degrees he or she will recruit less EMG activation than a patient who rotates his or her head 30 degrees. Not only will the activation level differ with different degrees movement, but so will the nature of the muscle synergies. During rotation, a separation of function for the right and left side may not occur until the patient has rotated adequately to mandate that the concentric contraction exceed the eccentric contraction. Forward flexion is another good example. During forward flexion of the torso, the muscles of the low back should shut off at approximately 45 degrees of flexion. If you do an assessment on a low back pain patient, ask them to do a forward flexion and their EMG levels do not shut off, perhaps it is due to the fact that they engaged in only 25 degrees of flexion. If they were able to go further, beyond the 45 degree mark, would their muscles then turn off as they go out onto ligament support ? What stops them from going further ? Fear of pain and protective guarding, muscle shortness, splinting, etc. ? In any event, a major moderating variable in EMG recruitment patterns will be the degree of movement. Hopefully, the instrumentation of the future will allow us to easily monitor the degree of motion along with our EMGs. Until then, the therapist must split his or her attention and watch not only the EMG recruitment patterns, but also the movement pattern.

## SITE SELECTION FOR EVALUATION LOW BACK

The protocol described next involves the monitoring of the right and left aspect of the erector spinae muscles (dual channel recording). The active electrodes are placed in the region of L3. Refer to the verbal description and graphics of the scan sites if you are unfamiliar with this location. The iliac crest is used as the vertical landmark for placement of the lower active electrode, with the second active electrode being placed approximately 3 cm. directly above it. Both active electrodes are placed approximately 3 cm. out from the vertebral ridge on the same side of the back. The ground electrode may be placed directly above the upper active electrode. The right side of the electrodes is typically set up first and these are used to guide the placement on the left aspect. Symmetry of placement is important.

As is presented in the chapter in this book by Donaldson and Donaldson, one might want to monitor four channels of EMG activity, instead of only two. The additional two channels allows one to assess the muscle synergies or orchestration of the back in a more comprehensive fashion. For more information on multi channel recording strategies, I refer you to his chapter.

## PROCEDURE FOR EVALUATION OF LOW BACK MUSCLES

The procedure described below has seven steps and is one commonly used in clinical practice. Variations, such as lateral bending, stooping, walking, etc. are encouraged, depending upon the therapist level of expertise and the patient's presenting symptoms.

The "formal" evaluation is done with the patient in the standing posture and the patient's weight evenly distributed on each foot. The therapist sits behind the patient and places his hands on the patient's hips to stabilize the pelvic girdle during the rotational patterns. The data collection and display must be done in a dual channel monitoring basis. The visual displays of computer driven graphics greatly simplify this process. They clearly depicts the pattern and levels of activation which occur throughout the entire movement. The concepts of "splinting" or muscle irritability may be observed and easily described to the patient.

The sequence of the evaluation is as follows:

---

1. STAND QUIETLY: The therapist asks the patient to "stand quietly, with his weight balanced and his arms at his side". A two minute recording period is initiated.

---

2. FORWARD FLEXION: The therapist asks the patient to "lean forward as far as he can without hurting himself or causing high levels of pain". The pattern of muscle activity is observed throughout the forward motion. Careful attention is given to the level of symmetry observed during the "eccentric" activation phase of this movement. In addition, the therapist should note when the activation "turns off". Careful observation of the actual movement and posture associated with the EMG pattern is encouraged.

---

3. EXTENSION: The therapist asks the patient to "return to the standing posture" and the pattern of muscle activity observed. Again, the level of symmetry and activation during this "concentric" contraction are noted.

[Forward flexion and extension are completed within a 20 second period.]

---

4. ROTATE RIGHT: The patient's hips are stabilized in the midline position by the therapist. The therapist asks the patient to "continue to look straight ahead, while rotating his torso to the right (without causing undue pain)". The extent of rotation, along with the co-contraction or synergistic patterns of muscle activity are observed.

---

5. RETURN TO MIDLINE: The therapist asks the patient to "return to midline". The recovery, and symmetry in the EMG at midline is noted.

6. ROTATE LEFT: Again, with the hips stabilized, the therapist asks the patient to "look straight ahead, while rotating his torso to the left (without causing undue pain)". The extent of rotation, along with the co-contraction or synergistic pattern of EMG activity are observed.

7. RETURN TO MIDLINE: Finally, the therapist asks the patient to "return to midline". The recovery, and degree of symmetry at midline is noted.

[The right and left rotational patterns are completed in a 20 second time period.]

**INTERPRETATION OF POSTURAL EVALUATION**

During quiet standing, the level of EMG activity in the low back should be symmetrical in nature and at low levels of activation (3-4 microvolts or below on a narrow filter RMS amplifier). Excessive levels of activation on one or both sides suggests excessive bracing or guarding patterns. With computer graphics, level of activation and symmetry can clearly be seen.

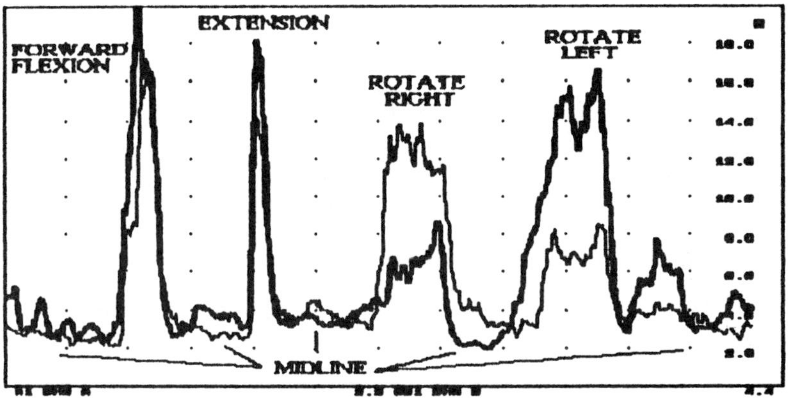


**FIGURE 17.** NORMAL FLEXION AND ROTATIONAL PATTERNS

During forward flexion the muscles should become activated in a symmetrical fashion through the first 45 degrees of flexion. Beyond that point of flexion, the muscles should essentially turn off as the back goes onto ligament support. If the pattern of activation is asymmetrical in nature a splinting pattern is suggested. If the EMG activity does not quiet down during the flexion, pay careful attention to the extent of flexion. As mentioned earlier, some patients will not conduct an adequate forward flexion to allow for the relaxation phase of the movement pattern. This may be due to a fear of inducing pain with such an extreme of movement. This type of restriction in movement is commonly termed "protective guarding". When restricted movement is observed the examiner should make an inquiry of the patient: " Can you go further ?" If so, try it again and see if the EMG is different. If he can't go further, what stops him? In addition to his fears, it may be that the forward flexion stimulates a muscle spasm. Or the muscles of the back or hamstrings may have an exceptionally "short" resting length, thus not allowing a normal flexion pattern.

During extension, a symmetrical pattern of activity should be seen. Again, an asymmetrical pattern may suggest a splinting pattern or an irritable muscle.

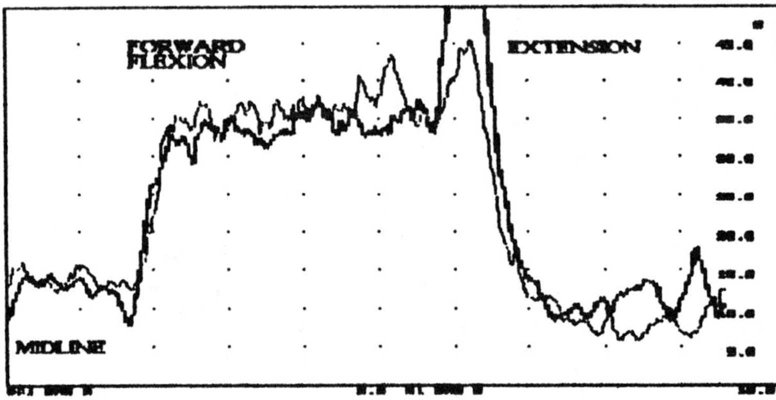


**FIGURE 18.** ABNORMAL FLEXION-EXTENSION PATTERN

During rotation to the right, the left aspect of the erector spinae should be more active than the right aspect. In approximately 80% of the population, the muscle group contralateral to the direction of the rotation becomes activated to stabilize this rotary motion in the low back region. It should be noted, however, that approximately 20% of the population activates the ipsilateral muscle group. If an ipsilateral activation occurs, this should be noted but its clinical meaning is uncertain at this time. The dysfunctional marker observed during a rotation pattern is a "co-contraction" of both the right and left aspect. Here, both the right and left side fire equally and simultaneous. Quantitatively, there is less than a 50% difference between the two activation patterns. If this happens during rotation to one direction, but not the other, pay careful attention to the extent of rotation for both directions. Differences in the degree of rotation could account for the differences in synergy. Be certain to explore any differences to understand the nature of the limited movement. Another common pattern observed is when one side (say the left) is briskly recruited no matter what the activity or direction of rotation. This would be identified as an "irritable" muscle. When co-contraction occurs, it places an undue torque on the spine and is clearly abnormal.

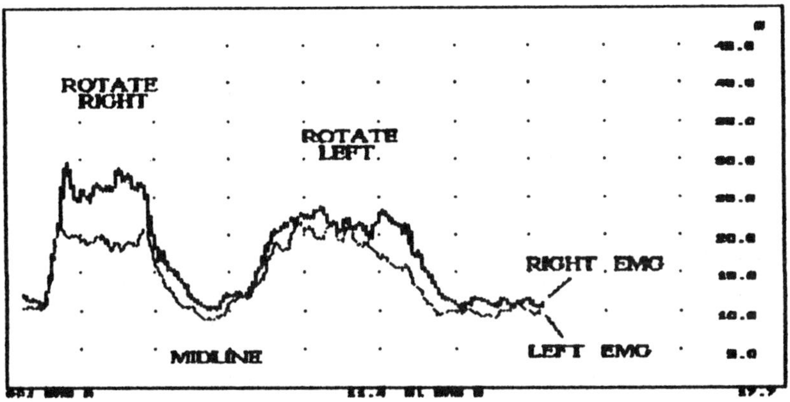


**FIGURE 19.** ABNORMAL PATTERNS OF ROTATION

During rotation to the left, the right aspect of the back muscles should primarily come into play. Again, this is found in 80% of the population. It is the co-contraction of the right and left aspects of the erector spinae which is considered abnormal. Remember to take into consideration the extent of actual movement as you begin to draw your conclusions.

When the rotational patterns are presented on a computer driven display, it is useful to have the patient rotate right, come to midline, rotate left and come back to midline during one screen (approximately 20 seconds). In this way, the therapist may examine (and explain to the patient) the manner in which the muscles are being orchestrated. One might see, for example, that the right aspect is clearly activated while turning both right and left. Or one might clearly see both the right and left side being activated during both right and left rotations. Following the time course of the action on one screen greatly simplifies this diagnostic endeavor.

This type of diagnostic assessment typically clarifies the nature of the postural disturbance and adds information about the dynamic component. Abnormal patterns may be detected and EMG biofeedback may be used to teach the patient about biomechanics and gravity, and how to correct dysfunctional patterns of muscle activation during dynamic movement.

## DYNAMIC EVALUATION OF THE HEAD, NECK, AND SHOULDER

The dynamic assessment of the head, neck, shoulder relationships provides a second variation of the dynamic assessment model. The astute and well studied therapist or practitioner should be able to abstract the concepts behind the assessment of these two regions and apply them to other regions. For example, one might study shoulder/arm relationships, scapular stability, etc.

Head, neck, shoulder relationships are commonly found to have dysfunctional muscle synergies in patient populations. Patients with headaches may clearly display poor or inappropriate "dynamic patterns" which play a significant role in their pain. Typically, these chronically poor "muscle synergies" are associated with a shortened resting length of the muscles which support the head. This abnormality may have developed as a function of poor posture, protective guarding, or emotional response in the anticipation to the experience of pain in the head, neck or shoulder regions. You may have observed some patients restrict their range of motion of the head and neck during a severe headache.

## SITE SELECTION FOR EVALUATION OF HEAD/NECK/SHOULDER

The site selection for electrode placement may vary depending upon the patient's presenting complaint, the findings from the muscle scan and the physical examination (palpation) of the muscles of the shoulder and neck. The placement strategy described in this section is commonly used in clinical practice and allows the examination of the neck/shoulder region in general.

Dual channel monitoring is mandatory for this procedure. The right and left aspects of the neck/shoulders are monitored independently. The electrode configuration for the left aspect are the same as for the right and a high level of symmetry of placement of electrodes is necessary. The electrode configuration for the right aspect will be described and should be replicated on the left aspect.

The first active electrode is placed over the belly of the right cervical paraspinals. Identify the muscle and cervical spine through palpation and place over the muscle mass. This should be at about the C4 level where you scanned this muscle. Place the ground electrode just to the right of the T1/C7 prominence. This is very easily found where the neck inserts into the torso. Place the second active electrode over the muscle belly of the upper fibers of trapezius. Palpate the upper crest of the shoulder and place the electrode directly over the center of the bulge of this muscle. This location is identical to where you scanned this muscle. Replicate the electrode configuration for the left aspect, keeping symmetry of electrode placement in mind. In my reports, I refer to this site as the "CT" site, indicating the Cervical and Trapezius sites of recording.

**PROCEDURE FOR HEAD/NECK/SHOULDERS**

The procedure described below takes the patient through six major planes of motion for the head/neck, and four major planes of motion for the shoulders. Midline is used as a common reference position. The experienced therapist may want to add some variations to these movement patterns to help clarify the nature of the dysfunction. This formal evaluation is conducted with the patient in the sitting posture. Usually the back is supported by the back of the chair. However, I occasionally request the patient to sit away from the back of the chair to see if it makes a difference. One might play around with the postural support concept. The data is monitored on a dual channel system. As with the postural evaluation, the computer driven graphics are ideal in that it provides a clear picture of how the muscles are recruited during the different planes of motion.

The sequence for the dynamic evaluation is as follows:

---

1. SIT COMFORTABLY. The therapist asks the patient to sit in a comfortable position with his hands in lap. Resting values are observed.

---

2. HEAD FORWARD FLEXION. The therapist asks the patient to let the head fall forward towards the lap (without causing undue pain), letting the head hang. Particular attention is paid to the symmetry of activation, along with the level of recruitment. Pay close attention to the extent and rate of the forward flexion. Ask the patient to do it again more slowly, and to go further if little change or movement occurs.

---

---

3. RETURN TO BASELINE. The therapist asks the patient to bring the head back to midline. Attention is paid to the symmetry of this concentric recruitment patten.

---

4. HEAD EXTENSION. The therapist asks the patient to lean the head backwards (without causing undue pain), letting it hang. CAUTION should be used here, this position has a modest probability of causing pain. I do not use this with all patients because of this. Attention is paid to how these muscles shut off at the end of this movement.

---

5. RETURN TO MIDLINE. The therapist asks the patient to return to midline. Attention is paid to the symmetry of the recruitment patten.

---

6. LEFT LATERAL BENDING. The therapist asks the patient to tilt the head sideways to the left, moving the ear towards the shoulder. This should be only a head movement, not a shoulder movement. While it is not mandatory, it is not a bad idea to place a hand on the shoulder contralateral to the direction of movement. If the therapist notices any movement of the shoulder, the EMG recording needs to be redone this time with the hand on the shoulder to stabilize it. Particular attention is given to the initial recruitment pattern of the ipsilateral side and the possible later recruitment of the contralateral side. Notice the extent of movement. Small movements may not show much in the way of EMG activation patterns.

---

7. RETURN TO MIDLINE. The therapist asks the patient to return to midline. Attention is given to the recruitment pattern of the contralateral side, and how the movement has effected the new resting values.

8. RIGHT LATERAL BENDING. The therapist asks the patient to tilt the head to the right, moving the ear towards the shoulder. Again, this should be a head movement only. Particular attention is placed on the initial recruitment of the ipsilateral side and possible latter recruitment of the contralateral side. Visual observation of the extent of lateral movement is strongly encouraged.

9. RETURN TO MIDLINE. The therapist asks the patient to return to midline. Attention is given to the recruitment pattern of the contralateral side and how the movement has effected the new resting values.

10. LEFT HEAD ROTATION. The therapist asks the patient to rotate the head to the left, as if looking over his shoulder. The therapist carefully watches the extent of the rotation pattern. If the patient does not rotate the head far enough to activate a strong recruitment pattern, the therapist asks the patient to turn the head further (without causing pain). If the patient rotates with his shoulders as well as his head, the movement is redone with the therapist placing one hand on the contralateral shoulder to stabilize its movement. Attention is paid to the initial recruitment on the ipsilateral side, and the possible later recruitment of the contralateral side.

11. RETURN TO MIDLINE. The therapist asks the patient to return to midline. Attention is given to the EMG on the contralateral side and how the movement has effected the new resting values.

12. RIGHT HEAD ROTATION. The therapist asks the patient to rotate the head to the right, as if looking over his shoulder. The therapist carefully watches the extent of the rotation pattern. If the patient does not rotate the head far enough to activate a strong recruitment pattern, the therapist asks the patient to turn the head further (without causing pain). If the patient rotates with his shoulders as well as his head, the movement is redone with the therapist placing one hand on the contralateral shoulder to stabilize its movement. Attention is paid to the initial recruitment on the ipsilateral side and the possible later recruitment of the contralateral side.

13. RETURN TO MIDLINE. The therapist asks the patient to return to midline. Attention is given to the EMG on the contralateral side and how the movement has effected the new resting values.

14. SHOULDER ELEVATION: The therapist asks the patient to raise both shoulders to hold them there briefly. Upon request, to let them go. Attention is given to the symmetry and levels of activation and the impact this movement has on resting levels. Visually observe whether or not one shoulder is lifted higher than the other.

15. SHOULDERS FORWARD. The therapist asks the patient to bring the shoulders forward, hold them there briefly and upon request let them go. Attention is given to the symmetry of the recruitment patterns and the impact of this movement on resting levels. Visual observation of the symmetry of the actual movement is also made.

16. SHOULDERS BACK. The therapist asks the patient to bring the shoulders back, hold them there briefly, and upon request to let them go. Attention is given to the symmetry and levels of activation, as well as the impact of this movement on the resting levels. Careful observation of the symmetry of this movement and whether or not the patient also elevated the shoulders simultaneously with retraction should be note.

17. SHOULDERS DOWN. The therapist asks the patient to pull the shoulders down, hold them there briefly, and upon request to let them go. Attention is given to the symmetry of the recruitment pattern and the effect of the movement on the resting levels.

**CLINICAL INTERPRETATION**

The midline position is utilized for comparison to all other planes of movement. This is a true example of the idographic approach to analysis mentioned earlier. While in the midline position, the EMG activity should have fairly low levels of activation and be fairly symmetrical for the right and left aspect. If you note a radical imbalance in the EMG, stand either in front or behind the patient and look at the symmetry of the shoulder elevation and head position. If one shoulder is higher than the other, this may explain the nature of the imbalance and suggest a direction for your dynamic assessment. If the head is tilted or rotated, this may explain the nature of the imbalance. In either event, these positional factors should be noted. If the symmetry of resting levels becomes more normalized as the patient moves through the different planes of movement, reassess these positional clues.

The resting levels of EMG activity should be fairly low. With a J&J EMG set at the narrow filter setting (100-200Hz), these should be easily below 5 uv. With the J&J wide filter (25-1000) or instruments which use a wider filter (i.e., 30-300), these values should be well below 8 uv.

During forward flexion of the head, the recruitment patterns should be fairly symmetrical. If one side is recruited twice as much as the other side, one would want to consider a shortened resting length on the side with the excessive recruitment. The shortened resting length creates an environment in which the muscle become more irritable, tends to fire more often and at higher than necessary levels. A clear eccentric activation pattern will probably not be seen unless the patient is requested to go into the forward flexion in a slow fashion. As the patient allows the head to "hang", this should place head support onto the ligaments and the muscles should shut off. One would not see the flexion relaxation, however, unless the eccentric activation pattern was clearly seen first. If the EMG values actually go up as a function of this movement, one might consider protective guarding. The continued activation of these muscles might reflect that the patient has learned not to allow adequate forward flexion, perhaps due to fear of pain. The alternate hypothesis is that of muscle shortness, and the resultant irritability.

During the return to midline from this position, one would like to see a fairly symmetrical recruitment pattern. If excessive recruitment occurs, it is usually located on the shortened side and is a sign of muscle irritability.

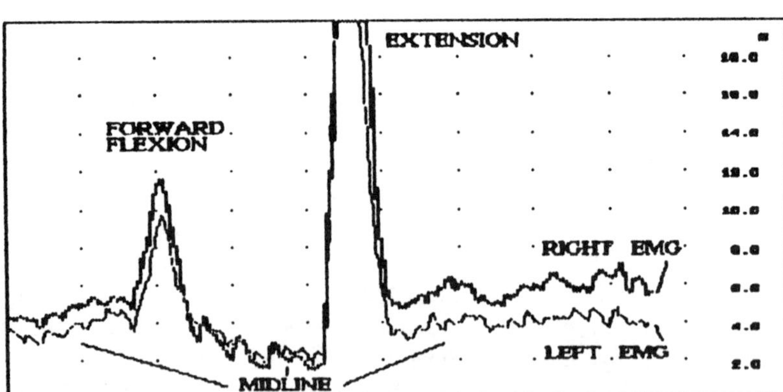


**FIGURE 20.** NORMAL FLEXION/EXTENSION PATTERN

During head extension, the initial recruitment pattern of the muscles is typically symmetrical and may turn off as the patient lets the head backwards on the anterior ligaments and muscles. Asymmetries are a clue at to muscle irritability. Continued activation while going into and staying in this position may represent a protective guarding pattern.

During lateral bending, the muscles on the same side of the "tilt" become active and shorten to pull the head over. The initial recruitment pattern then, should be seen on the ipsilateral (or same) side as the tilt. As the patient tilts more, the contralateral (opposite) side comes into play to stabilize the head as these muscles are placed on stretch. If a clear, strong co-contraction occurs during the initial movement, one would want to consider muscle shortening and irritability on the opposite side.

During the return to midline the side opposite the tilt contracts to pull the head back into the midline position. An excessively large recruitment pattern on that side may suggest a muscle irritability associated with the shortened muscle. A large recruitment on the ipsilateral side during return to midline, may suggest irritability within that muscle.

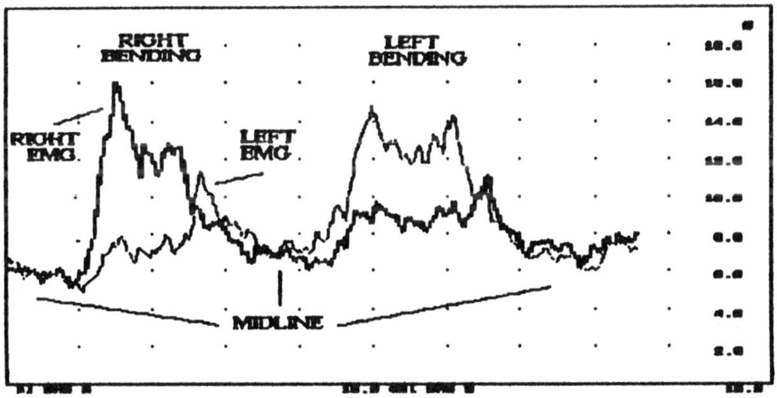


**FIGURE 21.** NORMAL PATTERNS FOR LATERAL BENDING

With head rotation, a lack of a clear recruitment typically reflects an inadequate amount of head rotation. Occasionally, the therapist will need to instruct the patient to rotate a little more in a given direction in order to make an adequate assessment. Pain associated with head rotation needs to be considered in making this type of request.

With adequate rotation and recruitment, the therapist would expect to see larger initial recruitment on the ipsilateral (same) side as the rotation, compared to the contralateral (opposite) side. If the therapist sees a clear and strong co-contraction pattern initially, either muscle shortness or protective guarding is suggested. If a strong contralateral contraction occurs towards the end of the rotation, the notion of muscle shortness is strengthened.

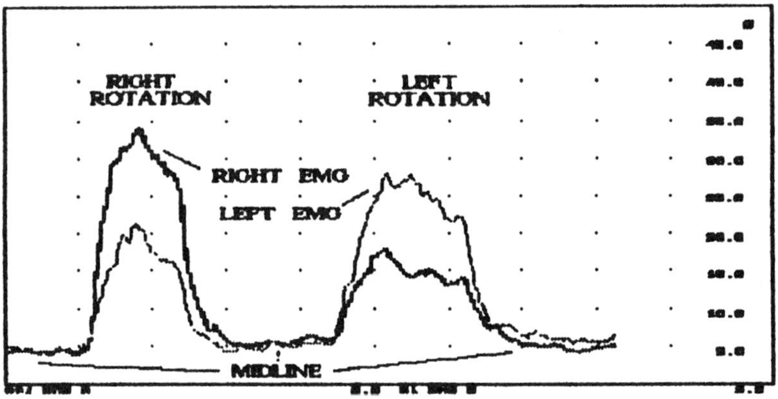


**FIGURE 22.** NORMAL ROTATIONAL PATTERNS

When the patient comes back to midline from a rotation, both muscles essentially turn off as the head returns to its neutral position.

During shoulder elevation, the therapist should expect to see a fairly symmetrical and strong recruitment pattern. Radical asymmetries during this motion typically suggest an irritability on the side with excessive recruitment. The alternate hypothesis is that the side with lessor recruitment may reflect protective guarding. In addition, if there was an initial asymmetry in the resting levels and this disappears as a function of the shoulder elevation, one would want to suspect a "chronic" protective guarding pattern on the elevated side. This may be on the side opposite of the pain, as the patient "moves away" from the nocioception.

When the shoulders are moved forward the muscles in the neck and upper back come into play due to being placed on stretch and to stabilize the shoulder. The recruitment pattern should be symmetrical and moderate in nature. If the pattern is asymmetrical or larger than shoulder elevation, one would want to suspect some muscle shortness on the overactive side.

When the shoulders are drawn back the therapist should expect to see a fairly symmetrical and strong recruitment pattern. Again, radical asymmetries observed during this motion typically suggest an irritability on the side with excessive activity or protective guarding on the side with lessor activity. This may occur on the same side as the pain ("splinting") or on the side opposite the pain ("protective guarding"). If the patient reports pain located on the head, the muscle irritability may participate with other muscles to irritate the greater occipital nerve or other more distal soft tissues.

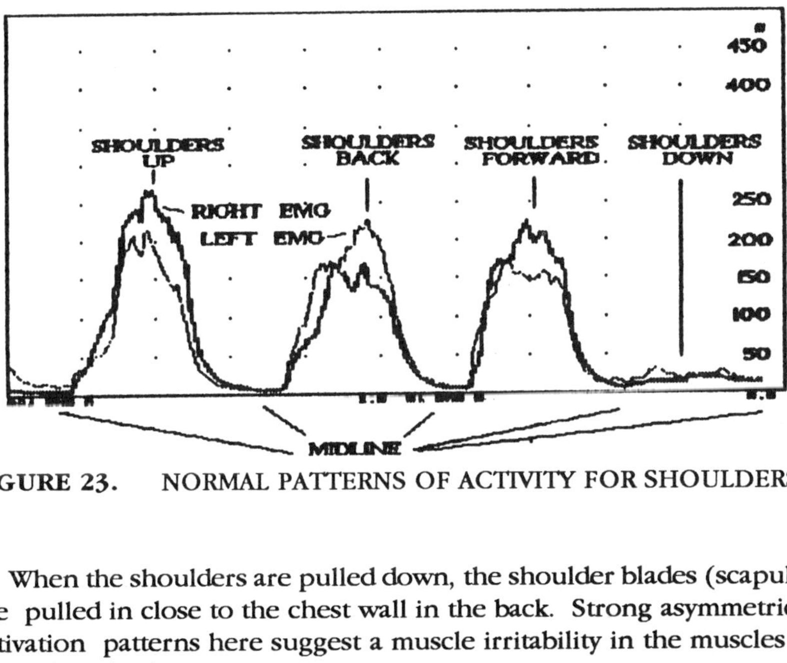


**FIGURE 23.** NORMAL PATTERNS OF ACTIVITY FOR SHOULDERS

When the shoulders are pulled down, the shoulder blades (scapula) are pulled in close to the chest wall in the back. Strong asymmetrical activation patterns here suggest a muscle irritability in the muscles of the upper back.

When the study is conducted using computer driven graphics display the therapist may easily observe asymmetrical patterns, as well as the levels of activation. If the sweep time is set up correctly, the therapist may view the recruitment patterns for two opposing planes of movement. For example, lateral bending to the right, midline, lateral bending to the left, and return to midline can be viewed on one screen (approximately 20 seconds). In this way, the therapist may examine and compare the two recruitment patterns along with their effects on the resting baseline. If a dysfunction pattern is present, the graphic display can be frozen and reviewed with the patient.

Once the study is completed, the therapist examines the results and looks for muscle shortness, irritability, and protective guarding. Conclusions concerning muscle shortness are drawn from those activities in which there is asymmetrical and/or excessive recruitment of EMG when the muscle is placed on stretch or plays a stabilizing role during an eccentric contraction. Conclusions concerning muscle irritability are drawn primarily from asymmetrical and/or excessive recruitment seen during activities requiring concentric or activation of the muscle. Conclusions concerning protective guarding are drawn from excessive recruitment patterns on the side opposite of the pain, as well as during limited movement in a plane of motion due to pain avoidance. Treatment protocols are then created to correct the dysfunctional patterns. These might entail stretching and strengthening exercises, stretch to relax or postural change.

## SECTION 8. SURFACE EMG TREATMENT CONSIDERATIONS

As was noted at the beginning of the Overview of Diagnostic Considerations section, DIAGNOSIS SHOULD LEAD TO TREATMENT PLANNING. We have just reviewed several common diagnostic protocols and will now consider four types of treatment protocols. Once you have systematically conducted a diagnostic assessment, you are ready to consider your treatment options. In this section, we will briefly review how EMG biofeedback may be used as a treatment tool. We will consider relaxation training, transactional psychophysiology, postural training, and stretch to relax.

### RELAXATION BASED TREATMENT

EMG biofeedback is rarely used in an unattended fashion, where the patient is "hooked up" to the instrument and left to relax on his own. Typically, the patient works with a therapist who acts as a coach or guide, assisting the patient in achieving the desired physiological state. The therapist, then, augments the information from the instrument with relaxation protocols. The feedback helps to guide and direct the therapist's relaxation inductions, while informing the patient how well he or she is doing. Such protocols typically include progressive relaxation, autogenics, guided imagery, etc. Much has been written about these ancillary techniques, and the reader should consult the literature for more information on these. Suffice it to say, the greater and richer the therapist's "requisite repertoire" of these techniques, the better the clinical service he or she will provide to the patient.

EMG biofeedback provides information about the neuromuscular system in a variety of ways. Visual information may come from a computer screen or a light bar. There may be several types of audio signals. Thresholds may be set for both the visual and auditory channels. One of the most important tasks of the therapist is to assist the patient in deciding which form of information he can learn best from. Some patients thrive on audio feedback, while others find it very distracting. Others encode information best through the visual channel. Whichever modality of feedback the patient chooses, the therapist needs to educate the patient so that he understands what the feedback means. This is particularly true in that many visual and audio feedback displays utilize a scaling which is "totally arbitrary".

The nature of the visual display should be demonstrated to the patient. The therapist indicates that the position of the visual stimulus (feedback) indicates the patient's level of muscle activity. If the visual display has a "zero point", the therapist clarifies for the patient where a relaxed level of muscle activity would be located. In computer driven graphics, it is not uncommon for the scaling of EMG to have a zero point. The therapist would then indicate that moving in a downward direction would be highly desirable and that reaching 1 microvolt, for example, would be ideal. With light bar displays, it is common for the scaling to be totally arbitrary. Here, the therapist adjusts the light bar display to approximately the "center" position of the display and points out to the patient that this represents the current level of muscular tension. The patient's goal then, is to lower the level of activity on the light bar. Movement in this direction indicates becoming more relaxed. If the patient is successful in lowering the light bar into the lower third of the light bar display, the therapist comments that he is resetting the display back up to the center position and that the patient is now at "x" microvolts.

Another way to orient the patient to a visual display is to explain the information given above and then have the patient tense the muscle being monitored and then let it relax. As the display moves upwards and rebounds downward, the therapist allows the patient to verify the relationship of the display to the level of muscle tension he or she is experiencing. This is also an excellent way for the therapist to verify that the EMG instrument is working properly.

Threshold lights, symbols, and tone may be selected and utilized in a shaping procedure. With the light bar display adjusted correctly, the light bar display should hover somewhere around mid-scale. The patient is instructed that as he relaxes, the pitch of the tone will decrease and the light bar will fall. When he becomes so relaxed that the light bar drops below the threshold indicator, the tone will turn off. His goal is to learn how to keep the tone off and the light bar moving in the desired direction. This way, the patient may close his eyes during the relaxation attempts. As he becomes more and more successful at this (i.e., keeping the tone off 80% of the time), the patient is given plenty of reinforcement for his success and the light bar is re-adjusted so that it is placed back into the middle of the range. The patient is then encouraged and guided to deeper levels of relaxation.

Dovetailing the relaxation procedures with feedback displays is not always easy. Many patients (and therapists) like to have the eyes closed in order to facilitate the procedure. In this case, the visual displays are useless and one must learn to rely upon and utilize audio feedback. Again, some patients find the audio displays distracting or even aggravating. At first glance, this may suggest that no audio display should be used. This is indeed an option, and the visual feedback display becomes solely the therapist's guide. The therapist may choose, on the other hand, to set the audio threshold as described in the above paragraph such that it acts as a negative reinforcer. Only when the patient relaxes to a preset criterion does the audio feedback (stimulation) turn off. In addition, the therapist should consider keeping the volume at a low level so that it becomes more of a background sound.

With computerized biofeedback programs, a nice combination of audio and visual feedback may be utilized. Here, the patient has the option to close the eyes and access only the audio information. At the end of the session, however, the computer may be programmed to display the muscle activity of the entire session on one screen. The therapist would then review the changes in EMG activity from the monitored site with the patient. This sense of history may be an invaluable aid to the patient's understanding and commitment to the biofeedback process.

Visual feedback may be the essential ingredient for some relaxation procedures. During progressive relaxation, for example, the visual display may easily be used to guide the level of contraction as well as the rapidity with which the activation pattern is turned off ("letting go"). In addition, the patient can easily acquire the notion of learning to control and discriminate large versus small contractions or "efforts".

If a generalized state of relaxation for a region is desired by the therapist, monitoring more than one site of muscle activity is highly recommended. This may be done by simultaneously monitoring two channels of EMG or by sequentially monitoring one muscle site and then another. As was pointed out in an earlier section, there appears to be a high level of specificity to the effects of EMG biofeedback and that one must cultivate a generalized effect rather an expect it. By monitoring more than one site, the therapist may verify the effects of generalization. If generalization is not occurring, one should consider switching the site for which the feedback is being given.

In addition, EMG feedback training may become locked into the status of the eyes or even the posture in which the training was done. Some patients lose their cultivated low arousal as soon as they open their eyes. In this case, the EMG levels would return to the pre-relaxation levels immediately upon opening the eyes. This phenomenon has meaning. It means that the therapist may want to eventually have the patient practice the relaxation technique in the eyes open state to help with generalization. The training effects may also be specific to posture. The patient trained in the reclining position may lose the cultivated low EMG when he or she moves to a sitting upright or standing posture. When this occurs, the therapist may want to consider conducting the relaxation procedure in different postures to promote generalization.

## TRANSACTIONAL PSYCHOPHYSIOLOGY

Transactional Psychophysiology is a term coined by James Lynch in his book The Language of the Heart. While Lynch's concepts are associated with cardiovascular reactivity, talking, and psychotherapy, his concepts may easily be generalized to work with the neuromuscular system. The essential concept behind Lynch's work is that of monitoring the patient's physiology while doing psychotherapy, orienting the patient to dysfunctional response patterns when they arise, and posing the question of "I wonder what that means?" As this form of therapy evolves with a patient, the emotional issues associated with the dysfunctional pattern of physiology are identified and, hopefully, resolved.

The neuromuscular system is very rich and well known for the role it plays in emotional displays. The muscles of the face, for example, are one of the strongest cues as to a person's emotional disposition. The shoulders are another strong emotional display system, being elevated when the person is "up tight" and slumped when the person is "depressed". If one recalls the way in which the cat arches it's back during a threat display, it's not too terribly hard to consider that the back musculature in humans has an emotional capacity. These emotional display capabilities of the neuromuscular system are the meat of transactional psychophysiology for EMG oriented work. Once you have gone through your diagnostic evaluation, observed a very strong emotional reactivity pattern during the stress profiling and have concluded that there is a strong emotional component, the therapist might want to consider this model of therapy in place of or in addition to the general relaxation model.

The ease of conducting this model of therapy is greatly facilitated by computer driven graphics. This is because the computer leaves its sense of history on the screen. Both the therapist and the patient may clearly observe a sudden or slow rise in the levels of activation of the muscle being monitored. When this occurs, the therapist can reach over and freeze the display, commenting "I wonder what this is about?". Implicitly, the patient is asked to pay attention to the way in which the mind, body and emotions interact. The patient, then, begins to access the proprioceptive cues (feelings) and associate these with the thoughts, images or content of the discussion. It is up to the skills of the therapist to assist the patient in "working through" these issues on a psychological level. In addition, the therapist might encourage the patient to "let go" of the emotional tone expressed in these muscles. Techniques such as the Quieting Response or "Breath-Blow-Let Go" might be considered.

The electrode placement sites for monitoring for this type of training are usually suggested from the muscle scanning or stress profiling procedure. The most common sites are those of frontalis, temporalis, masseter and trapezius. Some recent evidence, however, strongly suggests that the L3 muscles should be considered in the low back patient. In addition, one should consider monitoring just the right or left aspect of a muscle group, particularly if asymmetries were clearly noted during the diagnostic investigation.

## POSTURALLY BASED TREATMENTS

Posturally based training strategies becomes relevant when a strong postural component is discovered during the diagnostic work-up. Postural training is typically applied to the neck, upper or lower back. I will describe the treatment of a patient with low back pain and dysfunctions as an example.

The electrode configuration commonly used in the training of the low back muscles is identical to those described in the postural evaluation section. Please refer to that section for specific details. Essentially, what one wants to do is independently monitor the left and right aspect of the erector spinae (or other groups).

The general approach to this type of training is a "socratic" one, the development of a sense of self discovery. Rather than instructing the patient to take on a particular postural stance, say to the patient "I wonder what would happen if you did "X" ?" Suggest the patient try it out and let the feedback guide his performance. By letting the patient discover what it is he did, the patient tends to retain the principles he learned for longer periods of time.

The initial goal of postural training is to teach the patient about the biomechanics of posture. They should learn, for example, that raising their hands out in front of their chest brings the muscles of the low back into play in order to counteract the forces of gravity. They should become aware of the fact that when they lean to the left, the muscles of the right come into play to stabilize the torso over a new center of gravity. Once the patient has come to appreciate this fact, the next goal is to assist the patient in balancing the level of neuromuscular activity of the right and left aspects of the back. This is commonly referred to as "equilibration" training. The third goal is to assist the patient in lowering the levels of activation of the muscles in the back in a symmetrical fashion. The final goal is to assist the patient in attempting to do activities of daily living in such a way that he uses his back muscles appropriately.

When conducting this type of training a dual channel light bar or computer graphics display is mandatory. The computer graphics are ideal, in that they leave a trace of history so that the pattern of muscle recruitment for a movement can be studied and reviewed with the patient. A dual channel light bar allows the patient and therapist to view only the moment by moment fluctuations of EMG activity on the right and left side of the low back.

Let's pretend, for a moment, that you are working with a dual channel light bar display. The electrodes have been attached to the patient, the instrument is turned on and the patient is standing. The light bar display on the left is in the center position and the light bar display on the right is significantly higher. The therapist suggests to the patient that we want to see how moving his body around in space effects the low back muscles. You then might suggest some of the ideas listed below and see if both you and the patient can come to an understanding of the biomechanics of the back. Next, you inform the patient that you would like him to bring both light bar displays into the center position, thereby balancing the muscle activity in his back. Sometimes, the patient will stand there and attempt to "think" the light bar to the center point. If this happens you might say "I wonder what would happen if you moved your body in such and such a way?" When the patient complies with your request, the light bars moves and the connection between movement and the feedback display is made. The feedback from the light bar, then, tells both you and the patient whether or not that movement helped or hindered in balancing the muscle activity in his back.

Since one of the underlying goals of your treatment is to teach the patient as much as you can about "body mechanics" and posture, there are a few principles you might want to keep in mind. These principles, listed below, may be suggested to the patient in the "I wonder what would happen if. . ." format.

---

1. WEIGHT DISTRIBUTION. What happens when you place all of you weight on the right foot? The left foot? The balls of your feet? The heels of your feet? Equally on both feet? Sometimes two bathroom scales help clarify the meaning of weight distribution.

---

2. KNEES. What happens when you lock your knees? Unlock your knees slightly? Bend your knees?

---

3. PELVIC TILT. What happens when you rotate your pelvis up? Rotate it down? Shift it from side to side?

---

4. STOMACH. What happens when you hold your tummy in? Let it out?

---

5. LATERAL BENDING. What happens when you lean to the right? Lean to the left? Which side of your back is active when you lean?

---

6. EXTENSION/FLEXION. What happens when you lean forward? Lean back?

7. ARMS. What happens when you hold your arms out in front of you? At your sides? Behind your back, as in a "parade rest"?

8. CHEST. What happens when you hold your chest out? Drop it down?

9. SHOULDERS. What happens when you raise both shoulders up? Just the left shoulder? Just the right shoulder? Pull your shoulders back? Push them forwards?

10. HEAD/NECK. What happens when you hang your head forward? Lean it back? Lean it to the left? Lean it to the right?

As the patient begins to move his body about, he begins to notice how each movement and how movement of other parts of his body may effect the low back muscles. Essentially, what he discovers is that when he is over his center of gravity the back muscles tend to be the most relaxed and balanced. When he rotates his pelvis up, balances the weight on his feet, allows his knees to be unlocked, etc., the muscle activity during quiet standing is improved. Whenever the patient takes on an "absurd posture" which results in the light bars being centered, this needs to be noted by saying, "Well, that seems to work, but its not very useful". For example, if the patient leans way back (hyperextends the back), the light bar usually become balanced and show low levels of activation. While this posture appears to bring about the desired state of low and balanced EMG, it is awkward and places undue strain on other parts of the body. Once the patient has "mastered standing," take a look at sitting. Then work on going from sitting to standing and visa versa. Finally, have the patient engage in activities which he finds difficult or painful to do. Have him practice these while keeping the muscle activity from his back within reason.

## STRETCH TO RELAX

Stretching has been around for a very long time. If you are unfamiliar with stretching as a therapeutic concept, you should consult the vast literature on this topic. One of the more popular lay volumes on stretching is Stretching, by Bob Anderson.

The stretch to relax concept is presented in this chapter is a suggested method for lengthening shortened muscles discovered during the dynamic evaluation for the head-neck-shoulder. As was mentioned in earlier sections, if a muscle has a shortened resting length, traditional EMG assisted relaxation will do absolutely nothing to help correct this problem or the aberrant muscle synergies associated with it. However, when stretching is couched within a relaxation framework, the therapeutic effects may be dramatic. A slow gentle stretch facilitates a lengthening of the muscle tissue, while concurrent relaxation feedback helps the CNS reorganize the way in which it orchestrates the muscle. In some ways, it is my belief that doing a relaxation based EMG biofeedback procedure while the muscle is on stretch is the clearest way to training the Gamma motor system. The gamma motor system, as you might remember, essentially calibrates the sensitivity of the stretch receptors, gives us our balance and helps to determine our overall resting muscle tone.

The stretching procedure utilized in conjunction with feedback entails a slow gentle stretch. The stretch should never be so strong as to cause pain. Just enough stretch so that the patient can feel it. The stretch should be held for approximately 30-60 seconds. The intensity of this intermediate stretch is guided by the pull of gravity. Short stretches, 10 seconds or so, do not appear to be adequate to bring about the reduction in EMG activity you want to see. As the person puts the muscle on stretch, he is instructed to "breathe into the stretch", taking a large enough breath to feel it in the stretched muscle. As the patient exhales, it is suggested that he relax into the stretched muscle and the muscles around it. In essence, to relax a little bit more with each breath.

With computer graphics, the therapist may clearly observe the muscle go onto stretch by an increase in EMG activity. As the patient breathes into the stretch, the rhythmic oscillation of the respiration can easily be seen. During the inspiration phase, there is a modest to intermediate increase in EMG activity should be seen. During the expiration phase, a phasic decrease in EMG activity is expected along with a lowering of the tonic levels associated with the stretch (see figure 23). This phenomenon may also be seen on a dual channel light bar display in a moment to moment fashion.

In doing stretching exercises, it is usually wise to stretch the two opposing muscle groups. If you do a lateral stretch to the left, you should do a lateral stretch to the right. Pain would be the major factor allowing you to violate this rule. The following is a series of suggested stretches for the neck. The therapist should use their own clinical experience, knowledge and judgement as to whether or not these stretches are appropriate for their patient. In addition, one should not necessarily limit themselves to these stretches, for there may be others which would work well for the particular patient.

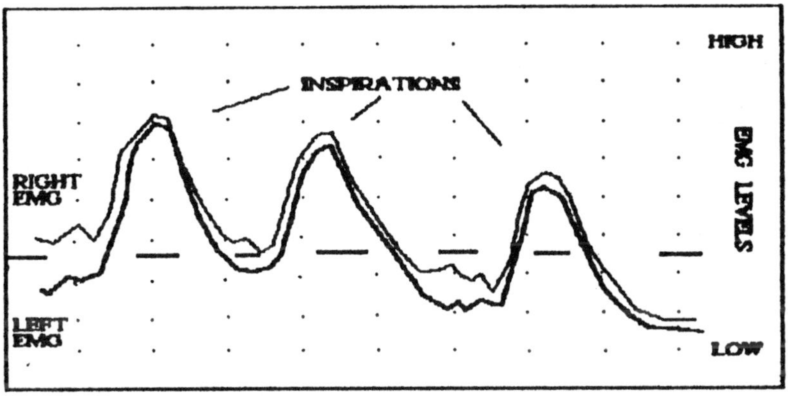


**FIGURE 24.** BREATHING INTO A STRETCH

## SUGGESTED STRETCHES FOR THE NECK

---

1. FORWARD FLEXION STRETCH. The therapist asks the patient to let the head hang forward towards the lap. A stretch sensation should be felt around the T1 region, up onto the neck and down towards the shoulder blades. Some patients will report feeling the stretch clear down to the L3 region. Suggest

that the patient breathe into the stretch, feeling it more clearly with each inspiration. As the patient exhales, encourage that the patient let the muscles relax a bit more with each breath. The stretch is held for 30-60 seconds.

---

2. EXTENSION STRETCH. The therapist asks the patient to let the head hang back over the shoulders. The stretch should be felt on the front of the neck and underneath the jaw. The patient may want to open and close the mouth to get a better sense of these muscles. Breathe into the stretch and relax a little with each exhalation. The stretch is held for 30 to 60 seconds. CAUTION SHOULD BE USED WITH THIS STRETCH. IF ANY PAIN IS FELT WITH THE EXTENSION, DISCONTINUE IT IMMEDIATELY.

---

3. LEFT LATERAL BENDING STRETCH. The therapist asks the patient to let the head hang to the left over the shoulder. The stretch should be felt on the right side of the neck and down into the right shoulder. Breathe into the stretch, relaxing a little with each exhale. The stretch is held for 30 to 60 seconds.

---

4. RIGHT LATERAL BENDING STRETCH. The therapist asks the patient to let the head hang to the right over the shoulder. The stretch should be felt on the right side of the neck and down into the right shoulder. Breathe into the stretch, relaxing a little with each exhale. The stretch is held for 30 to 60 seconds.

---

5. LEFT ROTATION STRETCH. The therapist asks the patient to rotate the head to the left gently. The stretch is felt in the right aspect of the neck and into the right shoulder. Breathe into the stretch and relax a little with each exhale. The stretch is held for 30 to 60 seconds.

---

6. RIGHT ROTATION STRETCH. The therapist asks the patient to rotate the head to the right gently. The stretch is felt in the right aspect of the neck and into the right shoulder. Breathe into the stretch and relax a little with each exhale. The stretch is held for 30 to 60 seconds.

One interesting finding observed from my initial work in this area, was the impact of stretching upon asymmetrical patterns found during the resting, midline position. Before I began using stretching for the neck, I would treat resting asymmetries in the neck in the same fashion as I treated asymmetries in the low back. This entailed the postural change model described above, generalized to the head/neck/shoulder region. As I began using some very simple stretches, I was amazed at how these could bring about an equilibration of resting values.

Stretching is a very nice place to begin the therapeutic process. Compared to the postural change model, the patient appears to more clearly understand and practice the stretching exercises at home on a regular basis. During an office visit, we will monitor the muscle activity while he practices the stretches and modify the home practice as needed. Figure 11 provided a case study of a stretch to relax session. Across sessions, the patient begins to notice a more normalized recruitment pattern during the stretches. Through this he can see the importance of home practice. As the muscles begin to take on a more normalized resting length and recruitment pattern, the theme of the EMG biofeedback training changes to postural training. Now the patient is muscularly ready to learn the biomechanics needed to conduct the activities of daily living.

## REFERENCES

Ahern D.K., Follick M.J., Council J.R., Laser-Wolston N. (1986). Reliability of lumbar paravertebral EMG assessment in chronic low back pain. *Arch Phys Med Rehabil*, 67, 762-765.

Alexander, A.B., & Smith, D.D. (1979). Clinical applications of EMG biofeedback. In R.J. Gatchel & K.R. Price (Eds.), *Clinical Applications of Biofeedback: Appraisal and Status.* New York: Pergamon Press.

Anderson, B.J.G., Jonsson, B., & Ortengren, R. (1974). Myoelectric activity in individual lumbar erector spinae muscles in sitting. A study with surface and wire electrodes. *Scandinavian Journal of Rehabilitation Medicine,* Sup 3, 91-108.

Bancroft, H., & Millen, J. (1939). The blood flow through muscle during sustained contraction. *Journal of Physiology, 97,* 17.

Basmajian, J.V. (1976). A fresh look at the intrinsic muscles of the back. *American Surgeon 42,* 685-690.

Basmajian, J.V. (1983). *Biofeedback: Principles and Practice For Clinicians.* Baltimore: Williams and Wilkins.

Blanchard, E.B., Andrasik, F., Ahels, T.A., Teders, S.J., & O'Keefe, D.M. (1980). Migraine and tension headache: A meta-analytic review. *Behavior Therapy,* II, 613-631.

Bonica, J.J. (1977). Introduction to symposium on pain. *Archives of Surgery, 112,* 357.

Budzynski, T., & Stoyva, J. (1974). Cultivated low arousal: An antistress response. In L. DiCara (Ed.), *Limbic and Autonomic Nervous System Research.* New York: Plenum.

Cahn, T.S., Cram, J.R., & Powell, J. (1990). Relationship between muscle regions and average total muscle activity: Support for the back. *Perceptual and Motor Skills,* in press.

Calliet, R. (1977). *Soft Tissue Injury and Disability.* Philadelphia: F.A. Davis.

Cannon, W. J. (1932). *The Wisdom of The Body.* New York: Norton.

Carmichael, S.W., & Burkart, S.L. (1979). Clinical anatomy of the lumbar complex. *Physical Therapy, 59,* 966-968.

Cram, J.R., & Engstrom, D. (1986). Patterns of neuromuscular activity in pain and non-pain patients. *Clinical Biofeedback and Health, 9*(2), 106-116.

Cram, J.R., & Freeman, C.F. (1985). Specificity in EMG biofeedback treatment of chronic pain patients. *Clinical Biofeedback and Health, 8*(2), 101-108.

Cram, J.R., & Steger, J.C. (1983). EMG scanning in the diagnosis of chronic pain. ***Biofeedback and Self-Regulation, 8,*** 229-241.

Cram, J.R., Lloyd, J., & Cahn, T.S. (1990). The reliability of EMG muscle scanning. ***International Journal of Psychosomatics,*** in press.

Davidson, R.J., & Fox, N.A. (1982). Asymmetrical brain activity discriminates between positive and negative affective stimuli in human infants. *Science, 218,* 1235-1237.

De Luca, C.J. (1979). Physiology and mathematics of myoelectric signals. *IEEE Transactions on Biomedical Engineering, BME 26,* 313-325.

Dolce, J.J., & Raczynski, J.M. (1985). Neuromuscular activity and electromyography in painful backs: Psychological and biomechanical models in assessment and treatment. *Psychological Bulletin, 97,* 502-520.

Donaldson, C.C.S.(1989). The effect of correcting muscle asymmetry upon chronic low back pain. Unpublished doctoral dissertation, University of Calgary, Calgary.

Duffy, E. (1957). The psychological significance of the concept of "arousal" or "activation." *Psychological Review, 64,* 265-275.

Ekman, P., & Friesen, W.V. (1972). *Unmasking The Human Face.* Englewood cliffs, NJ: Prentice-Hall.

Flor, H., Turk, D.C., & Birbaumer, N. (1985). Assessment of stress related psychophysiological reactions in chronic back pain patients. *Journal of Consulting and Clinical Psychology, 53*(3), 354-364.

Floyd, W.F., & Silver, P. (1955). The funtion of the erector spinae muscles in certain movements and postures in man. *Journal of Physiology, 129,* 184-203.

Goldstein, B. (1972). Electromyography: A measure of skeletal muscle response. In N. Greenfield & R. Sternbach (Eds.), *Handbook of Psychophysiology.* New York: Holt, Rinehart & Winston.

Hoyt, W.H., Hunt, H.H., DePouw, M.A., Bark, D., Shaffer, F., Passias, J.N., Robbins, D.H., Runyon, D.D., Sermrad, S.E., Symonds, J.S., & Watt, K.C. (1981). Electromyographic assessment of chronic low back pain syndrome. *Journal of the Osteopathic Association, 80,* 57-59.

Hudzinski, L., & Lawrence, G. (1988). Significance of EMG surface electrode placement models and headache findings. *Headache, 28,* 30-35.

Jacobson, E. (1938). *Progressive Relaxation.* Chicago: University of Chicago Press.

Jones, A., & Wolf, S. (1980). Treating chronic low back pain. *Physical Therapy, 60*(1), 58-63.

Kiesler, D. (1966). Some myths of psychotherapy research and the search for a paradigm. *Psychological Bulletin, 66,* 110-136.

Lacey, J., & Lacey, B. (1958). Verification and extension of the principle of autonomic response-stereotype. *American Journal of Psychology, 71,* 50-73.

Lader, M.H., & Mathews, A.M. (1968). A physiological model of phobic anxiety and desensitization. *Behaviour Research and Therapy, 6,* 411-421.

Loeser, J. (1978). Setting a chronic pain treatment plan. *Patient Care, 12*(4), 75-102.

Lynch, J. (1985). *The Language of The Heart.* New York: Basic Books.

Malmo, R.B. (1959). Activation: A neuropsychological dimension. *Psychological Review, 66,* 367-386.

McKenzie, R. (1985). *Treat Your Own Neck.* Lower Hut, New Zealand: Spinal Publications.

Nunnally, J.C. (1967). *Psycometric Theory.* New York, McGraw-Hill.

Patton, H. (1976). Muscles and neuromuscular transmission. In H. Patton, J. Sundsten, W. Crill, & P. Swanson (Eds.), *Introduction To Basic Neurology.* Philadelphia: W.B. Saunders.

Phillips, C. (1977). The modification of tension headache pain using EMG biofeedback. *Behaviour Research and Therapy, 15,* 119-125.

Price, J.P., Clare, M.H., & Ewerhardt, R.H. (1948). Studies in low backache with persistent muscle spasm. *Archives of Physical Medicine and Rehabilitation, 29,* 703-709.

Prudden, B. (1980). *Pain Erasure: The Bonnie Prudden Way.* New York: Ballantine Books.

Rodbard, S. (1975). Pain in contracting muscles. In B.L. Crue (Ed.), *Pain: Research and Treatment.* New York: Academic Press.

Sargent, M.M. (1946). Psychosomatic backache. *New England Journal of Medicine, 234,* 427-430.

Spector, B. (1979). Surface electromyography as a model for the development of standardized procedures and reliability testing. *Journal of Manipulative and Physiological Therapeutics, 2*(4), 214-222.

Sternbach, R. (1966). *Principles of Psychophysiology: An Introductory Text and Readings.* New York: Academic Press.

Thompson, J.M., Erickson, R.P., Madson, T.J., Offord, K.P. (1988). EMG muscle scanning: Comparison with attached surface electrodes. Poster presented at the Amer Acad Rehab Med, Seattle.

Traue, H., & Bischoff, C. (1983). Myogenic headache. In K. Holroyd, B. Schlote, & H. Zenz (Eds.), *Perspectives In Research On Headache.* New York: C.J. Hogrefe.

Travell, J. (1976). Myofacial trigger points; Clinical review. In J.J. Bonica & D. Albe-Fassard (Eds.), *Advances In Pain Research, Vol. 1,* p. 919, New York: Raven Press.

Travell, J., & Rinzler, S. (1952). The myofascial genesis of pain. *Postgraduate Medicine, 11,* 425-434.

Turk, D.C., & Flor, H. (1984). Etiological theories and treatments for chronic back pain. II. Psychological factors and interventions. *Pain, 19,* 209-233.

Turk, D.C., Meichenbaum, D.H., & Berman, W.H. (1979). Application of biofeedback for the regulation of pain: A critical review. *Psychological Bulletin,* 86, 1322-1338.

Turner, J., & Chapman, D. (1982). Psychological interventions for chronic pain: A critical review (I,II). *Pain, 12,* 1-46.

Ulmer, H.V. (1977). Arbeitsphysiologie-Umweltphysiologie. In R.F. Schnidt & G. Thews (Eds.), *Physiologie des Menschen.* Berlin: Springer.

Van Boxtel, A., & Goudswaard, P. (1984). Absolute and proportional resting EMG levels in chronic headache patients in relation to the state of headache. *Headache Journal, 24*(5), 259-265.

Whatmore, G., & Ellis, R. (1959). Some neurophysiologic aspects of depressed state. *Archives of General Psychiatry, I,* 70-80.

Whatmore, G., & Kohli, D. (1974). *The Physiopathology and Treatment of Functional Disorders.* New York: Grune and Stratton.

Wolf, S., & Basmajian, J. (1978). Assessment of paraspinal electromyographic activity in normal subjects and chronic back pain patients using a muscle biofeedback device. In E. Asmussen & K. Jorgensen (Eds.), *International Series On Biomechanics* (Vol. 6B). Baltimore: University Park Press.

Wolpe, J. (1958). *Psychotherapy By Reciprocal Inhibition.* Stanford: Stanford University Press.

# CHAPTER 2

# *Multi-Channel EMG Assessment and Treatment Techniques*

by

Stuart Donaldson, Ph.D.
and
Mary Donaldson, M.Ed.

---

The advent of computers with their capacity to process, analyze, and store massive amounts of information and data has facilitated the development of new assessment and treatment techniques in the field of electromyography (EMG). These techniques, while progressing logically as the next step from single stand instruments, require further training of the biofeedback practitioner. This chapter is directed towards the biofeedback practitioner who has a working knowledge of EMG techniques and wishes to add to his repertoire of skills.

## ELECTROMYOGRAPHY AND MUSCLE PHYSIOLOGY

Before exploring some of the recent developments, examination of relevant concepts will be undertaken in order to ensure a common conceptual frame of reference. Electromyography, as used in this chapter, is a form of biofeedback in which the myoelectric signal of the striated muscle is the physiological system under study. Striated muscles are composed of numerous muscle fibres ranging in diameter from 10 to 100 microns (Guyton, 1971). Each muscle contains several hundred to several thousand myofibrils, which are composed of myosin and actin filaments. The muscle fibre is innervated by a motor unit which comes from the ventral horn of the spinal cord (Basmajian, 1988). The number of fibres that are innervated by a motor unit varies from a few in the smaller muscles (e.g., muscles involved with eye movement) to a great many in the larger muscles of the limbs. The fibres that are innervated by a single motor unit are spread throughout the muscle and are intermingled with fibres innervated by other motor units. Muscle contraction occurs when an electrical impulse, which is transmitted along the axon of the motor unit arrives at the muscle, activating a biochemical process which causes the filaments to contract. The size of the contraction is determined by (a) the rate of the impulses transmitted by the motor unit, and (b) the number of motor units which are active at that moment.

While muscle contraction itself is not accessible, the electrical aspects of the contraction may be studied (Peek, 1987). The surface electrode picks up the electrical aspects of the contraction including the number of spikes, amplitude, rise time, and amplitude-rise time (Komi & Viitasalo, 1976). This electrical signal which is measured in microvolts (one millionth of a volt) is thought to represent the amount of force generated by a muscle. Bigland and Lippold (1954) originally proposed that the EMG signal corresponded with the generated force at $r = 0.9$ level in a linear manner. Since then, various studies have supported this position with some qualifications. Bouisset (1973), and Fujiwara, Miura, Numasaki and Hasue (1979), and Vredenbregt and Rau (1973) found a linear relationship between the EMG readings from isometric contractions and force generated. Further studies have demonstrated that the mathematical relationship varies somewhat depending upon the size of the muscle studied (Janda, 1983), load (Vredenbregt & Rau, 1973), fatigue (Mulder & Hulstijn, 1985; Stulen & De Luca, 1979; Vredenbregt & Rau, 1973). Movement also affects the relationship, for as movement is initiated there is a burst of electrical activity creating a curvilinear relationship (Vredenbregt & Rau, 1973). However, as movement continues the relationship becomes linear.

In summary, the EMG signal (within certain limitations) appears to adequately represent the amount of force generated by a muscle. The signal is a composite of various factors reflecting the electrical properties of the muscle.

## THE VALIDITY AND RELIABILITY OF EMG RECORDINGS

The criteria of validity and reliability are employed as the determinants to scientific acceptability. The validity of EMG techniques may be established by determination of face and concurrent validity. Face validity is established by rigid adherence to standardized procedures when applying surface electrodes (Basmajian and Blumenstein, 1983; Cram, 1986). These procedures include (a) application along the belly of the muscle, (b) monitoring striated muscles which are close enough to the surface to permit pick-up of the signal, (c) avoidance of adipose and thick tissue, (d) utilization of a standard sized electrode and inter-electrode distance, (e) a universal reference, and (f) correct abrasion, application of gel, and good firm adherence. Concurrent validity is established by comparing the item under study (in this case surface electrodes) to a known standard (needle or wire electrodes). Bouisset and Maton (1973) compared surface electrodes to wire electrodes measuring the activity of the right biceps brachii. While performing an isometric contraction they found a relationship of r = .98 between the electrode recordings. The authors concluded that surface electrodes are at least as representative as needle electrodes. Anderson, Jonsson, and Ortengren (1974) compared the myoelectric activity recorded by surface electrodes to that from wire electrodes while studying the erector spinae at the L1, L3, and L5 levels with subjects in numerous positions. The results showed a strong congruence in the amplitudes recorded, but the wire electrodes displayed a higher standard deviation. The authors concluded that surface electrode were more precise in their sampling of the activity.

Reliability of EMG scores have been established by various authors in test-retest studies. Jonsson and Komi (1973), utilizing the brachioradialis, compared surface to needle electrodes at 10-minute and at 3-day intervals under various isometric load conditions (20 - 100 units of tension). Studying 12 subjects they found that the reliability coefficient for the surface electrodes for the 10-minute interval was r = .88 and .69 for the 3-day interval. Komi and Buskirk (1970) studied the test-retest reliabilities of surface electrodes assessing isometric and isotonic contractions (both eccentric and concentric). They found the coefficients for 3 and 55 day intervals ranged from r = .91 to .97 for all conditions. Cram (1986) studied 12 chronic pain patients examining 11 different muscle sites. He found a median coefficient of r = .83 for all sites when assessed and then immediately reassessed. A high degree of variation amongst sites ranging from r = .57 to .99 was reported.

Examination of the procedures utilized in the various studies may account for the differences in reliabilities. Komi and Buskirk (1970) and Jonsson and Komi (1973) utilized tightly controlled positions in which the limb was repositioned in a manner identical to the initial assessment. Cram's (1986) subjects were sitting and standing in unsupported position possibly allowing for slight variations in posture. Biedermann and Monga (1985) also found a high variation between EMG scores on repeated measures. They suggested that the wide variation they found was attributable to barely noticeable shifts in patient positions.

In summary, there appear to be a number of factors which may significantly reduce the validity and reliability of the EMG data.

Comparison of data from one patient to the next becomes difficult for muscle size, muscle fatigue, load, and posture all will affect the outcome.

## WITHIN SUBJECT ANALYSIS

In view of the problems of high individual variability and artifact due to postural shifts practitioners have started using within subject data analysis procedures. Traditionally practitioners are taught to collect data and compare this data to the norm (usually established by collecting data on hundreds of subjects and differentiating the normal from the pathological). The patient is then assigned into the pathological or normal group. EMG data with its tremendous variance makes this assignment difficult. Within subject analysis utilizes the subject as their own control. Data may be compared at the start of the session to the end of the session, or as in more recent research, by comparing the activity of one muscle to another. For example, Cram (1986) and Wolf, Nacht and Kelly (1982) have started studying the activity of the paraspinal muscles comparing the activity of one group (i.e., the left side) to the activity of the other side (i.e., the right side). Until recently, most of this type of work was undertaken in static positions (e.g., sitting, standing). However, recent developments, particularly the interfacing of computers with biofeedback equipment, have led to the rapid development of dynamic (movement) assessment techniques.

## WITHIN SUBJECT MOVEMENT ANALYSIS

In movement analysis, the subject is asked to perform various movements while the EMG monitors the firing patterns of the various muscles. The implications to the practitioner are numerous. First, obviously more than one EMG instrument is needed. It is unreliable to monitor the activity of one muscle group, then set up on the next muscle group and have the subject repeat the movement, for there is no guarantee that the movement will be performed in an identical manner the second time. Second, the practitioner's knowledge of anatomy and physiology will need to be expanded. To ensure valid readings it is important that the belly of the muscle be located and the electrodes be placed over this. Third, it is important the practitioner know the primary function of the muscle to be assessed. For example, the primary function of the extensors of the neck is to draw the head towards the back. It makes no sense to ask the patient to bend from the waist while assessing these muscles. Before the reader gets discouraged and assumes they have to be an occupational or physical therapist, there are some very good books available which outline these procedures and how to perform them.

Finally, there are two equipment issues. Movement analysis cannot be performed without computers. The data occurs at a rate which cannot be collected by human means, then stored and displayed for subsequent analysis. Computers are a must for this. The second issue is that of a universal reference. When more than one EMG channel is utilized and the data will be compared, it must be established that the reference is the same for all electrodes. The universal reference solves this problem. This is particularly important for legal work as diagnoses based upon EMG analysis may be attacked in court! In clinical work where absolute precision is not so necessary this is not as critical. The margin of error between two referenced electrodes is approximately 1% (Dr. J. Lowenbregh, personal communication, August 8th, 1989). To avoid introducing this source of variance into the assessment it is recommended that whenever possible universal references be employed.

## MOVEMENT ANALYSIS - EXAMPLES

Movement analysis represents one of the more exciting developments to emerge in the last several years. The material which follows is based upon data collected in clinics and laboratories throughout North America on approximately 1000 patients (see footnote at the end of the chapter for locations). To start the focus of attention will be the cervical paraspinals (C5-7) of the neck outlining assessment procedures utilizing two channels of EMG data. Then the sternomastoids will be added to the assessment and the EMG data studied from four channels.

The setting up of the electrodes follow the protocols as recommended by BCIA. The skin in the cervical areas on either side of the spine is scrubbed with rubbing alcohol and abraded in the standard fashion. Four electrodes are attached (using double sided adhesive collars) 2.5 cm. from the spineous processes on either side of the spine, 3 cm from center to center. For references on the appropriate set up locations see Basmajian (1983) and Cram (1986). A universal reference is located on the spineous processes in a convenient place. It is important to remember that at this stage it cannot be said that the electrodes are monitoring the activity of a particular muscle, but are monitoring the activity of a number of muscles which are located in this area. The subject, while seated in a supported position, is asked to bend the head forward as far as is comfortable, then lift the head back to neutral and continue back extending the head backwards. This procedure must be completed in a symmetrical position in order to prevent artifact from rotation occurring. The activity from each side is recorded and displayed on a graph. An example of a non-pain subject's movement is outlined in Figure 1.

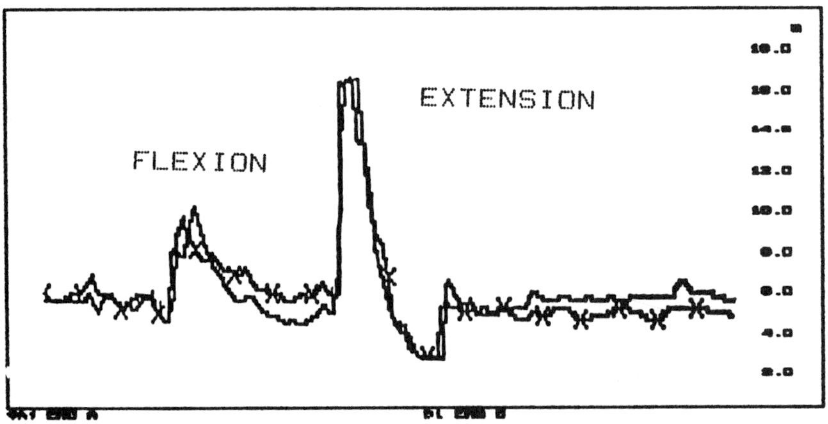


**Figure 1.** Flexion/Extension Pattern of the Cervical Paraspinals - Non Pain Subject[1]

Visual inspection of the graphs reveals a number of things. First, the patterns follow a predictable course. As the subject flexes (bends forward) the activity of the muscles increases then goes quiet. As the subject draws the head up to neutral and then into extension the muscle activity increases to a much greater degree than during flexion. This is due to the force of gravity which must be over come during extension. Second, what is also apparent is that both muscles contract (the amplitude as recorded in mvs) at approximately an even rate producing graphs that are symmetrical in nature. This has led to the development of two rules as first suggested by S. Wolf (J. Cram, personal communication, June 1988). Rule #1 reads as follows: **In non-pain subjects symmetrical movement produces symmetrical patterns**.

Rule #1 may be applied to all movements which involve symmetry of movement (i.e., the body is not rotated in any manner). Some of the more common symmetrical movements are flexion/extension of the waist and head, bilateral abduction of the arms, bilateral drawing back of the scapula, bilateral pressing of the elbows into the rib cage (assesses latismus dorsi activity), bilateral tightening of the gluteus maximus, bilateral abduction of the legs which assesses gluteus medius activity, and chewing when the food is located on midline in the front teeth.

When applied to pain patients Rule #1 becomes: **In pain patients symmetrical movement produces asymmetrical patterns**. In Figure 2, it is readily apparent the right side is not generating as much force as the left side during extension. The exact reasons as to why this occurs is not known. Current thinking suggests that one or several factors may cause the reduced recruitment, including: a) hyperextension of the muscle myofibrils specifically tearing some of the myosin and actin, b) joint displacement, or c) a dysfunction of the gamma motor circuit in which the response bias of the motor circuit is involved. Recent research (Ahern, Follick, Council, Laser-Wolston, & Litchman, 1988; Donaldson, 1989) indicates that these procedures are highly correlated with the reported pain and discriminate between pain and non-pain sufferers. It must be emphasized that to ensure valid results the subject must be symmetrically positioned and the movement must be completed in a symmetrical manner.

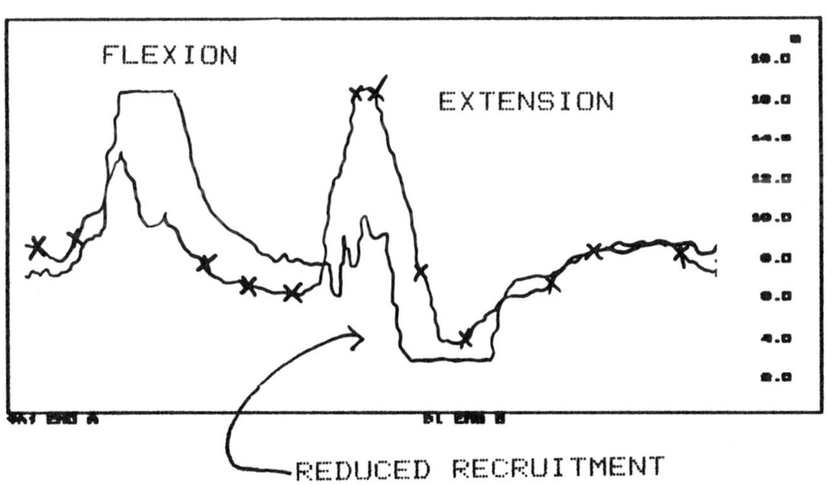


**Figure 2.** Flexion/Extension Pattern of the Cervical Paraspinals - Pain Subject. Note: the activity of the left cervical paraspinals is marked with x's and for graphing purposes is reduced by a factor of 4.

After completing the flexion/extension pattern to the therapist's satisfaction (four or five repetitions completed with clear and consistent results obtained for the particular group of muscles) the data is recorded and stored for future analysis. The subject is then asked to turn the head to the left and then the right. This produces an asymmetrical pattern which is predictable and consistent. Rule #2 states: **In a non-pain patient an asymmetrical movement produces an asymmetrical pattern.** In Figure 3, it may be seen that as the subject rotates the head to the left, the left paraspinals fire (increase in magnitude) until rotation is completed. The right paraspinals are essentially quiet with only a little burst of activity seen. As the head rotates to the right, activity of the left paraspinals decreases while that of the right increases, until rotation is completed. As seen in the third burst of activity (second rotation to the left) a partial rotation of the head produces only a partial firing pattern, so it is imperative that the subject performs a consistent and complete rotation every time.

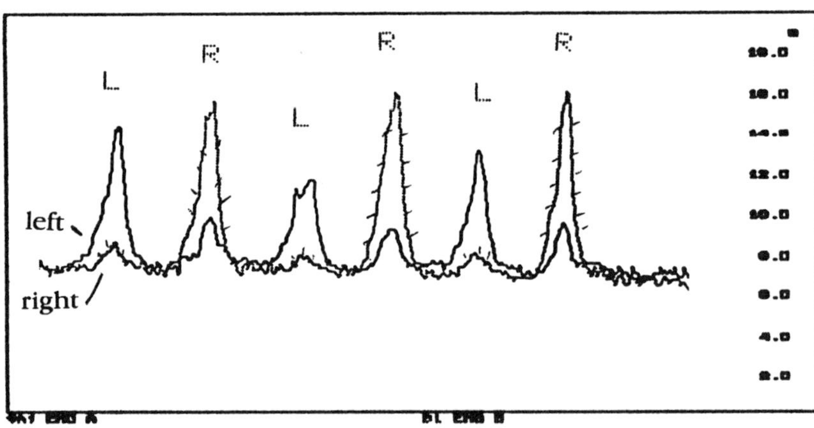


**Figure 3.** Rotations of the Head Showing Cervical Paraspinal Activity - Non-Pain Subject

As in the first rule, the opposite holds true for pain patients. For pain patients, Rule #2 becomes: **An asymmetrical movement produces a symmetrical pattern.** In Figure 4, rotation of the head to the left produces very little activity of the left side paraspinal processes.

Rotation of the head to the right produces the appropriate activity as evidenced by the firing pattern of the right cervical processes. Often the dysfunctional pattern (in this case the rotation to the left) produces an even greater symmetrical pattern than is evidenced here. The activity of the functional muscle (in this case the right cervical paraspinals) may rise as high if not higher than that produced by the dysfunctional muscle. If one muscle may be conceptualized as not producing adequate force, then it may be speculated that the other muscle assists in completing the movement, producing the inappropriate patterns seen in Figure 4. It should be noted in Figure 4 that the inappropriate pattern is produced only during the rotation to the left as the right cervical paraspinals produce adequate activity during movement to the right.

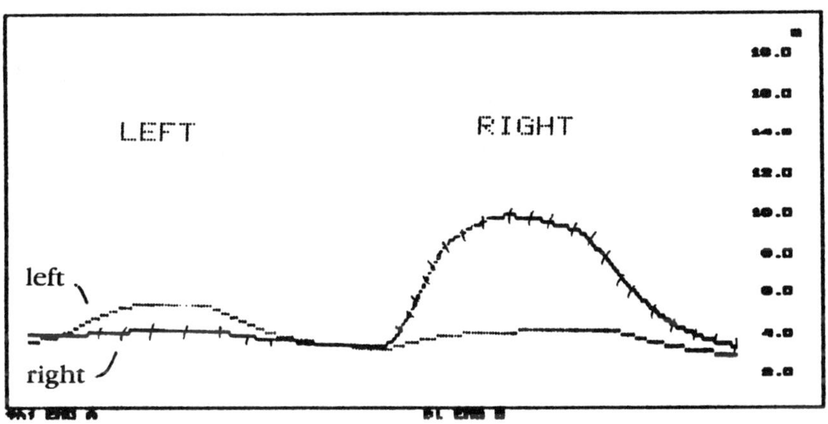


**Figure 4.** Rotations of the Head Showing Cervical Paraspinal Activity - Pain Subject

As most movements involve more than one set of muscles, it becomes important to be able to assess all involved muscles as possible. For example, the sternomastoids are actively involved with the cervical paraspinals in movement of the head. Utilization of systems that allow for the simultaneous assessment of several muscle groups is important. Our own clinical work utilizes four channels, for we have found that this allows for examination of two sets of muscles which more often than not accounts for most of the muscle activity. In trying six and eight channels of data it was found to be too difficult to simultaneously process so much data. Most of the present systems are set to handle four channels with some development presently occurring on eight and sixteen channel systems.

The rules as outlined above apply to four channel assessments exactly in the same fashion as for two channel assessments. In our hypothetical case, the sternomastoids (SCMs) will now be added to the assessment procedure. The electrodes are attached in the standard fashions for both the cervical paraspinals and the SCMs. Again, it must be emphasized that adherence to rigid clinical procedures is important, for as more electrodes are attached there becomes a greater possibility of error (particularly inexact or unequal electrode placement). The assessment proceeds in a standard fashion as outlined above. The subject is asked to perform a flexion/extension movement with data being recorded from all four sites simultaneously as seen in Figure 5. For the non-pain subjects, symmetry of activity may be seen bilaterally between muscles. In Figure 5, the flexion activity is poorly defined but the symmetry of recruitment for the extension part of the movement is evident. It should be noted that flexion/extension of the neck is not the primary function of the SCMs so less activity (32 mvs as compared to 64 mvs in the cervical paraspinals) is expected.

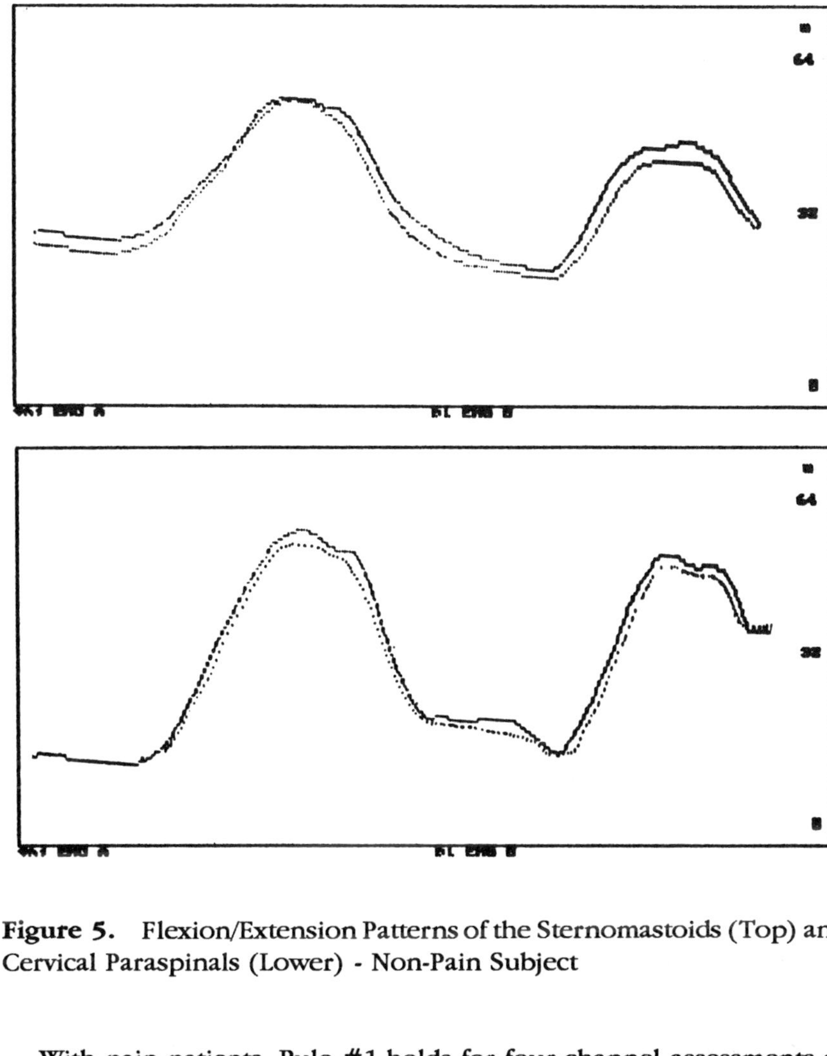

**Figure 5.** Flexion/Extension Patterns of the Sternomastoids (Top) and Cervical Paraspinals (Lower) - Non-Pain Subject

With pain patients, Rule #1 holds for four channel assessments as well. In the next example, the pain patient is performing a flexion/extension routine. The reduced recruitment may be seen primarily in the left cervical paraspinals and to a lesser extent in the right SCM as demonstrated in Figure 6. This movement is repeated twice in this graph with the second repetition interrupted in mid-motion so as to capture the absolute differences in magnitude.

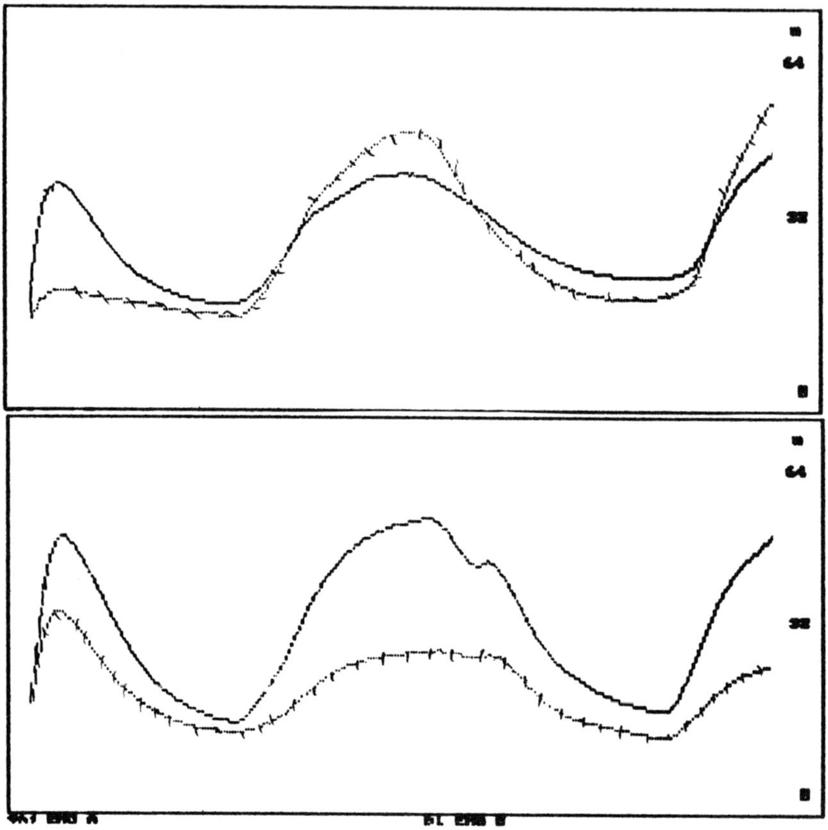

**Figure 6.** Flexion/Extension Patterns of the Sternomastoids (Top) and Cervical Paraspinals (Lower) - Pain Subject

The procedure then requires the subject to rotate the head as per the two channel assessment. The activity of the SCMs is apparent here. In Figure 7, for our non-pain subject, as the head rotates to the left, the left cervical paraspinals fire as do the right SCMs. This may be conceptualized as the cervical muscles pulling the head in the desired direction while the SCMs push. This process is then reversed for the rotation of the head to the right. In Figure 7, there are six rotations, alternating rotation to the left, then to the right. **The primary advantage of the four channel system is the assessment of the dynamic interplay between muscle groups.**

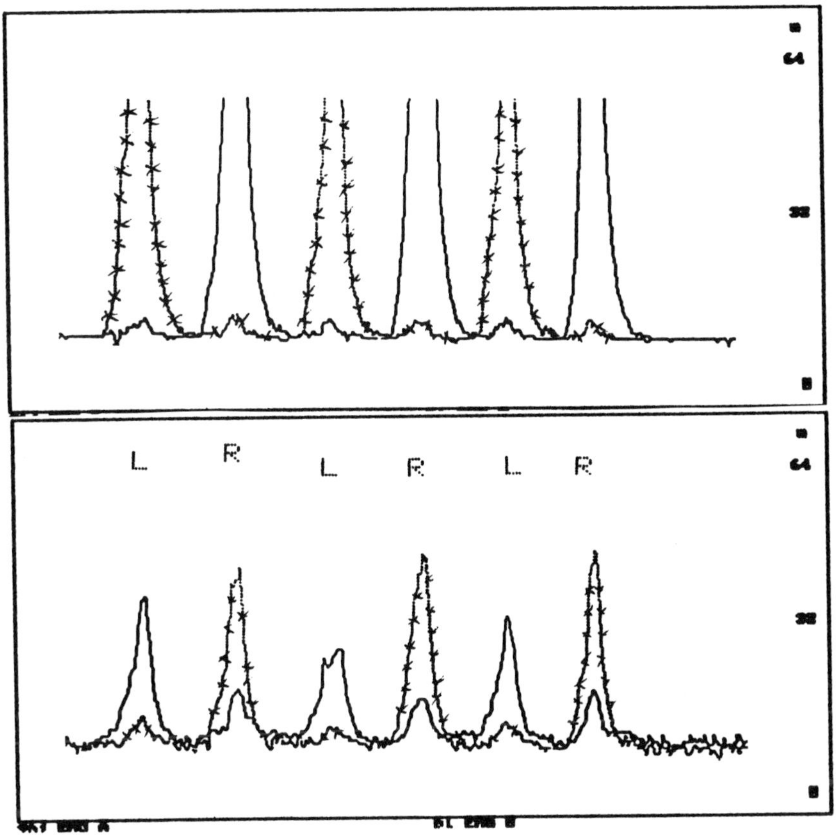


**Figure 7.** Rotations of the Head Showing Sternomastoid (Top) and Cervical Paraspinal (Lower) Activity - Non-Pain Subject

Figure 8 represents the rotation of the head by a chronic pain patient. In this case, as the head turns to the left the graphs appear abnormal. There is reduced recruitment of the left cervical paraspinals and the SCM. However, as the patient turns the head to the right both the right cervical paraspinals and the left SCMs show normal recruitment.

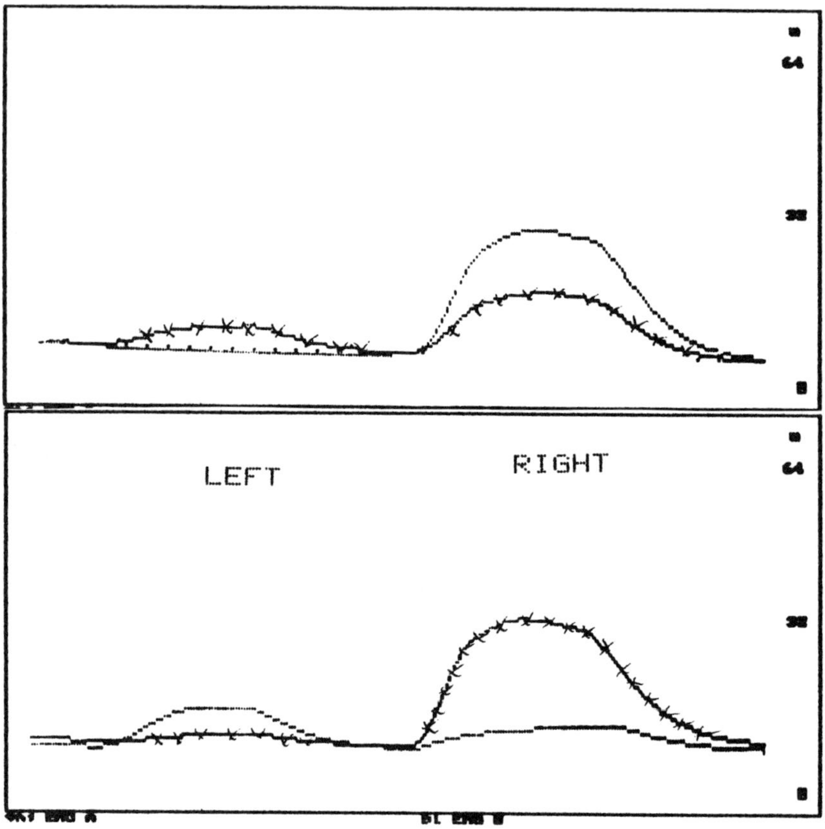


**Figure 8.** Rotations of the Head Showing Sternomastoid (Top) and Cervical Paraspinal (Lower) Activity - Pain Subject

The above pattern is characteristic of a person who was involved in a rear end collision. The data tells that the head was slightly rotated to the left at the moment of impact. As the head is initially snapped backwards, the right SCM is hyperextended; as the head is then thrown forward, the left cervical paraspinals are also hyperextended. This process produces a pattern of reduced recruitment which is characteristic of motor vehicle trauma patients. This pattern of reduced recruitment may be documented for other muscles and forms a rather consistent pattern of injury for this type of accident.

As noted earlier, placement of the electrodes over the belly of the muscle does not guarantee that the activity of that muscle is being exclusively monitored. Often crosstalk between muscles affects the validity of the readings. In placement of electrodes over the cervical processes it is possible to monitor the activity of both the flexors/extensors of the neck and the rotators of the neck. In situations where more than one muscle may be assessed, it is important that all possible movements that could be driven by the muscles be performed. Then the assessment becomes that of logical deduction. For example, if in the above example the rotation movements showed no significant differences, but the flexion/extension movement was abnormal, then it is concluded that the flexor/extensors are hyperextended and the rotators are normal. If the patterns are reversed, then the opposite is concluded. If both patterns are abnormal, then it is concluded that both sets of muscles are hyperextended.

## OTHER ASSESSMENT PROCEDURES

In addition to the points listed above, it is possible to document inappropriate muscle activity representing other problems. Often in examination of movement patterns the patterns will be normal in respect to symmetry and rotation, but reveal inappropriate activity during movement. For example, the cervical paraspinals should essentially remain quiet during chewing. In TMJ patients it is often found that one or both of the cervical paraspinal muscles co-contract with the masseters. In Figure 9, the right cervical paraspinal muscles are in perfect co-contraction with the masseters. It is suspected that this is a learned phenomena that at one point was compensatory for a dysfunctional muscle elsewhere, but did not extinguish with correction of the problem. Figure 9 portrays this phenomena of co-contraction which we have labelled chewing with the neck.

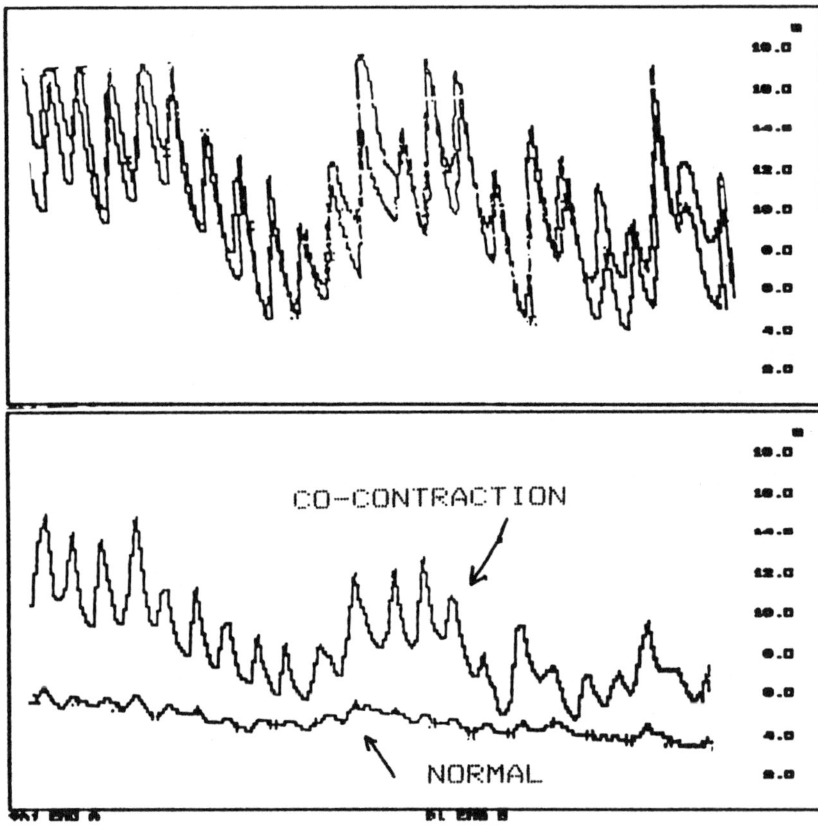


**Figure 9.** Co-contraction of the Right Cervical Paraspinals (Bottom) With Normal Chewing Action of the Masseters - Pain Subject

Finally, it is possible to document inappropriate muscle activity occurring as a form of compensation. In this case, muscles are found to be working when they should be quiet. For example, in bilateral arm abduction the muscles of the neck serve to stabilize the head during abduction. The SCMs should remain quiet during this activity. In Figure 10, the activity of the cervical paraspinals is quite appropriate, while there should be little or no activity of the SCMs. In this case it is suggested that incorrect posture, in which the head is thrust forward, causes this activity. When movement occurs, the cervical paraspinals shorten (contract) causing the SCMs to lengthen. It appears that it is the lengthening of the SCMs that is recorded.

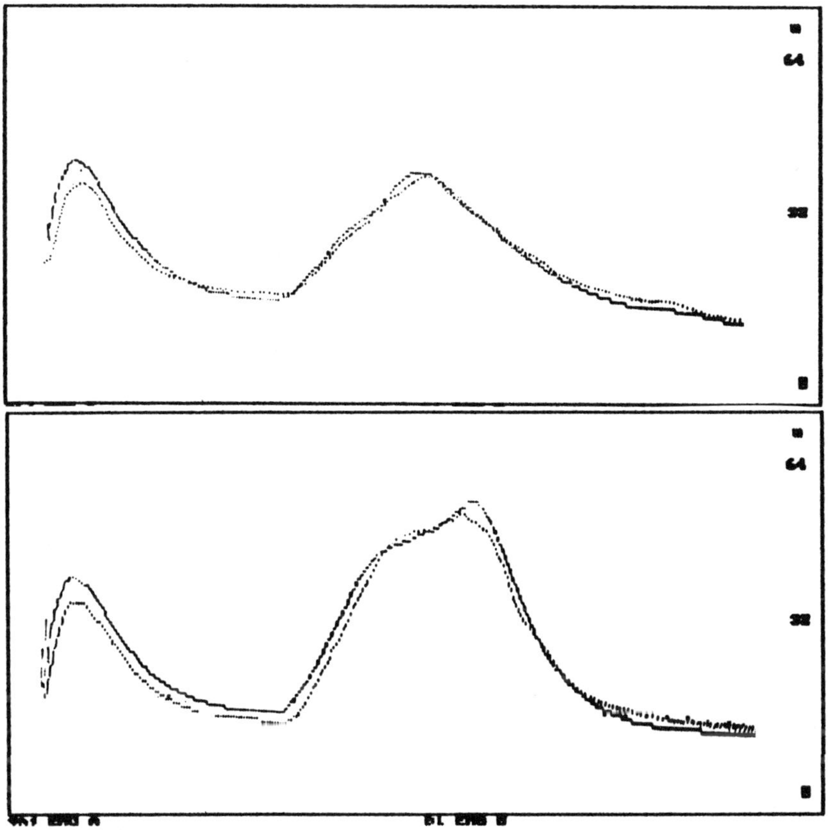


**Figure 10.** Inappropriate Recruitment of the Sternomastoids (Top) and Appropriate Recruitment of the Cervical Paraspinals (Lower) During Bilateral Arm Abduction - Pain Subject

In summary, multi channel EMG assessment procedures allow for the assessment of muscle activity that is impossible to otherwise assess. In particular, the amount of force that a muscle generates during a movement may be assessed at any particular moment in time. This activity may be compared to other muscles establishing the forces occurring about a joint. Evidence indicates that pain and non-pain sufferers produce differing patterns which are diagnostically significant.

## SUGGESTED THEORETICAL MODEL OF CHRONIC MUSCULAR PAIN

If as suggested, the EMG signal represents the amount of force that a muscle generates, it is then possible to monitor the force that a muscle generates during, and at certain points during, movement. Muscles work in conjunction with and in opposition to other muscles. A basic principle of physics states that an unequal force about a point creates movement. Body movement is caused by creating an imbalance of force about a joint which then produces movement about that joint. When working in balance and symmetry, muscles maintain a healthy level of activity which function in appropriate and consistent patterns. When muscles are hyperextended, they produce less force than their partner. In the situation where both paraspinals operate to produce movement, it is suspected that the loss of force by one partner is made up by the other partner. This shows up in increased EMG activity of the over worked muscle. Eventually, this over worked muscle goes into spasm, producing the pain-spasm-pain cycle and respective trigger points.

A word of caution is needed here. If the patterns discussed above are seen, do not automatically assume that the muscle showing the lowered readings is hyperextended. It is possible to get patterns of reduced activity when a joint is injured or dysfunctional. The material presented is based upon subjects and patients who were medically screened with joint problems eliminated as much as possible. The diagnosis of hyperextension is arrived at only when other medical factors have been eliminated.

## TREATMENT TECHNIQUES USING MULTI CHANNEL EMG PROCEDURES

The treatment of chronic pain and in particular muscular pain follows directly from the assessment. In general, there are two rules which apply to most problems. First, in problems in which there is unequal recruitment (amplitude) during movement (either symmetrical or asymmetrical patterns), the recruitment should be retrained to that of an equal amount of activity. Second, in problems of inappropriate activity, but an equal amount of activity between partners, the inappropriate activity needs to be extinguished. These will be discussed in detail later.

Much work has been undertaken in examining the issue of muscle asymmetry. The basic issue is that of which side should be treated: should the high side be reduced in activity or should the low side be increased in activity? Most present treatment programs focus upon the reduction of the activity of the high side, for most often, that is where the pain is reported. Physiotherapy typically utilized hot packs, stretching and massage to reduce the pain, while relaxation techniques (either focusing on relaxation of the body in general or specifically on a particular muscle) have been utilized by psychology. The results of limited research is equivocal with some studies indicating pain relief, while others indicate little relief (Turk & Flor, 1984). In particular, long term outcome studies show a return to baseline of the reported EMG activity and in most cases a return of the pain. However, it should be noted that these studies utilized EMG data obtained from static assessments as the independent variable.

Cram (1986) reports on a technique similar to those reported above, but slightly different in that movement analysis is involved. In this technique, labelled "stretch to relax," the subject is trained through EMG procedures to lengthen a contracted muscle, decreasing the spasm. In clinical trials (R. Clasby, personal communication, August, 1989) the stretch to relax procedure produces symmetry in the movement patterns after three sessions. This return to symmetry corresponds with a reduction in pain which has been maintained at follow up (three months).

Efforts have also been directed towards increasing the activity of the side that shows reduced recruitment during movement. There is precious little research presently in this area. Wolf, Nacht and Kelly (1982) trained a subject with low back pain (5 months duration) to increase the activity on the low side and to equalize the lumbar paraspinal activity during standing, flexion, extension and rotation. The pain level as reported on the McGill Pain Questionnaire dropped to zero during treatment and was maintained at follow up. Donaldson (1989) trained 12 chronic (7 years) low back pain subjects to increase the activity of the lumbar paraspinals (low side) using SMU (single motor unit) training procedures. There was a rapid decrease in pain which was maintained at follow up and corresponded to a decrease in EMG activity on the high side.

In the above studies, the EMG biofeedback was utilized as the tool of change. However, although preliminary, other studies indicate that symmetry may be obtained using other techniques. Donaldson (1989) trained 12 chronic low back sufferers (7 years duration) to obtain symmetry through the use of education techniques. The subjects were trained in concepts of muscle activity; how they made their backs asymmetrical, and how to move into symmetry. While the results were not as rapid as those of the biofeedback group, this group reported decreased pain at three months follow up which corresponded to a change in the EMG activity of the lumbar paraspinals. It is suspected that exercise programs would also move the back into more of a symmetrical position by making the muscles more efficient in their firing, reducing their degree of asymmetry. However, at present a study supporting this position is not available.

Figure 11 shows a flexion/extension pattern of the cervical paraspinals at pre treatment, while Figure 12 shows the same activity post treatment. The degree of asymmetry (pre treatment) in Figure 11 and the return to symmetry (post treatment) in Figure 12 is obvious. In Figure 11, hyperactivity of the lateral aspects of the trapezius is also evident. This appears to be some form of a compensatory pattern reactive to the hyperextension of the upper trapezius (cervical paraspinals). In this case, the movement data is treated as the independent variable. A decrease in reported pain corresponded to this change.

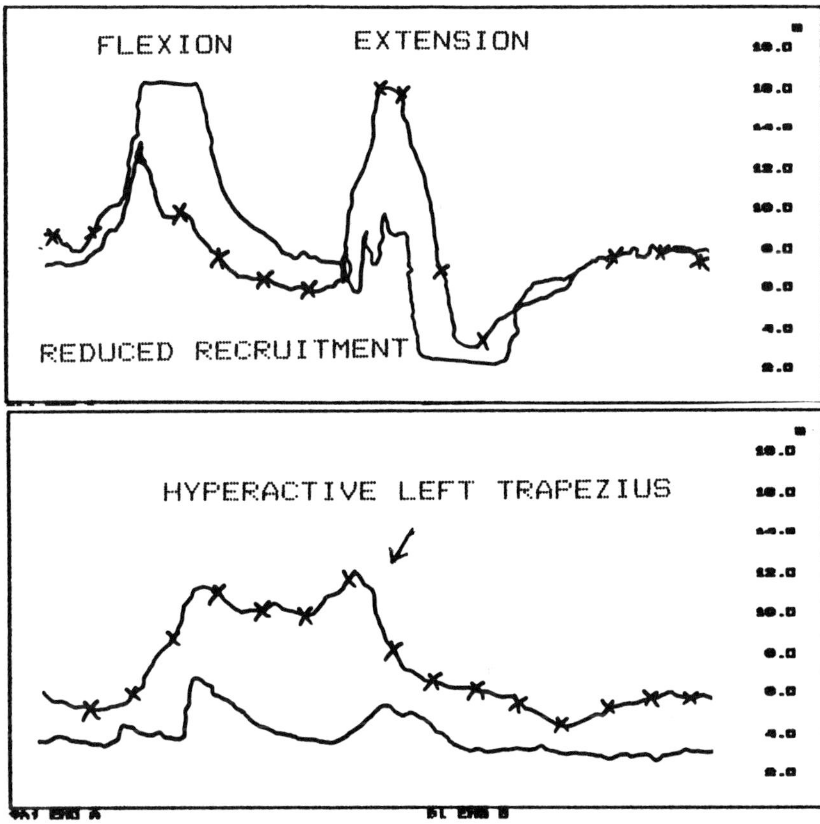


**Figure 11.** Pre-treatment Flexion/Extension Assessment of the Cervical Paraspinals (Top) and Trapezius (lateral) (Lower) - Pain Subject

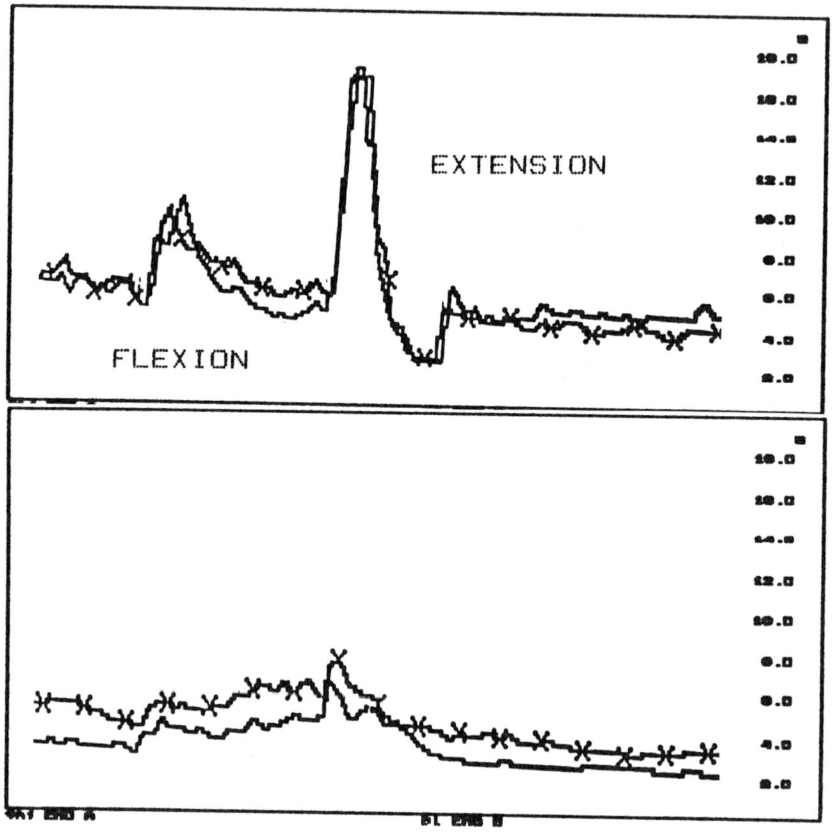


**Figure 12.** Post Treatment Flexion/Extension Assessment of the Cervical Paraspinals (Top) and Trapezius (Lateral) (Lower) - Pain Subject

Figure 13 shows the same cervical paraspinals assessed in a static position pre treatment, while Figure 14 shows the level of activity at post treatment. The decrease in spasm is obvious, dropping from an activity level of 41.08 (left side) and 10.43 (right side) (upper graphs Figure 13) to 2.62 and 2.63 respectively post treatment (upper graphs Figure 14). Remember in this case no treatment was completed involving static positions with the static activity treated as a dependent variable.

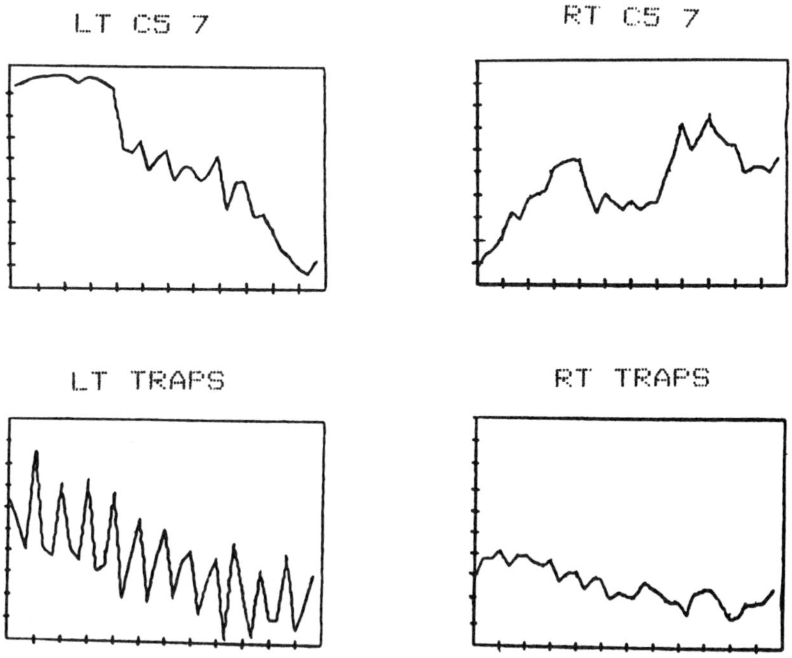


**Figure 13.** Pre-treatment Static Assessment of the Cervical Paraspinals (Top) and Trapezius (Lateral) (Lower) - Pain Subject

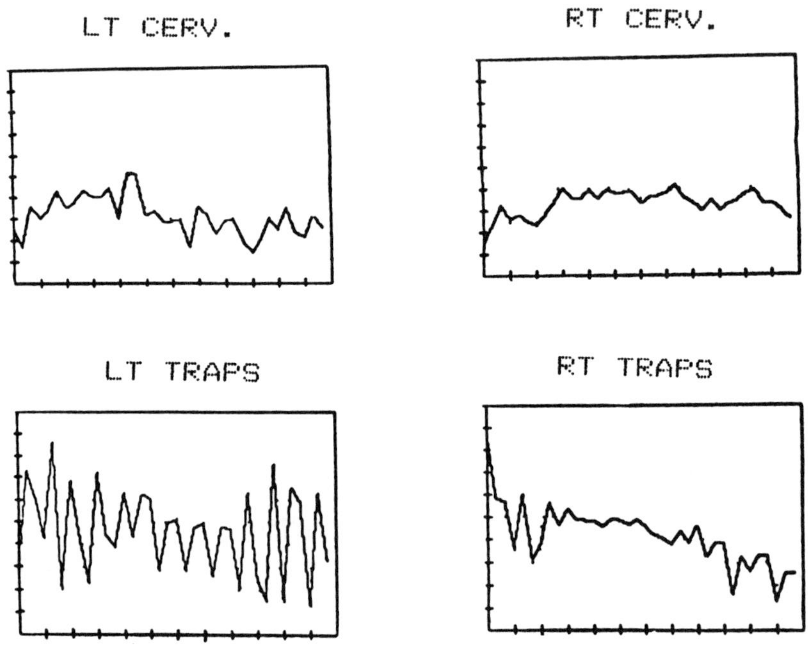


**Figure 14.** Post Treatment Static Assessment of the Cervical Paraspinals (Top) and Trapezius (Lateral) (Lower) - Pain Subject

Figure 15 shows a similar type of process occurring, but for an asymmetrical movement. The side showing the reduced recruitment was trained to increase its activity with the result that symmetry of amplitude was obtained for the two rotations. Concurrently, a reported reduction of pain occurred.

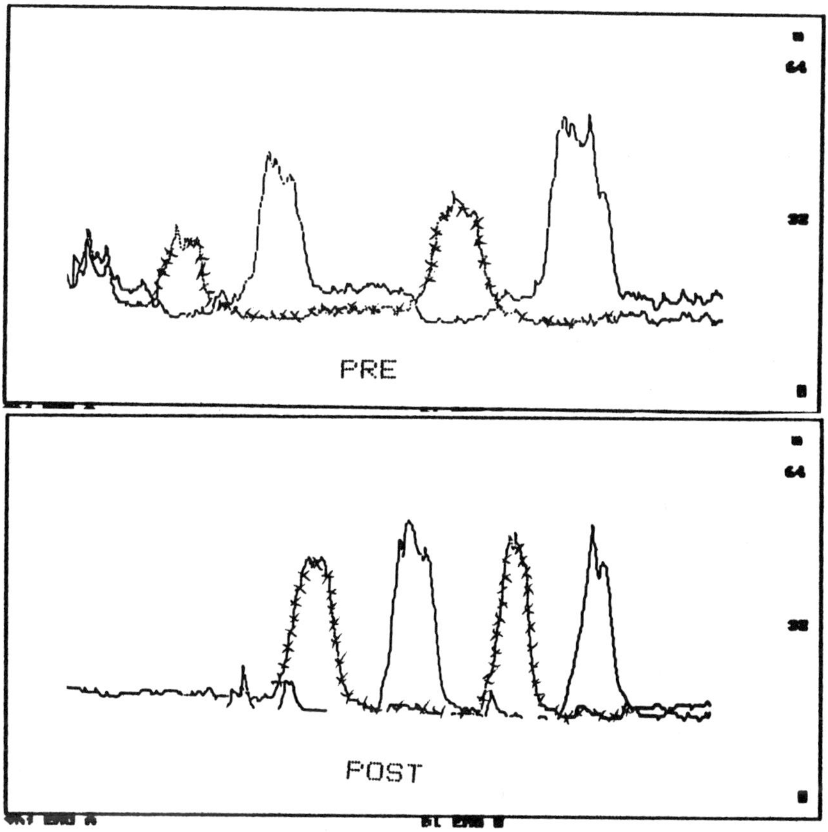


**Figure 15.** Pre and Post Treatment Graphs of Sternomastoid Activity During Rotations of the Head - Pain Subject

In summary, the return to symmetry (both sides producing an approximately equal amount of activity during movement) appears to be associated with a reported decrease in pain. It is suspected that this decrease in pain occurs in the muscles that are compensating for the lack of activity in the side opposite. As the muscle that shows the reduced recruitment strengthens, the demands placed on the other muscles decrease allowing these muscles to relax and breaking the pain-spasm-pain cycle.

The second use of the EMG biofeedback, the documentation of inappropriate muscle activity during movement, also produces results similar to those outlined above. The theory suggests that these muscles are in spasm due to their being fired inappropriately and therefore excessively. In these cases, the activity is extinguished by having the patient focus upon performing the desired movement while holding the muscle quiet. This procedure requires the use of an EMG during training, as often these patterns occur without the patient's awareness. The procedure is often lengthy, requiring patient compliance and continued motivation, but these problems may be substantially reduced by the use of a home trainer.

Finally, a comment about an area that is just starting to emerge in clinical settings. The EMG instrument provides information to the therapist which is otherwise impossible to obtain. For example, in treatment therapists have to rely upon their intuition and the patients' reports about the effect of the treatment. The EMG, when utilized in conjunction with standard treatments, can tell the therapist when the desired goal has been reached. This offers the therapist a tremendous advantage, as at all times it is known if the treatment is effective or if it should be modified.

## THE CRITICAL DIFFERENCE

It is important to recognize that a slight degree of asymmetry or imbalance exists in all people during movement. The question is at what point does the degree of imbalance become pathological and where in the movement should the imbalance be measured? If the EMG is, as suggested, a measure of force then it seems logical to measure the signal at the point of maximum contraction (for either side). Measurement at this point reveals the maximal load bearing aspects of the movement and appears to correlate ($r = .61$) very well with report pain (Donaldson, 1989).

Examination of data from over 200 clinical patients and 76 research subjects (all in pain) indicates that in 97.5% of the cases a 20% or greater imbalance was present. Examination of data from 40 non-pain subjects indicated that most subjects had an imbalance in the 5 to 10% range. In only one case has perfect symmetry been found. The degree of imbalance is established by locating the point of maximum contraction, then subtracting the reading of the opposite (lower) side to establish the difference. This score (the difference) is then divided by the score of the <u>low</u> side to obtain the percentage. If the difference is divided by the high side reading, the 20% rule shifts downward to approximately 18.5% needed for significance. Examination of the readings from Figure 6 shows the difference between the SCMs to be 11.4 mvs, which when divided by the low side reading (47.5 mvs) produces a score of 24%. The cervical paraspinal readings revealed a difference of 32.6 mvs, which when divided by the low side (23.0 mvs), produces a result of 141.7%. This great an imbalance is routinely seen in pain patients.

It must be emphasized that at this point in time, the 20% rule is speculative and applies to only the paraspinals of the back. The muscles of the body vary in their size and strength and consequently in the amount of force that each generates. It is logical to assume that the larger muscles of the body would require less of a degree of an imbalance to produce pain, while the small muscles would require more of a difference. In addition to the size, the tone of the muscle, the level of fatigue and the general state of health will all affect the readings. Finally, it should be noted that choice of the maximum contraction was based upon intuition and the high correlation between it and reported pain scores (Donaldson, 1989). There may be better ways to analyze the data (i.e., compare the areas under the curve), but this remains to be determined.

## APPLICATIONS TO OTHER SITES

The material presented focused upon the assessment of the sternomastoids and the cervical paraspinals. This was not to imply that other muscles could not be assessed in the same manner. The protocols as outlined may be utilized for any chronic pain problem which involves muscle pain. The data, the imbalances and the outcome (reduction in pain) follow similar patterns.

## CONCLUSION

The material presented in this chapter represents data gathered in clinical settings. As interesting and exciting as it may appear, much research remains to be completed. Protocols need to be developed and standardized, then subjected to scientific scrutiny before these procedures can be accepted as valid. These techniques require the practitioner to (a) modify his/her method of thinking to within subject procedures, (b) have a knowledge of anatomy and physiology, (c) utilize a computer with a minimum of two EMGs, and (d) be meticulous in the application of electrodes. Within these limitations, the techniques reported here appear to be promising, offering the biofeedback practitioner new techniques for the assessment and treatment of soft tissue injury.

---

Footnote 1. All graphs utilized in this chapter were obtained from patients in clinical settings throughout North America. The assessments and treatments were conducted by the author at the Calgary Pain Treatment Center (Calgary, Alberta), by Ms. Pat Bittermann of the Family Therapy Institute of Maine (Portland, Maine) and by Mr. Bob Clasby of the Biofeedback Institute of Cape Cod (Orleans, Mass.). A special thanks is expressed to them for letting me use their materials.

## References

Ahern, D. K., Follick, M. J., Council, J. R., Laser-Wolston, N., & Litchman, H. (1988). Comparison of lumbar paravertebral EMG patterns in chronic low back pain patients and non-pain patients. *Pain*, 34, 153-160.

Anderson, B. J. G., Jonsson, B., & Ortengren, R. (1974). Myoelectric activity in individual lumbar erector spinae muscles in sitting. A study with surface and wire electrodes. *Scandinavian Journal of Rehabilitation Medicine*, Supplement 3, 91-108.

Basmajian, J. V. (1983). *Biofeedback: Principles and Practice for Clinicians*. Baltimore: Williams & Wilkins.

Basmajian, J. V. (1988). Research foundations of EMG biofeedback in rehabilitation. *Biofeedback and Self-Regulation*, 13 (4), 275-298.

Basmajian, J. V., & Blumenstein, R. (1983). Electrode placement in electromyographic biofeedback. In *Biofeedback: Principles and Practice for Clinicians*. Ed. J. V. Basmajian. Baltimore: Williams & Wilkins. 363-377.

Biedermann, Heinz-Joachim, & Monga, T. N. (1985). Relaxation oriented EMG biofeedback with back pain patients: Evaluation of paraspinal muscle activity. *Clinical Biofeedback and Health*, 8 (2), 119-123.

Bigland, B., & Lippold, O. C. J. (1954). The relation between force, velocity and integrated electrical activity in human muscles. *Journal of Physiology*, 123, 214-224

Bouisset, S. (1973). EMG and muscle force in normal motor activities. In *New Developments In Electromyography and Clinical Neurophysiology*. Ed. J. E. Desmedt. Basel: Karger. 1, 547-583.

Bouisset, S., & Maton, B. (1973). Comparison between surface and intramuscular EMG during voluntary movement. In *New Developments in Electromyography and Clinical Neurophysiology*. Ed. J. E. Desmedt. Basel: Karger. 1, 533-539.

Cram, J. R. (1986). *Clinical EMG: Muscle Scanning and Diagnostic Manual for Surface Recordings*. Seattle: Clinical Resources

Donaldson, C. C. S. (1989). The effect of correcting muscle asymmetry upon chronic low back pain. Unpublished doctoral dissertation, University of Calgary, Calgary.

Fujiwara, M. Miura, H. Namasaki, K., & Hasue, M. (1979). Kinematic and electromyographic analyses of the strength of the abdominal and back muscles. Paper presented at the 4th Congress of I. S. E. K., Boston.

Guyton, A. C. (1971). *Basic Human Physiology: Normal Function and Normal Mechanisms of Disease*. Philadelphia: Saunders.

Janda, V. (1983). Pain In the Locomotor System: A Broad Approach. Unpublished manuscript.

Jonsson, B., & Komi, P. V. (1973). Reproducibility problems when using wire electrodes in electromyographic kinesiology. In *New Developments in Electromyography and Clinical Neurophysiology*. Ed. J. E. Desmedt. Basel: Karger. 1, 540-546.

Komi, P. V., & Buskirk, E. R. (1970). Reproducibility of electromyographic measurement with inserted wire electrodes and surface electrodes. *Electromyography*, 4, 357-367.

Komi, P. V., & Viitasalo, J. H. T. (1976). Signal characteristics of EMG at different levels of muscle tensions. *Acta Physiolica Scandinavica*, 96, 267-276

Mulder, T., & Hulstijn, W. (1985). The effect of fatigue and task repetition on the surface electromyographic signal. In *The Learning of Motor Control Following Brain Damage: Experimental and Clinical Studies*. Ed. T. Muler. Lisse: Swets & Zeitlinger. 54-66

Peek, C. J. (1987). A primer of biofeedback instrumentation. In *Biofeedback: A Practitioner's Guide*. Ed. M. S. Schwartz. New York: Guilford. 73-127.

Stulen, F. B., & De Luca, C. J. (1979). *Mean frequency of the myoelectric signal as a measure of localized muscular fatigue*. Paper presented at the 4th Congress of I. S. E. K., Boston.

Turk, D. C., & Flor, H. (1984). Etiological theories and treatments for chronic back pain. II Psychological models and interventions. *Pain*, 19, 209-233.

Vredenbregt, J., & Rau, G. (1973). Surface electromyography in relation to force, muscle length and endurance. In *New Developments in Electromyography and Clinical Neurophysiology*. Ed. J. E. Desmedt. Basel: Karger. 1, 607-622.

Wolf, S. L., Nacht, M. & Kelly, J. L. (1982). EMG feedback training during dynamic movement for low back pain patient. *Behavior Therapy*, 13, 395-406.

# CHAPTER 3

# *Dynamic EMG Biofeedback in Assessment and Treatment Using a Neuromuscular Re-education Model*

by

Will Taylor, M.D.

---

The use of EMG biofeedback in the assessment and treatment of pain syndromes has suffered from a lack of adequate conceptual models regarding the role of abnormal motor activity in the generation and maintenance of pain. The simplistic model of generalized excess activity present at rest, independent of posture, has been demonstrated experimentally and clinically to be a rare explanation for pain generation (Dolce & Raczynski, 1985). Adding site-specificity, such as paraspinal recording for low back pain, or wide-spread placement or muscle scanning for headache, results in a somewhat improved correlation between abnormal motor activity and pain (Hudzinski & Lawrence, 1988; Cram, 1986). The most robust demonstrations of abnormal motor activity in pain syndromes have been found in those studies that have employed ambulatory monitoring (Sherman & Sherman, 1989) or monitoring during movement in the clinical setting (Wolf, Nacht & Kelly, 1982; Jones & Wolf, 1980; Middaugh, 1982; Middaugh & Kee, 1987).

I would like to suggest that as the primary role of muscle is movement, its primary malfunction will be movement disorder; and that in investigating abnormal motor function in pain syndromes, it will be most fruitful to pay attention to disordered activity in movement. In a sense, abnormal activity at rest, and over various static postures, may be seen as a limited reflection of movement disorder, or the small "tip of the iceberg" of total muscle dysfunction. Individuals with abnormal resting activity invariably exhibit even more striking abnormalities of disorder in movement. More significantly, many pain patients with normal resting EMG activity demonstrate significant abnormalities during movement.

I find that consideration of several principles is helpful in interpreting normal and abnormal motor activity in movement:

(1) The concept of movement homeostasis,
(2) The clinical entity of myofascial pain and dysfunction (Travell & Simons, 1983),
(3) Decreased/distorted proprioceptive awareness in motor dysfunction,
(4) Sherington's reciprocal inhibition, and more inclusively, neurologic mechanisms of muscle antagonism and synergy (Basmajian & DeLuca, 1985), and
(5) Patterns of postural and phasic muscle disturbance (Janda, 1969, 1978, 1983; Lewit, 1985).

## MOVEMENT HOMEOSTASIS

Many physiologic and behavioral processes appear to be involved in the maintenance of normal patterns of movement. Important among these are proprioceptive awareness, neurologic mechanisms of muscle antagonism and synergy, and learning. Within certain limits, disturbances in motor activity may be corrected by these homeostatic processes; however, beyond these limits spontaneous correction may not be possible, and the same physiologic processes may serve heterostatically to maintain movement dysfunction. Significant to effective therapy is the possibility that interventions need not create perfect correction of dysfunction; return of motor activity to the *range of effective homeostasis* may permit the normal homeostatic mechanisms to restore functional movement patterns.

## MYOFASCIAL PAIN AND DYSFUNCTION

Functional pain of muscular origin is best interpreted within the model of myofascial dysfunction described by Travell and Simons (1983). Myofascial pain and dysfunction may occur as a primary process in response to muscular injury, overuse or disuse. It also often accompanies other pain syndromes such as intervertebral disc disease, shoulder impingement syndromes or tendinitis.

Myofascial dysfunction manifests as pain referred in characteristic patterns from the affected muscle(s). A single muscle may be affected, particularly in acute syndromes, but more often a group of biomechanically related muscles are involved. The characteristic lesion is the trigger point; a tender, palpable nodule within a taught band of muscular tissue, which behaves in characteristic manner (not all tender areas in muscle are trigger points - see Travell & Simons, 1983). The affected muscle(s) may be hyperirritable (prone to over-activity or spasm) or inhibited (see below). Stimulation of the trigger point(s) by pressure, stretch or muscular use results in pain, which may be referred a considerable distance from the trigger point in consistent patterns which have been well described for individual muscles (Travell & Simons, 1983; Simons & Travell, 1983).

## DECREASED/DISTORTED PROPRIOCEPTIVE AWARENESS

Although muscle tension and body-part position-in-space information are accessible to consciousness, they are not generally on the forefront of awareness. In chronic pain syndromes they may be severely distorted or inaccessible. Individuals with 30uV of "resting" activity in the upper trapezius may describe that muscle as "pretty relaxed," and may be unable to discriminate conditions under which that muscle exhibits 3uV vs 50uV of activity. Often the changes brought about with EMG biofeedback-assisted neuromuscular reeducation result in physical abreactions -- positions feel wrong ("my shoulders feel like they're down around my knees," "I feel like a giraffe") and tension levels may feel wrong. These distorted sensations of position and tension may be important factors in the perpetuation of dysfunction, preventing the spontaneous return of normalcy; addressing this is of crucial importance in effective neuromuscular reeducation.

## RECIPROCAL INHIBITION, PATTERNS OF ANTAGONISM AND SYNERGY

Muscles work together in any simple or complex movement pattern with carefully orchestrated patterns of interaction. The concepts of antagonism and synergy are relative ones, as the functional relationship of one muscle to another depends on the particular movement being performed. For example, the upper trapezius and lower trapezius (considered separate muscles) are antagonists in elevation/depression of the shoulder blade (scapula), but are synergists in scapular rotation. These patterns of interaction are determined in part by spinal and cerebellar reflexes, and in part by more complex movement stereotypes at the cortical level. Both of these levels are responsive to learning (Basmajian & DeLuca, 1985; Wolf & Binder-MacLeod, 1989).

Excessive activity in a muscle may result in excessive reciprocal inhibition of its antagonist(s); conversely, inhibition of a muscle by local factors (such as myofascial dysfunction) may result in disinhibition with resulting over-activity of its antagonist(s). Thus, myofascial dysfunction of a particular muscle, with resulting hyperirritability or inhibition of that muscle, may result in abnormal activity and abnormal participation in movement of other muscles to which the original is functionally related. The result is a complex disorder of movement.

## POSTURAL AND PHASIC MUSCLE DISTURBANCE

Muscles respond predictably to myofascial dysfunction by becoming either shortened and hyperirritable (postural muscles) or inhibited (phasic muscles). Both types of muscle are equally likely to harbor trigger points and generate pain.

The upper trapezius is a typical postural muscle. In dysfunction it becomes shortened, elevating the scapula and rolling it over the apex of the thoracic cage, with prominent contour of the neck angle and palpable firmness. It over-participates in scapular stabilization and movement and is prone to spasm in response to a variety of stimuli, including passive stretch, passive shortening, attempted limited contraction, or use of the muscles of the arm or forearm. This over-activity tends to reciprocally inhibit antagonist muscles, including the lower trapezius and the serratus anterior, resulting in global disturbance of scapular stabilization and rotation during use of the arm.

The lower trapezius is a typical phasic muscle. In dysfunction, it becomes inhibited, allowing the scapula to ride up and drift laterally on the thoracic cage. It does not feel tense, and tends to remain relatively inactive during movements in which it would ordinarily participate. Its relative inactivity tends to disinhibit the antagonistic upper trapezius, permitting excessive upper trapezius involvement in scapular stabilization and movement (Figure 1). Despite its lack of tenseness and lack of activity, it is just as likely to harbor trigger points and to be a source of pain.

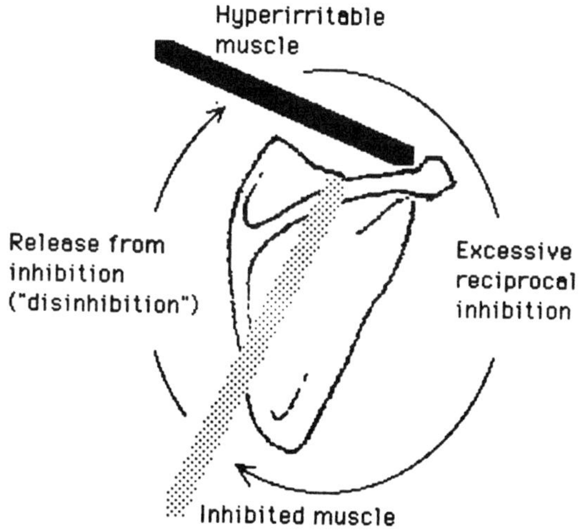


**Figure 1.** HETEROSTASIS DUE TO DISORDERED RECIPROCAL INHIBITION IN MYOFACIAL DYSFUNCTION

Vladimir Janda (1969, 1978, 1983) and Lewit (1985) have extensively investigated the role of postural and phasic muscle dysfunction in the development of movement disorders and musculoskeletal pain syndromes. A partial listing of muscles based on this classification is given in Figure 2.

| **Hyperactivity** ("postural muscles") | **Inhibition** ("phasic muscles") |
|---|---|
| *Lower extremity & pelvis* | |
| **Hamstrings** | **Vasti** |
| **Rectus femoris** | |
| Piriformis | **Glutei** |
| **Tensor fascia latae** | |
| Thigh adductors | |
| *Trunk* | |
| **Lumbar paraspinals** | **Oblique & transverse abdominals** |
| *Shoulder Girdle* | |
| **Upper and middle trapezeus** | **Lower trapezeus** |
| Levator scapula | **Serratus anterior** |
| Pectoralis minor | Rhomboids |
| Subscapularis | Supraspinatus |
| Pectoralis major (clavicular & sternal) | Infraspinatus |
| | Pectoralis major (abdominal) |
| *Cervical* | |
| **Scalenes** | Deep flexors of the neck |
| **Sternocleidomastoid** | |
| *Upper extremities* | |
| **Flexors** | **Extensors** |
| Pronators | Supinators |

**Figure 2.** MUSCLE GROUPS THAT SHOW A TENDENCY TO HYPERACTIVITY OR INHIBITION

## MYOFASCIAL DYSFUNCTION IN UPPER QUARTER PAIN SYNDROMES; DYSFUNCTIONAL STABILIZATION OF THE SHOULDER GIRDLE

Upper quarter (cervicobrachial, cervicocranial) pain syndromes are commonly encountered chronic pain problems, deriving from occupational cumulative trauma/repetitive motion and postural strain as well as from acceleration/deceleration ("whiplash") injury. Although there are many mechanisms of injury, the characteristic patterns of response of the regional musculature result in some rather consistent patterns of myofascial/movement dysfunction.

Figure 3 depicts the major muscles stabilizing and effecting movement of the scapula on the thoracic wall. The uppermost of these -- the upper trapezius, middle trapezius, levator scapula, and pectoralis minor (on the anterior chest wall) -- are postural muscles; in dysfunction they're prone to contracture and hyperirritability. The lower scapular stabilizers -- the lower trapezius, rhomboids and serratus anterior -- are phasic muscles, responding to dysfunction with inhibition. Significant shoulder girdle myofascial pain syndromes tend to involve several or all of these muscles to varying degrees; in general, one sees over-participation of the upper stabilizers, with under-participation of the lower stabilizers, in scapular movement and static scapular stabilization during use of the arm. This gives rise to observable disorders of movement and perhaps even resting posture, and to predictable patterns of abnormal EMG activity over the involved muscles during movement. Stabilization of the shoulder girdle by the upper stabilizers transfers stresses to the cervical spine, creating strain on these structures, which may extend myofascial/movement dysfunction into this area as well.

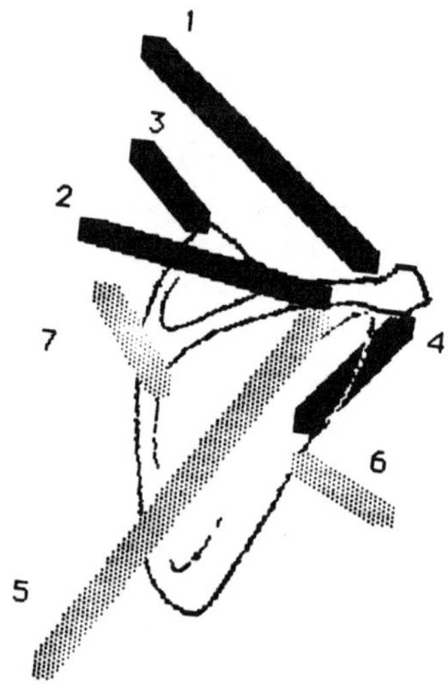


**Figure 3.** MAJOR MUSCLES OF SCAPULAR STABILIZATION: 1) UPPER TRAPS; 2) MID TRAPS; 3) LEVATOR SCAPULAE; 4) PERCORALIS MINOR; 5) LOWER TRAPEZIUS; 6) SERRATUS ANTERIOR; 7) RHOMBOIDS.

## RESPIRATOR PATTERNS IN PAIN SYNDROMES

Upper thoracic respiration is frequently seen in patients presenting in pain, and may be significant for several reasons. This pattern is often related to psychological distress (Jellinek et al, 1985), both as the result of emotional distress and as one factor in its perpetuation. Biomechanically, muscular fatigue in the upper thoracic/cervical accessory respiratory muscles may contribute to the development and perpetuation of upper quarter myofascial dysfunction and pain. The scalenes and sternocleidomastoid are most often involved in this manner. These muscles are both postural, becoming shortened and hyperirritable in dysfunction, thereby becoming further over-involved in elevation of the rib cage in respiration. These muscles are also involved in stabilization and movement of the neck, and as such are antagonists/synergists of the upper stabilizers of the shoulder girdle. Myofascial/movement dysfunction in one of these areas (shoulder girdle, cervical spine, respiratory pattern) may thus spread to, and/or be perpetuated by, dysfunction in the others. Respiratory dysfunction is often a very significant contributing and perpetuating factor in upper quarter pain syndromes, and failure to treat it is often a major reason for treatment failure.

## LOW BACK PAIN

Similar principles pertain in the muscular contribution to common low back pain syndromes. The paraspinals (multifidi, erectors spinae, quadratus lumborum), the psoas and hamstrings are postural muscles, prone to shortening hyperirritability; the glutei and abdominal oblique and transverse muscles (acting on the lumbodorsal fascia laterally; Gracovetsy et al, 1985) are phasic, prone to inhibition. In myofascial/movement dysfunction, the tendency is for pelvic and spinal stabilization to be shifted away from the gluteal and abdominal musculature, to the intrinsic musculature of the lumbar spine. This results in increased torque on lumbar structures with increased risk of mechanical fatigue, as well as pain of myofascial origin.

## EMG BIOFEEDBACK IN ASSESSMENT: GENERAL PATTERNS OF MOVEMENT DYSFUNCTION

### Postural muscles

Certain patterns of dysfunction seen on surface EMG monitoring are characteristic of postural muscles (such as the upper trapezius or the lumbar paraspinals). These are:

(1) Activity at "rest",
(2) Sustained activity on return to "rest" following use (failure to recover),
(3) Spasm during passive shortening (cramping),
(4) Spasm during passive stretch (exaggerated stretch reflex),
(5) Co-contraction with antagonists (rather than reciprocal activity),
(6) Activity during use of the contralateral muscle (exaggerated crossed reflex),
(7) Activity during "unrelated" movement (generalized activation),
(8) Over-involvement in regional moment and stabilization stereotypes, and
(9) Substitution for biomechanically related muscles in movement and stabilization.

Each of these general patterns may provide an opportunity for therapeutic intervention as outlined in the protocols to follow.

**Phasic muscles**

Phasic muscles (such as the serratus anterior and the glutei) typically respond to dysfunction with inhibition. On EMG monitoring, they tend not to exhibit activity at rest, and tend to be substituted for during stereotyped movement patterns. They are very important in pain generation and maintenance, as:

(1) They may harbor trigger points and generate pain directly, and
(2) They may, in their relative inactivity, be actively disinhibiting antagonistic postural muscles, enabling those postural muscles in their dysfunctional involvement in movement disorder and pain generation.

**Patterns of interaction**

The interactions between biomechanically related muscles are best seen in multiple-channel recording. Dysfunctional patterns most often seen are:

(1) Substitution of postural muscles for biomechanically related phasic muscles in stabilization and moment patterns; over-involvement of postural muscles, with under-involvement of phasic muscles,
(2) Co-contraction of antagonistic postural muscles, and
(3) Co-contraction of an antagonist postural muscle during movement involving use of a phasic muscle.

## PROTOCOLS FOR TREATMENT OF UPPER QUARTER DYSFUNCTION

**General Goals**

(1) Decrease hyperirritability/over-involvement of involved postural muscles (upper and middle trapezius, scalenes, sternocleidomastoid, masseter, temporalis),
(2) Restore normal resting length to postural muscles,
(3) Strengthen involved phasic muscles and retrain use in movement (lower trapezius, serratus anterior),
(4) Retrain moment and stabilization patterns to involve appropriate balance of postural and phasic muscles,

(5) Ergonomic and postural re-education to minimize upper quarter mechanical strain,
(6) Retrain respiratory patterns to eliminate upper quarter respiratory involvement, and
(7) Retrain psychophysiologic responses involving upper quarter and respiratory musculature.

**Electrode Placements**

(a) Bilateral Upper Trapezius
(b) Upper Trapezius / Lower Trapezius (unilateral, or 4-channel bilateral)
(c) Upper Trapezius / Serratus Anterior (preferably 4-channel bilateral)
(d) Bilateral Scalene
(e) Scalene / Abdominals
(f) Bilateral Sternocleidomastoid
(g) Masticatory muscles - Masseter, Temporalis

Basmajian (1983) provides a good reference for electrode placement. Travell & Simons (1983) is an excellent atlas for muscular anatomy, and is invaluable in its detailing of pain patterns generated by individual muscles.

**DETERMINING THE TREATMENT PLAN**

A guiding premise is that diagnosis should lead to treatment. The decision trees in Figures 4-9 outline the steps used in EMG assessment of the patient presenting with upper quarter dysfunction, and in assigning appropriate EMG biofeedback treatment protocols. Those stages marked with an asterisk (*) are often the most fruitful during assessment, and generally demand the most attention in treatment. While moving through the assessment phase, I prefer to think of the abnormal patterns identified as *therapeutic opportunities*; this additionally presents a much more positive image to the client than the amassment of ***abnormal patterns***. One could move through these decision trees, successively collecting assessment data prior to any treatment sessions; I often do this briefly in an initial session. However, I've found treatment most effective following an orderly sequence, addressing identified dysfunctional patterns in the order outlined in these decision trees. Thus, if elevated activity in the upper trapezei is seen at rest with arms in lap (see Figure 4), treatment begins with general relaxation protocols until performance criteria are met; then resting activity is assessed with arms hanging at sides, etc., proceeding to shoulder abduction and return only if (when) this is assessed normal. Phase 2 (Figure 5) is begun only when criteria are sequentially met for (1) resting activity with arms in lap, (2) resting activity with arms at sides, and (3) post-movement recovery from arm abduction, flexion and shrug.

Two protocols are "floated" throughout this sequence, and used to inject some variety during treatment sessions; these are respiratory retraining (Figure 9) and psychophysiological (stress) profiling (Cram, 1986). Protocols for assessment and treatment at other sites (e.g., low back, forearm) are similarly constructed.

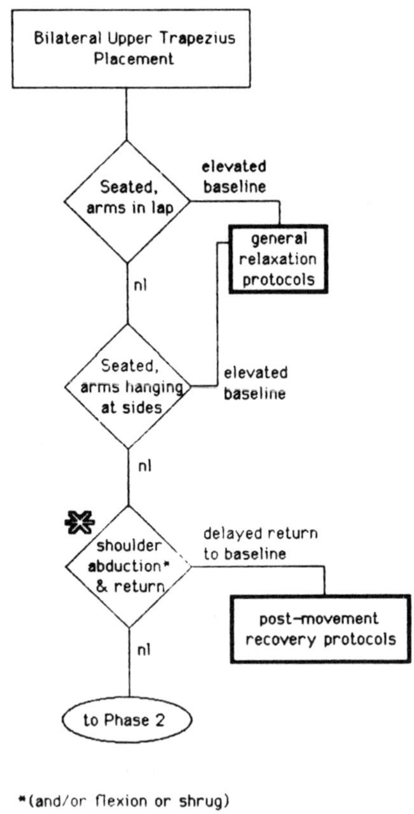


**Figure 4**. PHASE 1: POSTURAL MUSCLE RELAXATION

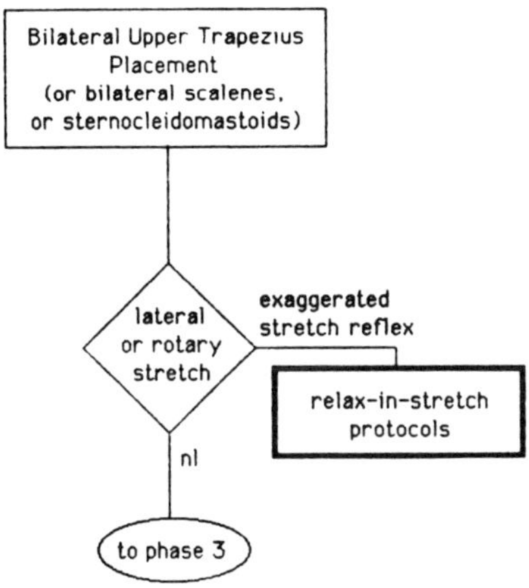


**Figure 5.** PHASE 2: STRETCH POSTURAL MUSCLES

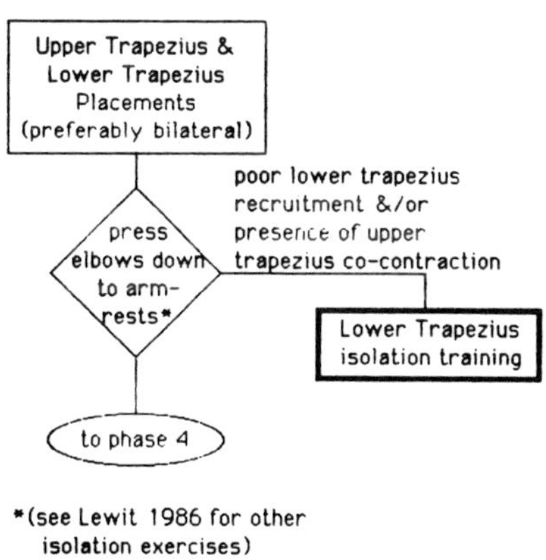


**Figure 6.** PHASE 3: ISOLATION OF PHASIC MUSCLES

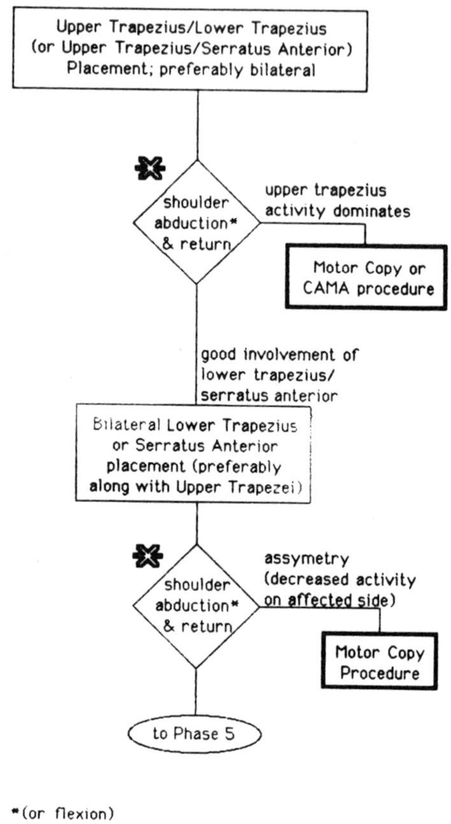


**Figure 7.** PHASE 4: POSTURAL PHASIC MUSCLE BALANCE

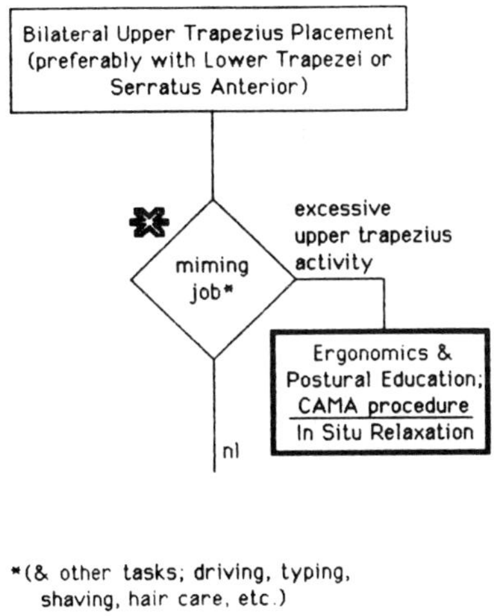


**Figure 8.** PHASE 5. STEREOTYPED MOVEMENT PATTERNS

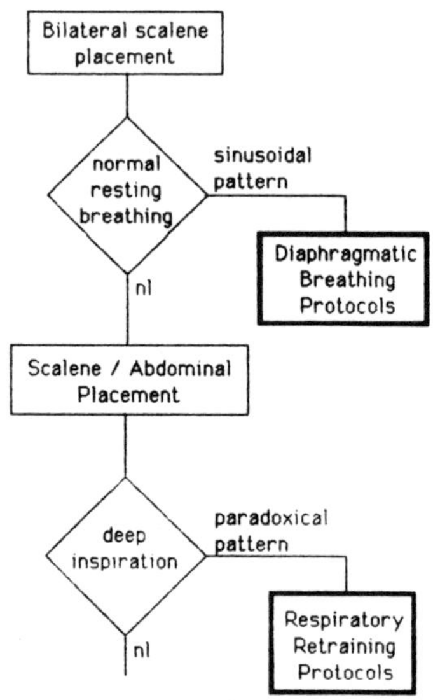


**Figure 9.** RESPIRATION (SIMULTANEOUS WITH PHASES 1-5)

## TREATMENT PROTOCOLS

### General Relaxation (Phase 1)

These are rather conventional, and so will not be detailed here. Be obsessive about performance criteria; this will vary according to equipment and bandpass setting. I find therapeutic opportunity here in less than 20% of clients.

### Post-Movement Relaxation (Phase 1)

This is often the workhorse in treatment. Over 80% of upper quarter pain patients demonstrate abnormality here.

- Electrode placement: Bilateral upper trapezius.
- Patient position: sitting, arms hanging loosely at sides.
- Action: variably,
  (1) abduction of the arms to 90° (begin with this);
  (2) flexion (forward elevation) of the arms to 90°; or
  (3) upward shrug of the shoulders, toward the ears

- hold position for 10 seconds, then return to the beginning, resting position.
- Look for: failure to immediately return to pre-movement baseline.
- To make easier: smaller movement, held shorter time (5 seconds). Return to lap or armrests, instead of hanging at sides.
- To make harder: hold longer time (20 seconds); add resistance.
- Treatment goal: more rapid return to resting baseline.
- Performance criterion: immediate return and maintenance below the general relaxation criterion figure.
- Strategy: initial -- gentle stretch ("press fingers toward the floor"), later substituting with general relaxation strategy.

**Relax-in-Stretch (Phase 2)**

This (along with respiratory retraining) is the treatment of choice for scalene or sternocleidomastoid dysfunction; it is also an important step in upper trapezius rehabilitation.

- Electrode placement: bilateral upper trapezius (generally start here), or scalenes, or sternocleidomastoids.
- Patient position: seated upright, or lying supine.
- Action: gently stretch the involved muscle (flex ear toward opposite shoulder, or rotate head toward the shoulder in question for upper trapezius; see Travell & Simons, 1983, or Lewit 1985, for more stretches).
- Look for: activity in muscle during passive stretch (exaggerated stretch reflex).
- Treatment goal/criteria: silence during passive stretch to full range.
- Strategy: general relaxation. Threshold audio signal may be helpful (e.g., relay to cassette recorder, with above-threshold-off setting); gradually shape threshold downwards as performance improves.

**Isolation of Phasic Muscles (Phase 3)**

The rationale here is to rehabilitate these muscles, as well as to use their increased activity to **reciprocally inhibit** biomechanically related hyperirritable postural muscles.

- Electrode placement: upper and lower trapezius, or upper trapezius and serratus anterior (preferably bilaterally, using 4 channels).
- Patient position: sitting, forearms resting on armrests of chair.

- Action: for lower trapezius, press arms downwards, by depressing shoulder blades. For serratus anterior, press arms forwards, against resistance (table edge).
- Look for: poor lower trapezius (or serratus anterior) activity, and/or any activity in the upper trapezei (co-contraction).
- Goal: no upper trapezius co-contraction; vigorous, symmetrical contraction of the lower trapezius (or serratus anterior).
- Strategy: "work from the lower shoulder blade." Hands-on direction helps.

**Postural/Phasic Muscle Balance (Phase 4)**

This is really the fun stuff. Using the preceding training as a base, the goal is retraining of functional movement stereotypes.

- Electrode placements: upper and lower trapezius, or upper trapezius and serratus anterior (preferably bilateral, 4-channel placement).
- Patient position: sitting upright, arms hanging at sides.
- Action: abduction -- i.e., sideways elevation (or flexion -- i.e., forward elevation) of the arm to 90° and return to starting position at rest.
- Look for: dominance of the upper trapezei, inhibition of the lower trapezei and serratus anterior; asymmetry of the lower stabilizers (affected side more inhibited).
- Goal: equal participation of the upper and lower stabilizers (in flexion, dominance of the serratus anterior); symmetry.
- Strategy: motor copy procedure (Wolf, LeCraw, Barton & Jann, 1988); use the "good side" as a model. Concurrent assessment of motor activity (CAMA) procedure (Wolf, Edwards & Shutter, 1986); hands-on guidance of movement with simultaneous biofeedback provided.

**Stereotyped Movement Patterns (Phase 5)**

- Electrode placements: bilateral upper trapezius, preferably 4-channel placement with lower trapezei or serratus anterior.
- Patient position: recreate job (good ergonomics, bad ergonomics, etc. -- "what happens if...").
- Action: mime performance of job or self-care activity; supply or ask patient to bring in props (keyboard, hair drier, welding torch, etc.).
- Look for: excessive activity in postural muscles, inhibition of phasic muscles.

- Goals: functional patterns of interaction (as for the phase 4 protocols, above).
- Strategy: motor copy or CAMA procedures, as for the phase 4 protocols above.

## SUMMARY

Attention to the dynamic aspects of myofascial dysfunction in musculoskeletal pain allows a novel perspective on this common clinical problem. EMG biofeedback becomes a powerful modality in assessment and treatment, allowing retraining of dysfunctional movement patterns in addition to its more traditionally established roles in psychophysiologic re-education.

## References

Bagg, Stephen D. & William J. Forrest (1986). Electromyographic study of the scapular rotators during arm abduction in the scapular plane. *Am J Physical Medicine* 65(3): 111-124.

Bagg, Stephen D. & William (1988). A biomechanical analysis of scapular rotation during arm abduction in the scapular plane. *Am J Physical Medicine* :238-245.

Basmajian, John V. & Carlo J. DeLuca (1985). *Muscles Alive*, 5th Ed. Williams & Wilkins, Baltimore.

Cram, Jeffrey R. (1986). *Clinical EMG: Muscle Scanning and Diagnostic Manual for Surface Recordings*. Clinical Resources, Seattle.

Dolce, J.J. & Raczynski, J.M. (1985). Neuromuscular activity and electromyography in painful backs: psychological and biomechanical models in assessment and treatment. *Psychological Bulletin* 97(3):502-520.

Fogel, Eric R. (1987). Biofeedback-Assisted Musculoskeletal Therapy and Neuromuscular Re-Education; chapt. 20 in Schwartz, M. S., *Biofeedback, A Practitioner's Guide*. Guildford Press, N.Y.

Fridlund, A., Fowler, S. & Pritchard, D. (1980). Striate muscle tensional patterning in frontalis EMG biofeedback. *Psychophysiology* :47-55

Gracovetsky, S., Farfan, H., & Helleur, C. (1985). The abdominal mechanism. *Spine* 10(4):317-324

Hudzinski, L.G. & Lawrence, G.S. (1988). Significance of EMG electrode placement models and headache findings. *Headache* 28:30-35

Janda, V. (1969). Postural and phasic muscles in the pathogenesis of low back pain. In: *Proc. XIth Congress of Intern. Soc. Rehabil.* pp. 553-554. Disabled, Dublin.

Janda, Vladimir (1983). *Muscle Function Testing.* Butterworths, Boston.

Janda, Vladimir (1978). Muscles, central nervous motor regulation and back problems. In Korr I. M. (ed.): *The Neurobiologic Mechanisms in Manipulative Therapy.* Plenum Press, NY 1978.

Janda, Vladimir (1983). On the concept of postural muscles and posture in man. *The Australian Journal of Physiotherapy* 29(3):83-84.

Jellinek, M.S., Goldenheim, P.D., & Jenike, M.A. (1985). The Impact of Grief on Ventilatory Control. *Am J Psychiatry* 142(1):121-123.

Jones, Ann L. & Wolf, Steven L. (1980). Treating chronic low back pain. EMG biofeedback training during movement. *Physical Therapy* 60(1):58-63.

Lewit, Karel (1980). Relation of faulty respiration to posture, with clinical implications. *J of AOA* 79(8): 525/75-529/79.

Middaugh, S. (1982). Muscle Training. In: Doley's, D.M.; Merideth, R.L.; & Ciminero, A.R. (eds.) *Behavioral Physiology in Medicine: Assessment and Treatment Strategies.* Plenum Publ. Corp, New York.

Middaugh, S.J. & Kee, W.G. (1987). Advances in electromyographic monitoring and biofeedback in treatment of chronic cervical and low back pain. In: Eisenberg, M.G. & Grzeslak, R.C. (eds.) *Advances in Clinical Rehabilitation.* pp. 137-172. Springer Publ. Co.

Sherman, Richard A. & Crystal J. Sherman (1989). Relationships between continuous, environmental recordings of posterior trunk muscle tension and patterns of low back pain and tension headaches. Proceedings, 20th annual meeting AAPB, San Diego.

Simons, David G. & Travell, Janet G. (1983). Myofascial origins of low back pain 1. Principles of diagnosis and treatment. *Postgraduate Medicine* 73(2):66-77.

Simons, David G. & Travell, Janet G. (1983). Myofascial origins of low back pain 2. Torso muscles. *Postgraduate Medicine* 73(2):81-92.

Simons, David G. & Travell, Janet G. (1983). Myofascial origins of low back pain 3. Pelvic and lower extremity muscles. *Postgraduate Medicine* 73(2):99-108.

Travell, Janet G. & Simons, David G. (1983). *Myofascial Pain and Dysfunction: The Trigger Point Manual*. Williams & Wilkins, Baltimore.

Wolf, S.L.; Basmajian, J.V.; Russe, T.C.; & Kutner, M. (1979). Normative data on low back mobility and activity levels. *Am J Phys Med* 58:217-279.

Wolf, S.L.; LeCraw, D.E.; Barton, L; & Jann, B.B. (1988). Comparison of motor copy and targeted training techniques for restitution of upper extremity function among neurologic patients. *Phys. Ther.* (submitted) 1988.

Wolf, Steven L., Nacht, M., Kelly, J.L. (1982). EMG feedback training during dynamic movement for low back pain patients. *Behavior Therapy* 13:395-406.

Wolf, Steven L. & Stuart A. Binder-MacLeod (1989). Neurophysiological Factors in Electromyographic Feedback for Neuromotor Disturbances. Chapt. 3 in J.V. Basmajian, ed. *Biofeedback: Principles and Practice for Clinicians*, 3rd ed. Williams & Wilkins, Baltimore.

# CHAPTER 4

# *Muscle Learning Therapy - A Treatment Protocol*

by

Dennis L. Ettare, M.A.
and
Ross C. Ettare, M.D.

---

A new treatment procedure, MUSCLE LEARNING THERAPY (MLT), has been developed to help the biofeedback therapist teach people the skills needed to reduce pain associated with certain BACK, NECK, HEAD and CUMULATIVE TRAUMA (CTD) disorders. Muscle Learning Therapy, based upon the same techniques used in training neuromuscular re-education, uses of biofeedback to facilitate the learning of specific skills which help reduce muscular components of pain. Cumulative Trauma Disorders, or repetitive motion syndrome, are not the focus of this chapter but they will be mentioned briefly due to the fact that it has recently become an industry-wide problem, and is related to musculoskeletal disorders upon which MLT has shown impact.

The advantage of a MLT procedure is that it trains the patient to recognize hypertonicity in all involved muscle (trapezius; cervical, thoracic & lumbar paraspinals; forearms and hands, etc.) and then trains the individual to relax those muscles specifically <u>without interruption of work.</u>

Many procedures exist today which measure or assess muscle involvement in pathology. Some are used for measuring stress, bracing or guarding. However, specific and structured protocols are also needed for the <u>treatment</u> of pain in the back, neck and head. In addition, new protocols should not be dependent upon relaxation tapes, imagery, meditation, self-hypnosis and other cognitively oriented therapy because the normal work day continuously inflicts internal and external stressors requiring constant attention. The work site is not the type of environment conducive to cognitive therapies. Most cognitive and behavioral interventions have had difficulty generalizing into the practical, working world for the following reasons:

1. High cognitive involvement which often requires time-outs,
2. A certain agility with imagery and insight,
3. Lay public resistance due to psychological overtones,
4. Legal resistance due to litigious considerations.

The therapist must be aware of these difficulties when looking for treatment options. The purpose of this chapter is to present a treatment option without such limitations.

The role of the neuromuscular - musculoskeletal systems in the above disorders is becoming more evident. Clinical evidence supports the fact that as less muscle involvement occurs, less pain is often experienced. However, the method in which the reduction of that muscle involvement is taught may be critical in determining if the muscular component of pain will become controllable and if that control will indeed be transferable as a skill into a practical environment. Therefore, control is the primary issue. The next most important issue is the generalization of that control into the working world.

Relaxation methods used in biofeedback have traditionally involved a cognitive and/or focused approach with the use of imagery, relaxation tapes, tense/release techniques, posture, hypnosis, and so forth. These methods require one's full attention, or at least a significant portion of it. Although these interventions do work at the time they are used, unfortunately, as a practical application, they become very difficult for anyone to use while working, studying, using dangerous or delicate equipment or simply negotiating rush-hour traffic. The following reasons explain why cognitive interventions are not as indicated for back, neck, head and CT disorders:

1. Cognitive training is difficult to quickly assess and then provide immediate feedback.
2. When feedback is not immediate, learning takes longer and is more easily extinguished.
3. Most cognitively based training requires concentration levels in excess of the practical availability during normal working and social conditions.
4. Cognitive work, within the area of back and neck trauma, alienates or irritates most people or at least causes needless resistance.

Through the general use of biofeedback, a person may learn how to reduce the muscular component of pain in one of two ways:

1. Cognitive skills (imagery,relaxation tapes, meditation, etc.), or
2. Proprioceptive skills (muscle learning).

The protocol used in MLT training is unique and has been developed, at this time, only for patients with back, neck, head and CT disorders. This approach reduces the muscular component of pain by directly accessing the muscles rather than engaging the mind. The skill is easily learned and is applied without interruption of activity. Also, since muscle activity affects the nervous system, MLT has proven to be extremely effective as a non-disruptive stress reducing technique.

Why is this process different from others in biofeedback? Simply because it does not involve relaxation tapes, imagery or other cognitive interventions. Instead, it involves the use of a highly structured muscle learning technique which applies a physical, non-psychological, short-term approach and which is easily applied while the person is at work, in the car or during social activities. Also, it does not involve exercise but is, in fact, used as an adjunct to other therapies.

The MLT protocol involves the use of a protocol used to train neuromuscular coordination wherein the patient learns to turn off muscle groups the same way they learn to turn them on - by repetition and reinforcement. A person is taught the process by watching and controlling their own Surface Electro-Myographic ($_{S}$EMG) graphs as they walk, sit, stand and simulate various work-related movements and conditions. Consequently, as they train to lower their EMG levels, pain may be either eliminated or reduced to more tolerable levels, thus allowing for a more rapid and permanent return to work.

It is the purpose of this chapter to present specific, short term, non disruptive protocols for the treatment of pain associated with back, neck, shoulder, head and cumulative trauma disorders. There are 3 types of pain in these areas which are effected by MLT. The last two are directly accessible to change through MLT. The first, however, is more refractory.

## THREE ETIOLOGIES OF PAIN HELPED BY MLT

1. PAIN ARISING FROM THE INJURY ITSELF
   The first etiology of pain which may be influenced by MLT originates from the pathogenic tissue itself. Pathology often causes pain, but although there is not a great deal that a relaxed muscle can do to remove pathology, there is still much it can do to minimize the effects of hypertonic contiguous muscle groups. A herniated disc is not going to become integral simply by relaxing adjacent muscles. However, if dysfunctional systems have a chance of healing on their own (with the help of all therapies combined), the softening of contiguous muscle groups will allow that process to occur more rapidly by reducing constant irritation. In addition, it may even have the potential of accelerating the process of healing. In fact, with soft tissue injuries, the relaxation of adjacent muscle tissue often provides the quickest and only method of pain relief.

2. INCREASE OF PAIN AT THE SITE OF INJURY
   The second type of pain is caused when muscles tighten around injured tissues and thereby irritate the original problem. This often causes the original pain to increase or, at times, even to reactivate a problem which had remained dormant. The continual irritation of injured tissue by contiguous hypertonic muscle groups is an important reason why people find themselves looping back into pain cycles.

DYNAMICS OF THE PAIN CYCLE

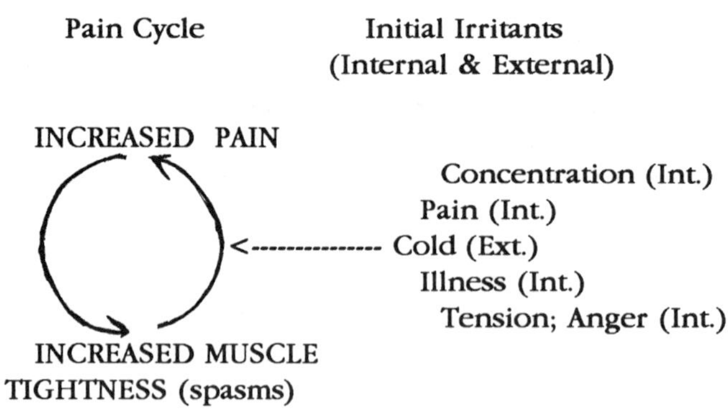

3. MUSCLE PAIN
   The third etiology of pain, muscle pain, is the result of constant, isotonic states which produce an over-worked, fatigued muscle. Muscle pain may only be indirectly related to the original dysfunction. It may instead be a parallel problem which increases to overwhelming proportions. It is a strategic mistake to underestimate the effects of this muscle pain as it may, in some cases, become even more severe than the original pain.

The following pages will describe the profiling process used to pinpoint where to begin this procedure and also an appointment by appointment summary of the main interventions used.
The MLT protocol is based upon two fundamental theories found in training procedures - repetition and reinforcement .

The following explains how those theories are used:

1. The purpose of repetition is simply to repeat a process so often that "muscle learning" occurs with or without reinforcement. It is not to help the person remember what to do as much as it is to render what is done as a habit. Muscle learning refers to the brain's recognition of correct proprioceptive cues and with adequate repetition, the conscious mind will require less and less involvement in order to perform a task. The same holds true with relaxing back and neck muscles. The more you relax, the less you need to try to remember and therefore the more it becomes a habit. As in sports, during intense situations, the trained person continues to perform well.

2. Positive Reinforcement occurs during repeated, induced experiences and in MLT it is facilitated by biofeedback in which the patient learns by watching his own EMG graphs. Negative reinforcement occurs when pain decreases as a direct result of lowered muscle involvement.

In MLT, repetition promotes the use of neurological pathways. For example,using an analogy from sports training, in learning to play tennis the player swings a racket repeatedly to increase muscle learning. A large portion of the conscious mind is actually used only during training to focus attention during the learning process. Afterward, the body performs its learned tasks while the mind is freed to attend elsewhere. As in another example, when first learning how to type, initial concentration is applied toward the goal of learning the position of the keys. Soon, however, fingers simply repeat the learned movement while the mind is freed to create. In the same way, muscles involved in bracing from pain, cold, stress, etc. will habitually learn to relax.

## E M G PROFILING

The EMG profile is an essential part of MLT training and helps decide the extent and direction of training. Its three main functions will be elaborated upon later. The process of applying the electrodes and going through the sequencing will be overviewed at this time in order to illustrate the procedure.

### ELECTRODE PLACEMENTS

A complete profile involves bilateral, dual channel placements. The term "bilateral" is used to indicate two ipsilateral and opposite placement sites, and "dual channel" means that each site is monitored by a separate channel. In other words, each side of the body is viewed independently but at the same time. The placements remain in the same position throughout both the profiling as well as the training . The following are three advantages in using the same sites for assessment as well as training.

1. It expedites the placement time needed for attaching the sensors;
2. It reduces the bother to the patient;
3. It lowers the cost of additional materials and wear.

If only one channel of EMG is available, each side of the body must be recorded independently for the profiling. In that case, to then conduct the training phase, both sides are combined (ambilateral placements) onto one channel of EMG.

Sensors are placed over two separate sites for the upper back with one on each and opposite side of the body, two for the mid-back (thoracic 8) and two for low back (lumbar 4, 5), making a total of 3 bilateral placements. Although each site requires a bilateral placement, set-up time is brief due to the economy of total sensors used. Only 8 electrodes are used for the entire profile because of the dual functions of the active and common leads.

Figure 1. RECOMMENDED ELECTRODE PLACEMENT

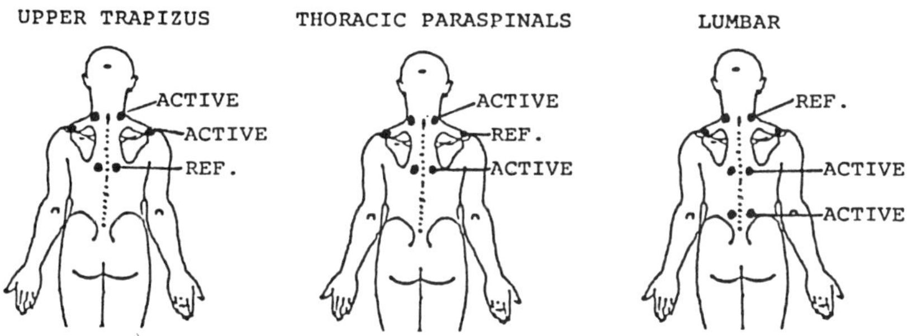


The placements are on the Left (L) and Right (R) cervical muscle groups, L/R upper trapezius , L/R T8 (equal to the lower edges of the scapulae), and L/R placements at an L4,5 . During the later part of the session, by simply unsnapping the electrodes and alternating the commons with the actives, all pertinent areas can be monitored with the least amount of trouble (see Figure 1).

It is important that the therapist not become compulsive about exact sensor placements. Accuracy is not necessary. Only the large muscle groups are of concern and even poorly placed electrodes will give adequate feedback for MLT training.

<u>SEQUENCE OF THE PROFILE</u>

A profile is begun by asking the patient to sit in a low back chair while the therapist attaches the sensors. Do not ask the patient to sit back or change position or relax. Unless he is sitting on the edge of the chair, assume he is sitting in his own normal position. You need to start with as normal a position as possible. However, all normative data have been collected with feet flat on the floor, hands in the lap and the back lightly supported. You will need to point out these minimum needs to the person.

"SITTING" (1st. period)

1. As soon as the person is seated and ready, begin the first 20-second period of the graph by turning on the instruments and again reminding the patient not to talk, move, or breathe deeply. Pause the graph when the 20 seconds are up (see Figure 2).

"STAND-UP" (2nd. period)

2. Upon completion of the previous 20 seconds, begin the graph again and then after 2 seconds of recording, ask the person to stand up and to continue facing the instruments with arms to the side. You will see a high spike during the movement used in standing up, and then the level will drop to the person's usual resting state. Whether it be low or high, no verbal feedback should be given until all segments are completed so that the profile is not influenced by the therapist. Stop the graph after 20 seconds.

"WALKING" (3rd. or middle period)

3. Ask the patient to start walking back and forth within the widest part of the office. Start the graph only when the person has walked for several lengths of the room. Be certain that there are no misunderstandings and that the wires do not tangle. Hands should be kept to the sides and avoid looking around while walking. Do not substitute "Walking in place".

"STANDING" (4th. period)

4. When 20 seconds have elapsed, stop the graph and tell the person to return in front of the chair, facing the instruments, and remain standing for another 20 seconds. As soon as the person is in front of the chair, start the graph at once and remind him/her to remain still for another 20 seconds with hands to the side. It is important to get a true recording of how quickly the muscles return to a resting state.

"SIT DOWN" (5th. period)

5. Finally, while the person is standing in front of the chair, start the graph, wait 2 seconds and then ask the patient to be seated as during the beginning. Again you will notice the upward EMG spike and then the final resting levels.

At this point a graph with all five 20-second periods will be displayed on the screen (see Figure 2). The graph is to be saved for later reference.

Also, with activity such as shown above, long leads are required. The best source for leads of 15 to 20 feet are standard EKG cables. These may be purchased either through a hospital or cable company. Typical cables supplied by biofeedback companies are often both too short and produce high levels of artifact.

## THREE FUNCTIONS OF THE PROFILE

There are three uses of the profile. These are:

(1) to evaluate in order to establish the person's "norms",

(2) to locate sites of elevated activity and,

(3) to reinforce resulting changes.

Data were gathered from over 400 back and neck patients using narrow filters on a J&J I330 computer interface. Graphs were 0-20 uv @100 seconds and were administered by Dennis L. Ettare, M.A.

### 1. EVALUATION

The first function of the therapist is to process the diagnostic and interpretive value of the profile. The profile indicates where elevated EMG levels are located and whether immediate or long term relief of symptoms can be expected. The following table provides the basic elements used for this interpretation. All levels refer to upper back (upper trapezius and cervical muscle groups); mid back (cervical and thoracic groups) and lower back (lumbar paraspinals). Please refer to diagram 1 for these sites.

Table 1. GUIDELINES FOR GENERAL INTERPRETATIONS OF UPPER, MID AND LOW BACK MICROVOLT LEVELS.

(All values are in microvolts.)

| Positions & Anatomical Sites | Normal Levels | Significant Levels | Poor to Very Poor Levels | Severely Hypertonic Levels |
|---|---|---|---|---|
| SITTING | | | | |
| Upper Back | 2 | 3-4 | 5-10 | 11+ |
| Mid Back | 2 | 4 | 5-8 | 9+ |
| Low Back | 2 | 4 | 5-6 | 7+ |
| STANDING | | | | |
| Upper Back | 2 | 3-4 | 5-10 | 11+ |
| Mid Back | 2-3 | 4-6 | 7-10 | 11+ |
| Low Back | 2 | 3-5 | 6-8 | 9+ |
| WALKING | | | | |
| Upper Back | 4-6 | 7-10 | 11-15 | 16+ |
| Mid Back | 5-6 | 7-10 | 11-15 | 16+ |
| Low Back | 5-6 | 7-12 | 13-16 | 17+ |

EMG levels which do not maintain a quiet pattern while the subject is motionless, may indicate a highly irritable nervous system and could also indicate where the person may be losing fine motor control. This irritability often accompanies chronic pain and is therefore also used in differential diagnosis.

As muscle activity increases above 2uv, as shown in the chart, the levels become more significant and indicates possible irritation of pre-existing problems. Muscles could become a significant part of the problem, or even become the predominant problem. This does not imply, however, that pathology exists simply because significant levels have been reached. It indicates only that muscles are exhibiting uncontrolled and unusually active behavior.

2. SITES OF MUSCLE ACTIVITY.

The second purpose of the profile is to demonstrate the <u>location</u> of increased activity. The importance of this increased activity is to point out the area of poorest control.

The site of poorest control is often the site where contralateral bracing occurs. The body tends to favor the side that hurts while it braces the other. This often produces pain in different areas at different times as bracing patterns change in order to accommodate painful areas. These changing patterns, sometimes diagnosed as referred pain, often confuse the patient as well as some health professionals.

3. REINFORCEMENT

The third purpose of the profile is to document and demonstrate the results of training for reinforcement and for reporting purposes. As therapists we are continuously aware of the benefits of demonstrating improvement over time and it is especially true here as patients learn to recognize increased muscle control and decreased pain. EMG biofeedback is especially helpful due to its immediate visual feedback of decreased muscle tension and increased subjective calming. The pain itself is often reduced more slowly and so is less helpful as an immediate feedback cue. Each incremental drop or rise of the visual or auditory feedback helps increase awareness of the physical self as well as increasing a sense of physical control.

The above purposes of profiling will help the therapist explain the origin for many of the symptoms felt by the patient. Often the patient experiences increased pain due to stress and does not understand the physiological connection between an external stressor and increased internal pain. After explaining bracing patterns associated with pain and stress, the patient comes to recognize that most of the problem lies not in the head, but in the muscles. It also affords a good opportunity for the therapist to begin providing informative and empathic data to people who have felt excluded by friends, relatives and health professionals, all of whom are often tired of hearing about the person's unrelenting pain.

## CUMULATIVE TRAUMA DISORDERS
## ("REPETITIVE MOTION SYNDROME")

The specific profiling for cumulative trauma disorders (CTD) will not be elaborated upon at this time as it is not the main focus of this chapter. Briefly, however, there are pertinent points which need to be mentioned about this emerging industrial problem. MLT profiles may still be used but only as an adjunct to other profiles taken of the arms and hands during work simulation (normative levels in CTD profiles are generally much higher). Training has proven to benefit participants inasmuch as MLT's focus on the musculature provides a significant learning protocol for all major muscle groups involved in CTD.

Recent studies have indicated that while millions of dollars have been spent on work site improvements and employee education, there is apparently another major factor which is influencing Cumulative Trauma Disorders. For even as workers sit with the correct posture, in the perfect chair, in front of the ideal monitor, and over the perfect keyboard, if at the same time they are also coiled like a tight spring due to the intensity of their job and/or due to internal and external pressures, underlying structures will be rendered more vulnerable to damage.

Pain and/or swelling in the hands or arms may indicate an irritation of various underlying structures (tendons, ligaments, etc.) due to repetitive movements and other causes. Specific MLT training for CTD is based upon an extended theory and protocol. It is accomplished by reducing the extremely high levels of muscle tension exhibited during intense working conditions and thereby reducing susceptibility to damage.

To reduce all the various types of internal and external "stressors" would be infinitely complex, if it were not for the fact that the direct goal of "stress reduction" within this narrow context is simply to reduce the muscular component of the problem. MLT accomplishes that goal by going directly to the muscles involved. This is accomplished without getting involved in the cognitive and emotional elements of stress. It renders the process much more simply, faster, is longer lasting and less costly.

## EXPLANATION OF THE MLT PROFILE AND GRAPHS

A total of three complete profiles are taken:

1) upper back
2) mid back
3) low back (only when appropriate).

Each profile will contain the five, 20-second segments (periods) mentioned earlier. A scale of 0-20uv is used when possible. Figure 2 shows a post-training profile where no bracing is evidenced. Figure 2 exhibits a typical guarding profile with high EMG levels and where pain and/or stress are present. Graphs of different sessions are compared as the patient progresses. In order to do this with the least confusion, however, all graphs should maintain the same scale. Changing a graph's scale in order to accommodate different EMG levels will invite confusion and a loss of visual meaning to the patient as well as to the referring physician. Clinically it has been found that 85% of all patients will have profiles which will remain within 20uv. Simply record levels by hand if they elevate greater than 20uv.

**Figure 2.** TYPICAL POST-TRAINING PROFILE (20 second epoch)

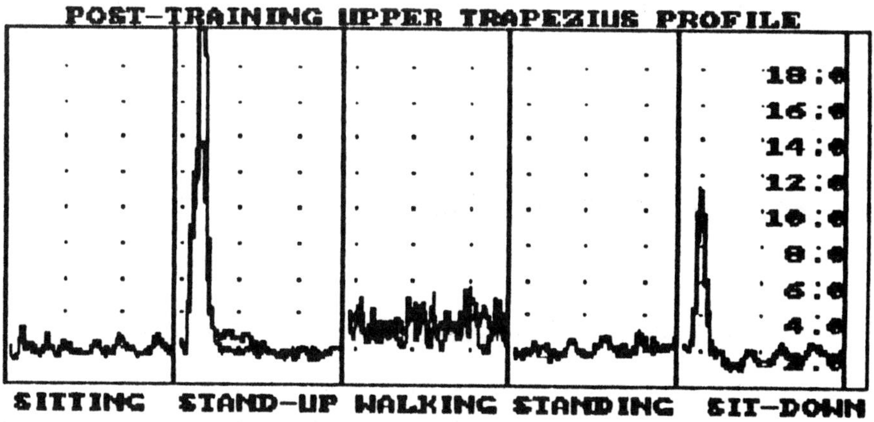

Figure 3. is an example of a typical guarding or hypertonic profile.

**Figure 3.** TYPICAL PRE-TRAINING GUARDING OR HYPERTONIC PROFILE (20 second epoch)

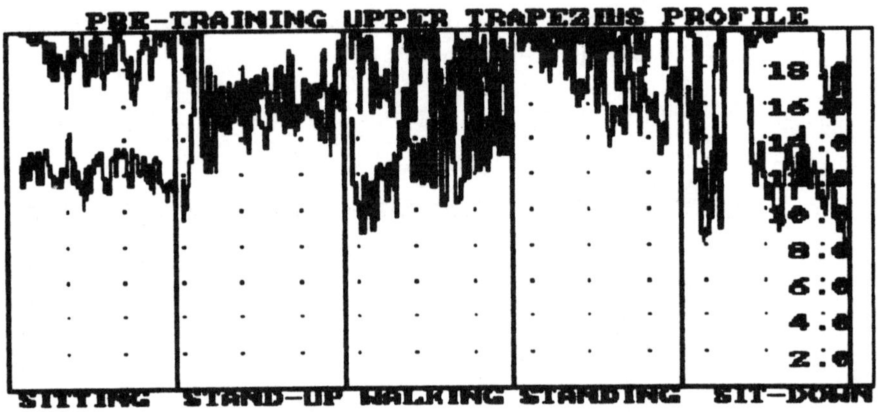


<u>INTRODUCTION OF THE INSTRUMENTS TO THE PATIENT</u>

Biofeedback therapists often assume that the lay person will readily accept and relate to instrumentation. However, even those who do will need to be given as much instruction as possible in order for feedback to become meaningful. This is accomplished with therapeutic skills but is especially necessary for people in physical pain. The patient needs to understand the following:

1. The <u>theory</u> of feedback
2. The clinical <u>significance</u> behind the feedback

After the therapist informs the patient about the basic theory of biofeedback using any of the available definitions in the literature, the most important point to impart is that EMG levels need to be associated with <u>active levels of control</u> rather than passive indicators of tension or pain. The person needs to quickly understand that surface electromyography (EMG) is not an indicator of deep, subconscious, uncontrollable reactions to life. EMG simply monitors <u>voluntary</u> muscle activity.

At the beginning of the first few appointments, patients will often collapse into a chair, look at their EMG readings and wonder out loud what "this crazy week did" to their muscles, as if they did not possess the slightest level of control. This verbalization should immediately indicate to the therapist that the patient has not yet understood the point. Skeletal muscles are voluntary. By definition, they will not be tight if they are trained to release. And releasing ones's muscles does not, after training, require a focused mind. It is accomplished simply by repeating the same sensations that were felt while their graph was lowered to normal levels. Through an MLT-type training, muscle systems are trained to relax quickly, even while the patient is under the worst of situations.

PROPRIOCEPTIVE AWARENESS

Proprioceptive awareness is defined as the ability of the mind to recognize internal cues originating from muscles, tendons, and other internal tissues. For MLT training, those cues are of tightness, soreness, pain, etc. A repetitive and structured training protocol is required in order to build or refine proprioceptive skills for the purpose of relaxing the muscles. The most efficient method of teaching voluntary muscle release is to induce a rapid change in the state of the muscle. By so doing the muscle is "felt" as it is quickly relaxed. In this way proprioceptive cues are soon learned, remembered and then each successive act of releasing becomes easier to do and to feel.

## TREATMENT INTERVENTIONS

After the baseline profiles are taken at the beginning of each appointment, other interventions are needed to teach the appropriate skills. Later, an appointment by appointment outline is presented where each intervention, (standing, sitting and standing, and walking, etc.) is introduced. There are many such positions and movements to facilitate training. Only the essential ones are presented at this time.

A basic ingredient in learning to control muscles is to be able to do so with the minimum of effort. The most efficient way to accomplish this is by repeating the process often, until the muscles themselves perform the task. This learning process is begun by releasing muscles rapidly so that the sensation of release is felt immediately. The act of sitting and standing (sit/stand) provides this opportunity. It also duplicates a movement often used during a normal day. Upon standing, however, muscles tend to remain contracted. Upon sitting, many people do not automatically release muscles. It is the goal of the therapist to teach rapid release of upper and lower back muscles. Learning to quickly release muscles has two functions:

1. It rapidly releases the structural irritation caused by tightened muscle tissue; i.e. the faster it is released the sooner its positive effect will be felt.
2. It allows the proprioceptive cues of released muscles to be felt more intensely due to the dramatic juxtaposition of a rapid "before" and "after" sensation. Learning time is decreased dramatically when that sensation is rapidly felt.

## INTERVENTIONS USED IN THE MLT TREATMENT PROTOCOL

SIT/STAND TRAINING

Figure 4 demonstrates a typical pre-training sit/stand profile wherein resting levels are only 5/9 respectively. Figure 5 demonstrates a typical profile, after training, of sit/stand levels lowered to 2/2. The purpose is to train for a controlled and rapid release or relaxation of muscle groups. This is first accomplished by having the patient sit comfortably in a chair. The therapist will then direct the person to sit and stand every time the graph indicates that the muscles have quieted to within the 2uv range. The theory of sit/stand training needs to be explained openly and clearly to the patient so that the three goals remain simple and easily obtainable. The therapist needs to indicate to the patient those three goals.

1. The first goal is to lower the graph as quickly as possible.
2. The second goal is to reduce the levels as far below 2uv as possible.
3. The third goal is to try to recognize the sensation of the released muscle as soon as the graph drops. (In this way the patient learns that the goal is to lower the graph by feel rather than by trying to understand what to do).

It is critical to the learning process that proprioceptive skills be emphasized. If time is taken to think, the process becomes disruptive and difficult to generalize. By feel it becomes possible to relax muscles even while talking, driving or working.

**Figure 4.** A Pre-Training Sit-Stand Profile (100 second screen)

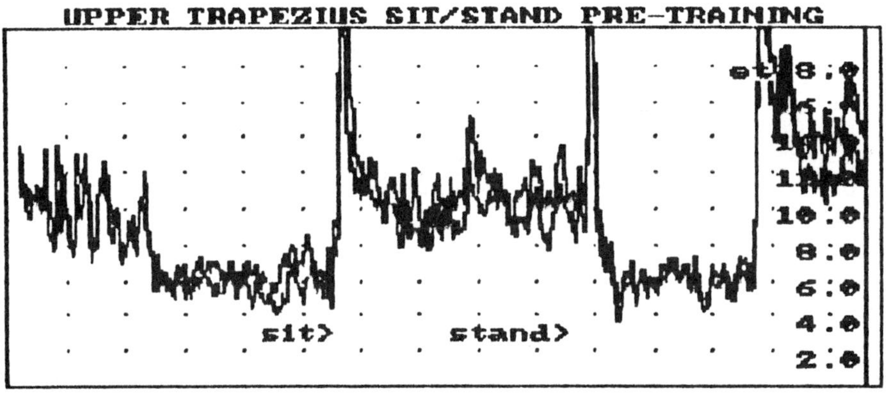


**Figure 5.** A Post-Training Sit-Stand Profile (100 second screen)

SIT/STAND USING AUDIO

Another routine called "sit/stand with audio", is the same as the above except that it incorporates the use of audio feedback in order to signal the patient when to sit and stand rather than relying on the therapist. The sit/stand routine continues as before but instead of the therapist signaling when to go to the next position, the audio feedback will signal with a tone. The threshold is set at 4uv with audio "on" when it is above threshold. The patient is instructed to intentionally release muscles quickly upon sitting or standing and to use the resulting cessation of sound as the cue to move to the next position. The person is also told that the therapist will be looking for the same objectives as before. In this way the muscles are trained for a repetitive and quick releasing pattern. The purpose is to further enhance neurological pathways into a recognized pattern of quick release responses.

SIT/STAND WITH REVERSE COUNT

Another routine using the same basic steps as in the "sit/stand" is the "sit/stand by reverse count". The goal of this intervention is to keep the person's mind occupied with cognitive tasks (listening to the audio, watching the graphs and processing information) while building upon only proprioceptive cues to specifically signal when to release muscles. The patient first must be introduced to the task as well as to the purpose and theory. As the name suggests, a person is instructed to attempt the same routine as with the "sit/stand with audio" but to also count backwards at the same time. It is best to begin with "reverse ones" (counting backwards from 100 by ones). Due to performance anxiety, the patient could become apprehensive. The therapist needs to point out that the nervousness will help add to later success. Patients must pay attention to the following three types of feedback:

1. visual descent of the graph,
2. sound of the audio,
3. internal proprioceptive cues.

These feedback cues are concentrated upon while at the same time they are rapidly sit and stand. Since this requires concentration, they will not be able to think about their muscles at the same time. In other words, they will feel when to release muscles while they are thinking of the other tasks. This procedure helps them to quickly separate the two processes - cognition and proprioception. This separation is critical for MLT or any physical training. The sooner the release of muscles is relegated to proprioception, the sooner it will generalize into work and play.

BLIND SIT/STAND

Another powerful intervention is the "Blind Sit/Stand". The following graphs (Figures 6 & 7) demonstrate typical pre- and post-training profiles. Vertical lines indicate decreasing time periods. The profile is initiated by turning off the visual and auditory feedback. The person is asked to begin by sitting normally. The person is then instructed to stand up and to quickly allow the muscles to go to their lowest level of the quieting response and to immediately signal the therapist when ready. At the exact moment that the person signals, the therapist quickly turns the graph on to record and verify if the EMG levels are indeed low. After the graph has recorded about 15 seconds of data it is paused (a vertical line marks the exact point on the screen), and the person is then told to be seated and to again signal when levels are "felt" lowest. This is repeated about 12-15 times, but each time the graph is allowed to run two seconds less until periods of only two seconds are reached and the entire screen is complete. The method of turning the graph on and off is actually accomplished by pausing the graph. The resulting profile is then used to demonstrate if muscle tension can indeed be quickly lowered to within the 2-3uv range without watching or listening to feedback. If a computerized graph is used, these trials are then left on the screen to inspect and are saved to the patient's disk. It would serve little purpose except to discourage the person. Finally, with the screen attesting to lowered levels of about 2-3 uv, it then becomes proof to both the therapist and the patient (as well as a referring physician or physical therapist) that the correct proprioceptive cues have been learned. Through continued reinforcement, patients soon realize they can relax muscle groups quickly and intentionally.

Figure 6. A Pre-Training, Blind Sit/Stand Profile (100 sec. screen)

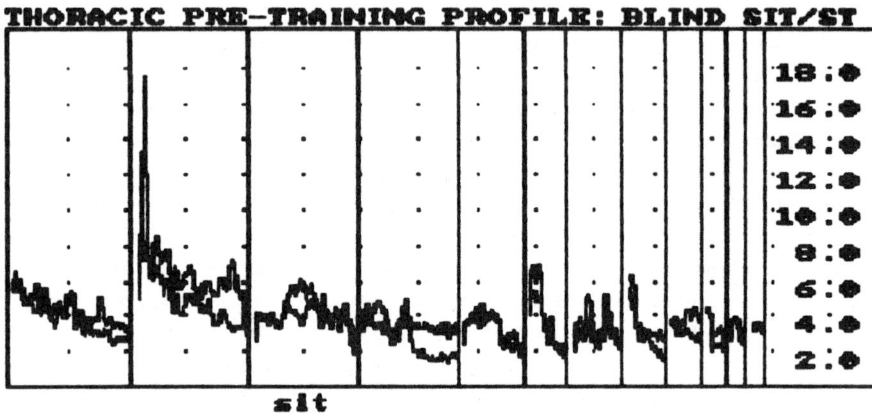


Figure 7. A Post-Training, Blind Sit-Stand Profile

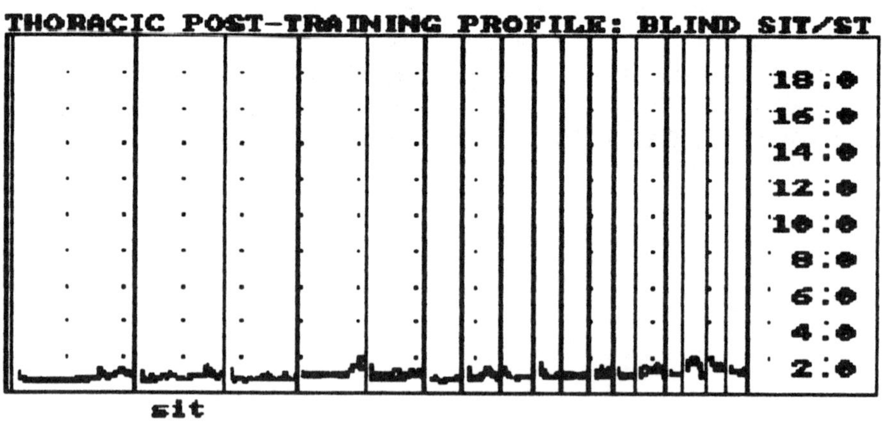


<u>THE TYPING INTERVENTION</u>

Another useful intervention is to simulate the working environment, in this case typing. In Figure 8 the graph starts high at levels of around

15uv and then quickly drops to 6uv as the person practices relaxing neck and trapezius muscles. Finally, at 3uv the training has succeeded in its first phase by allowing the person to begin to recognize the sensations of relaxed muscles while typing at the same time. In other phases, actual experience on a keyboard with EMG feedback would complete the training.

To begin the profile, the graph is started and the person is prompted to sit with forearms extended as if at a typewriter or keyboard. You will notice EMG levels typically rise 8-20uv. Ask the person to hold that position but to continue moving their fingers as if to simulate typing. As the person watches the graph, ask that the elbows be brought down and against the sides. Shoulders are then to be released while allowing the elbows to drop even further downward. The therapist needs to closely coach the person until EMG levels drop within 3-5uv while the person's fingers continue to simulate hitting the keys. The proprioceptive cues which are learned with this intervention need to be emphasized. The training needs to be repeated at each appointment if the patient works often on a keyboard or typewriter.

**Figure 8.** Training While Typing

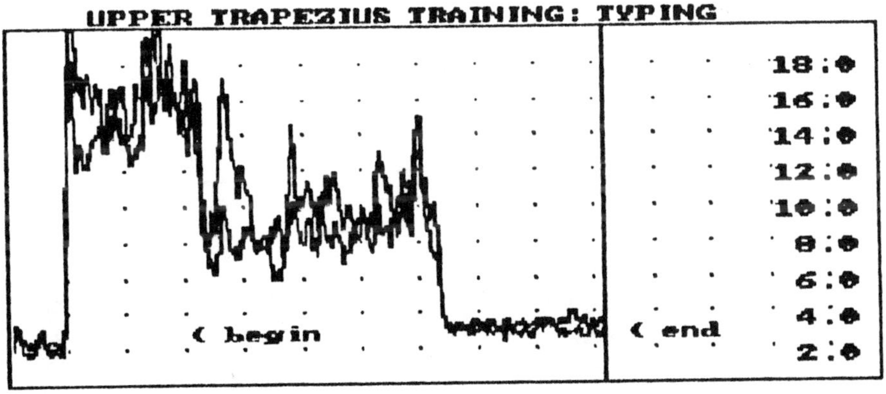

<u>WALKING</u>

Figure 9 shows the graph of a person walking with EMG levels recorded at 26uv (the graph does not allow levels above 20uv to be seen). As audio feedback is turned on and instructions are given to the patient, levels are seen to lower to 4uv (again, we look only at the lower portions of the spikes).

Many people take for granted the extreme levels of muscle tension which are present during walking. Most of this tightness is not used for actual movement. People tend to become intense in their determination to get to their destinations. EMG Biofeedback is a superb tool for teaching how to relax and walk at the same time. Begin by asking the patient to walk slowly about the room (do not turn on audio feedback yet). A word of caution - it is not appropriate to substitute "walking in place" for the following two reasons:

1. Walking in place does not 'feel' the same as actually moving oneself forward. Because we are teaching proprioceptive cues, it is essential that they be as close to the real sensations as possible.
2. Walking in place is not a typical event which the subject is likely to repeat in the practical world. Because appointments are very brief, it is better to use positions and movements which are likely to be repeated outside the office.

**Figure 9.** Training While Walking

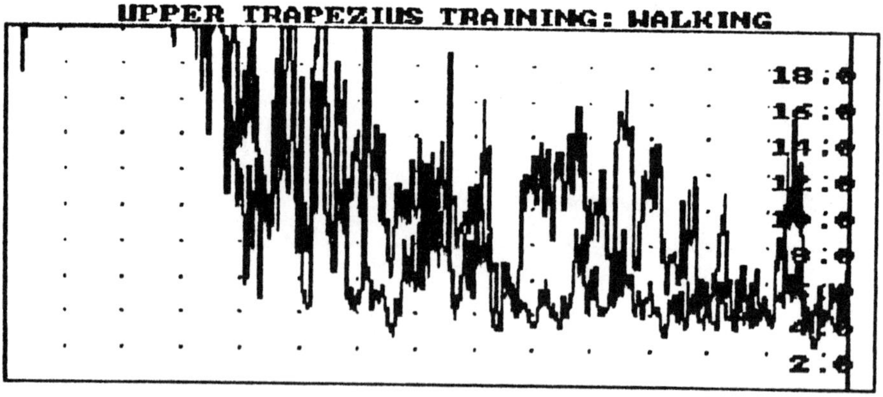

As the patient walks, set the threshold at a level which allows just enough audio to give the person a challenge. Look at Table 1 to obtain normal values for walking and then set the threshold about 4uv above that level. As you turn on the volume, warn them that they will hear sounds which signal that their muscles are too tight. They are to then release the muscles by allowing the heavy, pulling sensations to return again. The sensations of heaviness will be the same ones remembered while standing. Patients will also need to be warned that as they lower their muscle activity, they will not necessarily enjoy what they feel. It will begin to even feel quite awkward. Unless the person actually feels the same awkwardness while at work, muscles have not intentionally released. They are to be reminded that they have never really relaxed their back or neck muscles while walking and therefore their brain will not recognize the feeling as "normal". They remember the feeling so that it may later be repeated and incorporated into working conditions.

SIT/WALK/SIT

Figure 10 is a pre-training graph exhibiting the typical initial elevated activity associated with standing, horizontal levels which are unable to drop down to typical walking levels of 5-6uv and then a final elevation for sitting down. Figure 11 exhibits the trained profile where the graph will resemble a "U" or "H". The horizontal bar shows the EMG activity while walking at the same time that the vertical portions represent the sit/stand.

This procedure starts in the sitting position and incorporates part of the sit/stand as well as walking skills. The goal is to walk with the lowest EMG activity possible, as soon after rising from the chair as is possible. The therapist turns on the graph and asks the person to stand up and immediately start walking for about 5-10 seconds, return to the chair and then be seated.

Many people assume you need to hold all muscles tight while walking, especially those with low back pain. But patients often report feeling exhausted or unable to walk very far due to a continuous hypertonic state, which may also exacerbate pain. If they learn to selectively release muscles, they will benefit from the best of both worlds. The best way to accomplish this is by producing a lower abdominal, light "firmness" or "tone" while walking. At the same time, while reducing the lower abdominals, the upper and lower back muscles should be released. This takes time to practice and perfect but is definitely desirable and worth the effort.

**Figure 10.** SIT/WALK/SIT PRE-TRAINING PROFILE
(100 second screen)

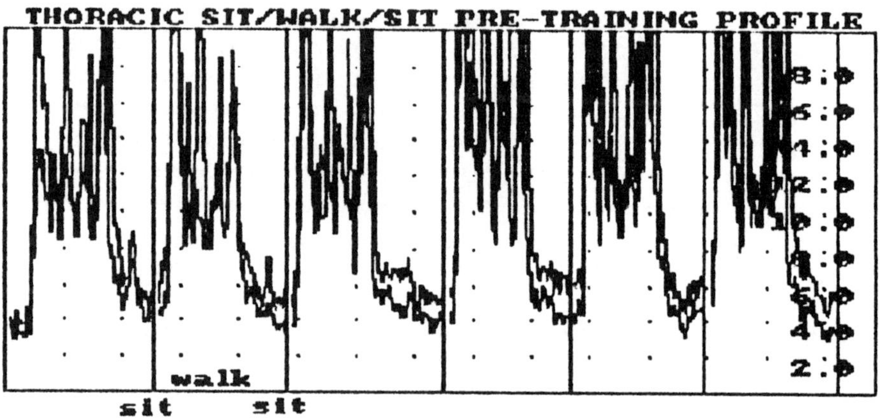


**Figure 11.** SIT/WALK/SIT POST-TRAINING PROFILE
(100 second screen)

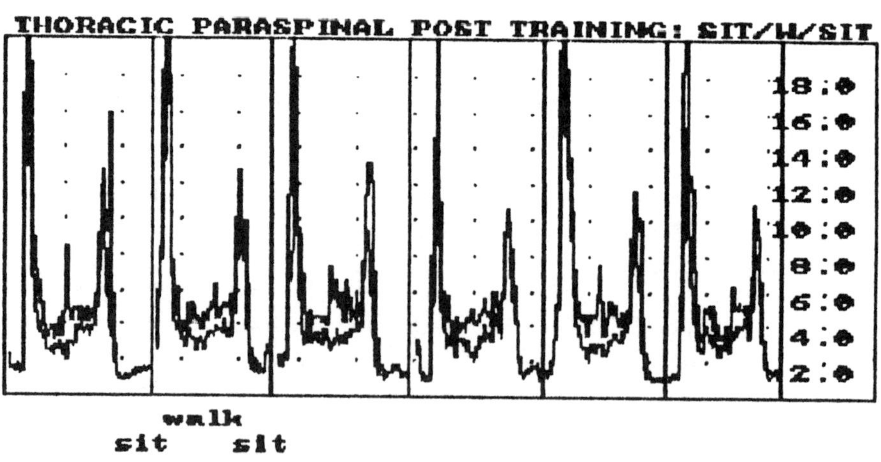

LYING DOWN

The purpose of the "lying down" intervention (Figures 12 & 13) is to teach people how to relax the back and neck muscles while lying down on them. For this training, a mattress or a carpeted floor may be used as long as there are pillows available. The main point to impart to the patient is that the direction of gravity is now no longer towards the feet and that the sensation of a relaxed muscle will therefore be perceived differently. Instead of feeling a dropping or heaviness toward the hips, as felt when sitting or standing, there will be more of a feeling of the body being pressed into the mattress.

Recordings should begin while the patient is still standing in order to record how quickly the quieting response is obtained after lying down. As the patient lies down, the graph (not shown here) will produce a fast drop into the 2uv range. Actually, levels of .5 and 1.0 are quite common for those who exhibit little hypertonicity.

After the patient has been lying down a moment, turn on the audio feedback using a threshold of about 2-3uv. You may then begin to talk with the person. As you continue to converse, remind them of the need to keep the audio off.

**Figure 12.** T8 PRE-TRAINING PROFILE: LYING DOWN

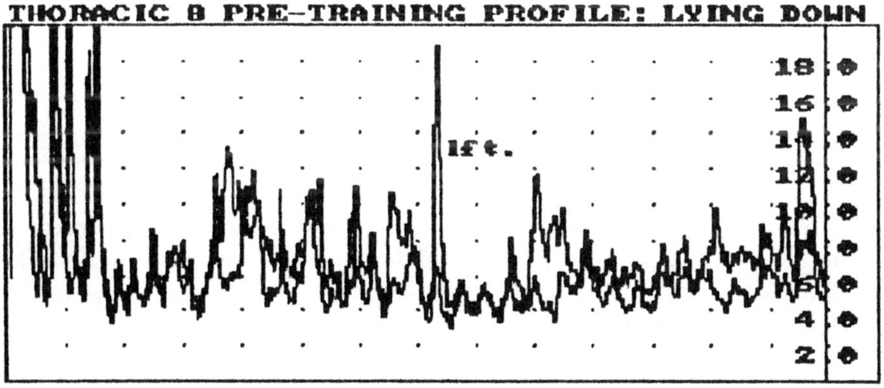

**Figure 13.** T8 POST-TRAINING PROFILE: LYING DOWN

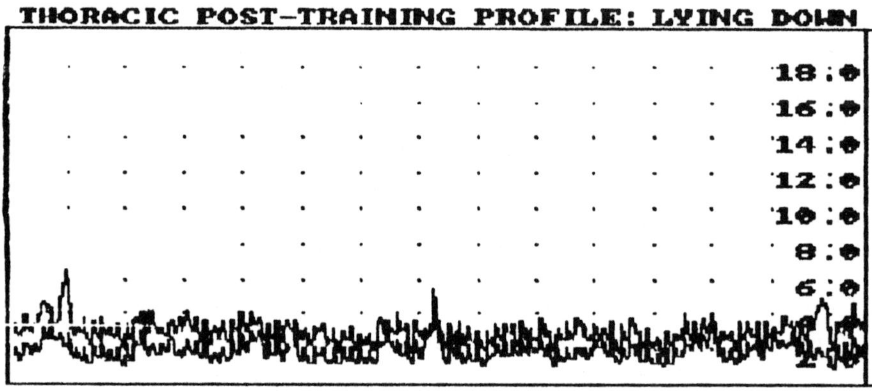


The lying down intervention often allows the therapist to be certain that the patient is indeed relaxing appropriately while in a prone position. If sleepless nights are reported, have the person practice this skill upon going to bed and awakening.

## THE LOW BACK (Lumbar paraspinals)

<u>THE LOW BACK</u>

The low back (lumbar paraspinals) is a region brought to our attention usually when pain is present. The area is so important to movement and posture, however, that when it produces pain it has the potential of affecting more than its own area. Low back dysfunctions, therefore, must be approached with caution as well as with respect. Caution is used because of the longer time frame needed for the prognosis and also because accessibility is more limited than with any other part of the back or neck. Respect is given to the low back due to that area's potential for extreme pain, strategic significance and pivotal influence on one's physical existence as well as one's emotional state.

LIMITATIONS OF WORKING WITH THE LOW BACK

Low back training is inherently the most challenging area to teach and in the future will be allowed its own chapter. The main difficulty stems from the fact that there are very few low back muscles to work with. This severe limitation has rendered that area apparently inaccessible, quite subtle and extremely elusive.

There are two frustrating aspects in working with the low back. First, is the length of time needed to produce change. Second is that when therapy reduces subjective pain levels a meager 20-30%, it is considered very significant for such a difficult area. On the brighter side, however, is the fact that MLT treatment also reduces stress-related neurological involvement and therefore the patient ends up feeling better, stronger, more in control, less irritable, and with the overall feeling of having made a positive first step toward recovery. The limited 30% change in pain levels eventually becomes very important, therefore, as a first sign that increased control has occurred and that some hope for further improvement is not too far fetched.

TREATMENT SUCCESS

Because there is less of an immediate feedback cue in the way of pain or soreness reduction in the low back, pain itself will obviously not be the first sign of choice to watch for in order to judge the extent of success. Initially, some or all of the following signs will be used instead:

1. Increased muscle control (quantifiable changes on the graphs)
2. The "dulling" of pain rather than its elimination
3. A more calm and accepting attitude reported by the patient
4. A more calm behavior pattern during appointments
5. Changing sleep patterns
6. Decreased depression
7. Increased energy (not that portion expected from less depression, but from a decrease in muscle activity due to less hypertonicity)
8. Decrease of interpersonal conflicts due to decreased tension
9. Increased sense of humor
10. Healthier interest in the future
11. Desire to bring parent or spouse into the appointments
12. Increased interest in learning during the appointments, outside work and as a general attitude

The time needed for reducing low back pain may be as little as two weeks and as long as three months depending upon the severity of the pathology and how quickly the patient learns.

The limitations of working with the low back need to be explained to the patient but preferably not at the beginning of therapy. The patient will be much more receptive after having already begun to notice a change and also when the nervous system has had a chance to regain a balance.

Although this chapter does not allow space for the complete and full treatment of such a complex, difficult and limited area, there are some major and strategic interventions which have been developed through MLT training which can have a major effect on the low back.

The traditional starting place for MLT low back training is in the upper back. In fact, the low back is profiled only in the first appointment and then is not monitored again until the upper and mid back muscle groups have been mastered. Limited low back instruction (without instrumentation) is given early in therapy, but biofeedback is not usually attempted. It is essential to first master other muscles groups for the following reasons:

1. It is difficult to pick up proprioceptive cues in the low back due to the latter's unobservable location, its continuous usage, and due to the meagerness of available muscle groups.
2. Adjacent muscle groups will tend to affect low back muscles. It is therefore important to control those adjacent, less difficult groups from the very onset of training.

It is essential to firmly establish in the patient's kinesthetic memory a standard sensation of how muscles feel when they are released. By working on control of the upper and mid back muscle groups first, it will allow the therapist the opportunity to establish that standard. Generalization into the low back, although slow and difficult, is accomplished more easily and quickly when the patient knows what sensations to repeat.

## TREATMENT OF THE LOW BACK

The goal in treating the low back is to quickly release adjacent low back muscle groups while being active and/or in pain. After initial introductions are given, it is explained to the patient that in order to really affect the low back, adjacent and more accessible muscle groups will have to be trained instead. The strategy will be to involve muscle groups which indirectly affect the low back but are more easily accessed. In other words, training will focus on the next most accessible muscles of the low back. These are the Gluteus and the External Anal Sphincter muscles. These muscles are trained independently of each other and are the most influential groups for the low back. Once they have been trained to relax while the patient is under pain or stress, the entire pelvic girdle will relax. This will reduce much of the hypertonic state surrounding existing Lumbar pathology. Also, by reducing firing of major muscle groups, patients will experience less fatigue and less irritablility. In fact, for those patients who exhibit very little low back muscle activity in the first place, the reduction of fatigue and irritability become critical areas of improvement.

## INTRODUCTION TO GLUTEUS MUSCLES

The function and location of the gluteal muscles are easily explained by simply placing your hands on your own, describing them as the "cheek" muscles and thereby showing what needs to be relaxed. The person must realize how easily they may be accessed and consequently controlled. A practice of tightening them and then allowing a quick release will help to teach recognition of the telltale "drop" as they relax. This tense-release teaching method is NOT to be used again. It is better to introduce proprioception at this early stage. It is important not to tense-release too often or more irritation may ensue. This muscle group is easily felt but still needs to be trained adequately so that the release occurs quickly and easily, even while the person is under pain, stress, cold etc.

## INTRODUCTION TO THE EXTERNAL ANAL SPHINCTER MUSCLE

After gluteus muscle control has been well trained and learned, the external anal sphincter muscle (EAS) is then trained. This is the most influential muscle of the entire low back but must obviously be introduced with some diplomacy. If the person is rather sophisticated or knowledgeable there is a good chance it will already be known but still must be introduced. However, in most cases it will not be known.

The EAS is described as that muscle, the "anus", which is relaxed in order to have a bowel movement and is tightened when one wishes to hold back. This description is usually an adequate explanation of its function as well as its location. But the person needs to be able to control the rectal sphincter and gluteus muscles independently of each other in order to assure the required level of control.

Teaching sphincter muscle control could also bring up social or cultural issues. There will be a temptation to make light of the subject, especially if the therapist is uncomfortable. Credibility will be sacrificed unless the therapist maintains professional control. If smiles, winks or jokes occur, the process will not be taken as seriously as it needs to be. It must be impressed that this muscle is the most important one of all the lower back. The EAS is more involved during voluntarily relaxing of the pelvic girdle than any other muscle. It is especially important when sciatica is present. Often sciatica may only be affected by a constant release of this muscle. The effect, however, will not be felt as quickly with sciatica as with a non radiating back pain.

## SEQUENCE OF LOW BACK INTERVENTIONS

As low back control is introduced and applied, the need for emphasis and diplomacy must be remembered. Uppermost in the therapist's mind should be the following four reasons for remaining professional and insistent:

1. The typical low back patient is usually irritable, lightly to moderately depressed, in constant pain, often under financial hardship and suspicious of all interventions due to historically poor results.
2. The muscle groups used in learning to control the low back (gluteus & sphincter) are unexpected.
3. The gluteus and sphincter muscles, due to their intimate locations, could cause minor compliance problems.
4. The gluteus and sphincter muscles are absolutely critical in learning to control the low back.

Ultimately, the therapist must be convinced and convincing when teaching all pain patients but especially with the low back group. Their pain is more centrally located and strategically placed and has inevitably affected their lives to such an extent that all activities are painful and all hope seems to be disappearing.

## FOLLOW-UP APPOINTMENTS

To best serve the interest of the patient and the MLT training procedure, appointments should never end abruptly simply because skills have been learned. Certain patterns will develop as time passes and need to be pointed out.

A routine will soon emerge in which the person will remember to use the new skills only when they are needed (i.e. when pain or muscle tightness is present). The patient needs to know that this is acceptable behavior and that the skills will always be there, ready to be applied.

Another important function of the follow-up is to allow time to pass so that more events will have had time to occur. These events (pain recurrences, illnesses, new injuries, vacations, etc.) will provide an opportunity for the person to try out the new skills under the most rigorous of conditions (all of which constantly occur ) and then bring the experiences back into the appointment for review.

The follow-up appointment is begun by telling the person that the purpose is to simply mark any significant patterns which might have developed over time and to reinforce all learned skills. No new work need be taught unless new interventions have been developed or the person is in need of experience regarding a new area of the back.

After baselines have been taken, further interventions (sit/stand; walking; sit/walk/sit) should be recorded and used to support the fact that skills have indeed been learned. If there are any problems visible from the baselines or interventions, the follow-ups would be the time to reinforce or review them and add any needed work.

## THE TREATMENT PROCESS

### INTRODUCTION

Three separate anatomical sites are used as the basis for the procedures which comprise MLT for back, neck and head pain:

(1) Upper Back
(2) Mid back
(3) Low back.

All three together are part of the complete protocol. However, space allows for only a general overview to be described at this time. Each area has its own specific type of intervention designed for its own specific needs and difficulties.

The first area to be trained will always be the upper back even when a low back injury is involved. Training continues down to mid back and finally down to the low back when indicated. Three criteria, when satisfied, dictate when one muscle group has been completed and the next is ready to be started. The first is objective and the last two are subjective. These three criteria are as follows:

1) Graphs must show a low EMG profile,
2) The person reports recognizing proprioceptive cues when muscles are released as well as when they are tight,
3) The person reports feeling better or reports that some symptoms have improved.

## OUTLINE OF APPOINTMENTS AND THE SEQUENCE OF INTERVENTIONS

The following is a typical appointment by appointment protocol, in an outlined, abbreviated sequence. Each intervention has been described previously in this chapter. The exact sequence will only occasionally vary. However, the <u>total</u> number of appointments may vary greatly depending upon the complexity of the case and the patient's level of proprioception and type of pathology.

### <u>APPOINTMENT 1: INTAKE AND BASELINE PROFILES</u>

The intake is the main focus of the first appointment. It asks questions which are specific to the diagnosis, fundamental to the neuromuscular system in general and emphasizes the functional level of muscles specifically. Typical questions concern symptoms of muscular distress, pain patterns, sleep behavior, daily irritants, etc. A basic part of the intake involves describing biofeedback in general and then emphasizing the uniqueness of MLT.

At the end of the intake appointment, baseline profiles are recorded but typically no attempt is made to explain results. The reason for not going into explanations with the patient during this first appointment is due to the limitations of time. It would be unfair to tease the person with what has been found in the initial profiles unless you had at least fifteen minutes to elaborate and entertain questions or possibly initiate training.

If a dual channel graphing capability is available, the normative data as described earlier are easily and immediately visible.

APPOINTMENT 1 PROCEDURE SEQUENCING

1. *INTAKE*
2. *BASELINE PROFILES*
3. *INTRODUCTION* (only if there is a full 15 minutes left in the appointment)

<u>APPOINTMENT 2: INTRODUCTION TO INSTRUMENTATION AND TO THE UPPER BACK.</u>

The second appointment is used to familiarize the person with the instruments and also to begin an introduction to upper back control training procedures.

APPOINTMENT 2 PROCEDURE SEQUENCING

1. *INTRODUCTION TO INSTRUMENTS*
   Also provide interpretation of results from previous baselines.
2. *INTRODUCTION TO THE THEORY OF THE SIT/STAND INTERVENTION*
3. *SIT/STAND*
   The appointment is begun by rapid, repetitive acts of sitting and standing. When the person begins to recognize the sensation of the released muscle, it is the cue that the person is ready for further interventions.
4. *INTRODUCTION TO SITTING AT A KEYBOARD OR TYPEWRITER.*

<u>APPOINTMENTS 3 & 4: CONTINUE BASIC TRAINING AND INTRODUCTION TO RESPIRATION</u>

By the third and fourth appointments, basic training should be underway. At this time an introduction to the low back must be made if the patient has a low back dysfunction. The patient must recognize and then learn that muscles will not spontaneously release. Only through specific training will that happen. Muscles must be actively released each time tightness, pain or cold is felt. The person should also be made aware of the fact that each time the muscles are released, the learning process will become easier and the area involved will be subjected to less and less physical abuse.

The third and fourth appointments may eventually take longer and could go to 6 appointments. The decision should be based upon the receptivity, skill and level of pain of each person.

There will be times when the therapist will need to provide more than physiological training. As an MLT therapist your primary objectives should be to provide psychological support while providing the skills needed to help lower pain levels as quickly as possible. You are not teaching pain control, you are teaching pain reduction and the patient needs to understand the difference. However, there are some very nice, physiological interventions for anxiety which we all know as "pursed-lip breathing" and which are a must for this patient population. This very effective intervention has been described by many and will not be elaborated upon at this time.

APPOINTMENT 3-4 PROCEDURE SEQUENCING

1. *SIT/STAND*
   Reinforce previous appointments by repeating the sit/stands.
2. *SIT/STAND WITH AUDIO*
   Audio feedback is introduced and then incorporated into Sit/Stand with audio.
3. *WALKING*
   Learn to walk while releasing muscles using audio feedback
4. *BLIND SIT/STAND*
5. *FASTER WALKING*
   By appointment 4 there should be an attempt to do faster walking and still keep the levels around 6uv (20% of your people will reduce them to 4uv or lower).
6. *SIT/STAND USING REVERSE 1s AND REVERSE 7s.*
7. *INTRODUCTION TO LOW BACK*
   General information and introduction to gluteus and external anal sphincter muscle control.
8. *RESPIRATION*
   Pursed-lip breathing.

APPOINTMENT 5: GENERAL REINFORCEMENT AND INTRODUCTION TO THORACIC PARASPINALS.

This appointment reinforces previous work, introduces the critical mid-back area and adds further interventions.

APPOINTMENT 5 PROCEDURE SEQUENCING

1. *REINFORCEMENT*
   Continue to repeat procedures from previous appointments until EMG levels indicate good control, until the patient definitely knows when muscles are relaxed and, finally, when the patient reports decreased pain.

2. *SIT/WALK/SIT*
3. *THORACIC PARASPINAL BASELINE PROFILES*
4. *INTRODUCTION TO THORACIC PARASPINAL CONTROL*
5. *BEGIN THORACIC SIT/STAND*
6. *REVIEW EXTERNAL ANAL SPHINCTER MUSCLE CONTROL*

APPOINTMENT 6: CONTINUE THORACIC PARASPINALS AND RESPIRATION

Session 6 repeats interventions or adds those not yet attempted due to time limitations.

APPOINTMENT 6 PROCEDURE SEQUENCING

1. *SIT/STAND*
2. *TYPING POSITION WITH THORACIC CONTROL*
3. *INTRODUCE WALKING WITH THORACIC CONTROL*
4. *LYING DOWN*
   Patient lying down on back while releasing muscles.
5. *RESPIRATION*

APPOINTMENT 7: CONTINUE THORACIC PARASPINAL TRAINING; UPDATE OF IMPROVEMENTS

The main focus of this appointment is the documentation of changes which the patient has experienced. The function of the update is to provide a vehicle of communication to the referring physician as well as to the patient. Patients do not often recognize improvements until they are pointed out. The elaborate patient update report has not been included due to space allowances.

APPOINTMENT 7 PROCEDURE SEQUENCING

1. *CONTINUE REINFORCEMENT OF WALKING*
2. *PATIENT UPDATE*
   The purpose of this is to establish what subjective and objective changes have occurred and to compare graphs of these changes.

APPOINTMENT 8: COMPLETE ALL THORACIC TRAINING PLUS REVIEW UPPER TRAPEZIUS.

APPOINTMENT 8 PROCEDURE SEQUENCING

1. *THORACIC SIT/STAND*
2. *BLIND SIT/STAND*
3. *SIT/WALK/SIT*
4. *SIT / STAND USING REVERSE 1s TO 7s*

5. *UPPER TRAPEZIUS BASELINE PROFILES*
   This is just a quick check of the upper back to make sure that control has not regressed.
6. *CONTINUE ANY NEEDED TRAINING OR REINFORCEMENT*

APPOINTMENT 9: 2-WEEK FOLLOW-UP

This appointment follows the traditional path of the MLT follow-up.

APPOINTMENT 9 PROCEDURE SEQUENCING

1. *THORACIC PARASPINAL BASELINE PROFILES*
2. *SIT/STAND REVIEW*
3. *WALKING REVIEW*

APPOINTMENT 10: 1-MONTH FOLLOW-UP

APPOINTMENT 10 PROCEDURE SEQUENCING

1. *UPPER TRAPEZIUS BASELINE PROFILES*
2. *THORACIC PARASPINAL BASELINE PROFILES*
3. *THORACIC SIT/STAND REVIEW*
4. *THORACIC WALKING REVIEW*
5. *RESPIRATION REVIEW*
6. *REINFORCEMENT*
   Only where indicated by objective evidence from graphs or reported by patient.

APPOINTMENT 11: 2-MONTH FOLLOW-UP

APPOINTMENT 11 PROCEDURE SEQUENCING

1. *SAME AS #10*
2. *UPPER TRAPEZIUS SIT/STAND REVIEW*
3. *UPPER TRAPEZIUS WALKING REVIEW*

APPOINTMENT 12: 4-MONTH FOLLOW-UP

APPOINTMENT 12 PROCEDURE SEQUENCING

1. *SAME AS #11*
2. *PATIENT UPDATE*

APPOINTMENT 13: 6-MONTH FOLLOW-UP

This length of time is not often available but is very important for long term control and reinforcement.

APPOINTMENT 13 PROCEDURE SEQUENCING

1. *SAME AS #11*
2. *PATIENT UPDATE*
3. *PATIENT RELEASED*

## CLOSING COMMENTS

There are several critical advantages to MLT training. The biggest advantage is its ability to generalize learned skills relatively easily and quickly. In the final analysis, a treatment procedure is only as good as its applicability into the practical environment of work and home. MLT also teaches life-long habits which promote a continuous awareness of change of hypertonic states. Patients no longer loop continuously back into the health system. Two distinct advantages which are due directly to the physical emphasis of MLT training are compliance and motivation. They become less an issue because training concentrates on lowering pain by building proprioceptive cues within muscle groups rather than on psychological issues.

A limitation of MLT training was found in working with people who had a severe difficulty in relating to their own proprioceptive cues. These individuals were rare but tended to need an uncommon amount of time to learn their skills. In that case, however, it was a matter of "when" they would learn rather than "if". Another limitation was found in working with depressed patients. Until pharmacologically or psychologically supported, this group could not muster enough energy or interest to really care about any intervention.

Muscle Learning Therapy is certainly not a cure for pathology nor is it a substitute for other basic medical interventions. Its goal is to teach an adjunctive skill which helps bridge the gap between traditional therapy and the realities of the working world. That skill is then used to help reduce the pain often caused by contiguous hypertonic muscle groups exacerbating injured tissues.

Any program concerned with back, neck, head pain or cumulative trauma disorders would be incomplete without the training of the voluntary muscles which so often are the sole cause for recidivism. MLT uses a uniquely structured training protocol which emphasizes brevity, cost effectiveness, precision and permanence.

# CHAPTER 5

# *Evaluation and Treatment of Myofascial Pain Syndrome Utilizing Biofeedback*

Barbara J. Headley, M.S., P.T.

The evaluation and subsequent treatment of soft tissue pain complaints have left much unanswered regarding the source of the complaints, directives for treatment, and treatment outcome. The identification of many of these complaints into specific syndromes, specifically myofascial pain syndrome and fibromyalgia, have been helpful in providing a model with which to view soft tissue pain complaints. It has been difficult to understand how patients suffering with myofascial pain syndrome or fibromyalgia, and have much greater levels of dysfunction than those who have other soft tissue complaints, do not seek treatment or report significant levels of interference in function.

The myofascial pain syndromes have been categorized recently as "sympathetically maintained pain." This may reflect the need to categorize these disorders by the referred pain which is the hallmark of the myofascial disorders, or it may be a reflection of the observable, and sometimes profound, autonomic responses of the trigger point areas to stimulation.

Biofeedback has been used by the author to investigate several of the components of myofascial pain complaints. The intent is, rather, to examine the referred muscle dysfunction, often with spasm, the autonomic involvement, and the restoration of function utilizing electromyographic (EMG) biofeedback to identify dysfunction and target treatment intervention. The first section will be devoted to an overview of myofascial complaints as they relate to autonomic dysregulation, referred muscle activity, and subsequent dysfunction. The components of autonomic dysregulation, observed by the author, will also be reviewed in the second section, as well as the proposed mechanism of the referred pain. In the final section, a methodology will be presented for evaluation of the trigger points as they relate to functional limitations. This will include the protocol for determining the primary components of the dysfunction, with the pain complaints being the secondary issue, and the ability of EMG biofeedback to target treatment in achieving more rapid resolution of the referred dysfunction.

## PATHOPHYSIOLOGY OF MYOFASCIAL COMPLAINTS

Simons (1988), a pioneer in the field, has identified myofascial pain syndrome by the signs and symptoms associated with the diagnosis, inclusive of one or many single-muscle myofascial pain syndromes resulting from "trigger points." The myofascial trigger point is defined as "a hyperirritable spot, usually within a taut band of skeletal muscle or in the muscle's fascia, that is painful on compression and that can give rise to characteristic referred pain, tenderness, and autonomic phenomenon" (Simons, 1988). While trigger points may exist in other tissue, such as tendons, joint capsules, ligaments, and periosteum, they do not generate patterns of referred pain as do those in the myofascial system. It is the referred pain and autonomic activity which have made the identification of myofascial complaints difficult and confounded the establishment of treatment techniques. Myofascial complaints are typically difficult to differentiate for the practitioner unaware of the patterns of referred pain and the associated autonomic dysfunction. The difficulty in establishing how these trigger points (which are prevalent in adults age 30 and over) can lead to marked dysfunction in some individuals has led to debate over whether the complaints are "real". Difficulty for some patients in obtaining treatment results in a delay in establishing effective therapeutic intervention.

Almost all adults have experienced the tight "knot" in the shoulder, or a brief interlude of back pain which is accompanied by a painful hardening in the muscle belly, but the majority of individuals recover and find no significant degree of functional impairment. It is not the intent of this chapter to review the literature in regards to the extensive theories regarding pathophysiology, or to discuss at length the discriminators of myofascial pain syndrome from fibromyalgia. One has only to work with a patient experiencing myofascial pain syndrome about the head and neck to grasp the autonomic dysfunction, which may be produced with palpation, or stimulation, of the painful trigger point.

The presence of referred pain has been well documented by many (S.B. Graff-Radford, et. al., (1987); Sherman (1990); Simons (1981, 1988); D.G. Simons & J.G. Travell (1983)). The detailed "mapping" of these referred pain patterns has been immeasurably helpful in assisting patients and practitioners in understanding that the cause of the pain is often not at the site of the pain. Validating these complaints to the patient often removes the major obstacle to treatment. The involvement of the autonomic nervous system has been identified (S.B. Graff-Radford, et. al., (1987); Simons (1988); D.G. Simons & J.G. Travell (1983)), but the dysfunction related to musculoskeletal involvement from the trigger points has been more difficult to pinpoint. Some patients can tolerate high levels of stimulation of the trigger points, while others suffer symptoms such as dizziness, nausea, lightheadedness, diarrhea and balance dysfunction with less intense stimulation.

Sarno (1981) has suggested that myofascial pain syndrome constitutes an autonomic myoneuralgia. While he introduces the concept of autonomic mediation of myofascial complaints, he concludes it is a "functional disorder" without significant tissue change. He rightfully points out the extent to which these patients have been educated and diagnosed to relate their complaints to herniated discs. The physical aspects of this disorder have been elucidated in more recent literature. More recent evidence has been reported to the effect that myofascial complaints may constitute as much as 95% of the low back complaints (Rosomoff, 1990). It has also been the experience of the author that, while degenerative disc disease may or may not be present, treatment of the soft tissue dysfunction alone may markedly reduce associated leg complaints, including numbness, tingling, and pain. The mechanism can be related to other research on referred pain. Myofascial pain complaints may well continue to be exacerbated by mechanical instability in patients who have had some injury to a disc, where healing has occurred without movement at a motion segment(s). However, the treatment of the myofascial pain syndrome can significantly alter the on-going pain complaints. The ability to decrease the pain experience of a patient with a bulging disc can facilitate the development of a self-management program.

The implications of myofascial pain syndrome, being a type of sympathetically maintained pain, led the author to correlate the autonomic dysfunction of these patients with other chronic pain patients treated over a two year period. The author has collected two years of data on low back pain patients regarding autonomic dysfunction. This data, being reported elsewhere, was compared with the autonomic profiles generated by patients identified as having myofascial pain syndrome. The criteria being used for this comparison is founded on theoretical and clinical work by Lacey (1958), Schwartz (1976, 1977).

The ability of active trigger points to elicit profound autonomic responses has been documented for points which can precipitate cardiac arrhythmia and the mimicking of the pain of gastrointestinal ulcers (Simons, 1988). Similarly, an active trigger point in the infraspinatus can present with a pain complaint of "bursitis" in the shoulder. These examples, and others cited, suggest that feedback loops exist between the central nervous system, trigger points, visceral structures, and their referred pain zones. The correlation between trigger points and acupressure points has been studied by Melzack (1986) and cannot be overlooked. The autonomic responses observed by the author reflected hyperactivity of the sympathetic nervous system. On several patients, the sympathetic point on the ear was treated after these patients demonstrated marked hyperactivity of the sympathetic nervous system to the point where the discontinuation of treatment was considered. In one patient, a single case study using an ABABC case study design was carried out, to evaluate the effect of the acupressure treatment and the possibility of a placebo effect. In each case, the patient responded favorably to the auricular treatment (see Table 1). When this treatment was believed to have been initiated, (in session number 10) which usually resulted in dramatic autonomic quieting, the treatment was, in fact, withheld; the patient responded with profound autonomic dysfunction. The reduction in sympathetic nervous system activity enabled treatment to be continued and increased patient compliance.

**Table 1.** Trigger Points and Possible Autonomic Responses

| SESSION | TRIGGER POINTS TREATED | AUTONOMIC RESPONSES | AURICULAR TREATMENT |
|---|---|---|---|
| 5 | L/R Upper Trapezius | Nausea, nose, histamine wheels | No |
| 6 | L/R Upper Trapezius | Nausea,head-ache, dizzy, histamine | No |
| 7 | L/R Upper Trapezius | None | Yes |
| 8 | L/R Upper Trapezius | None | Yes |
| 9 | L/R Upper Trapezius | None | Yes |
| 10 | L/R Upper Trapezius | Runny nose, nausea, dizzy | No |
| 11 | L/R Upper Trapezius | Nausea, dizzy, runny nose | No |
| | | Relief of Symptoms | Yes |

The profile of muscular activity involved in the trigger point itself is generally one of inhibition. Direct placement of wire EMG electrodes into the trigger point have demonstrated that these areas are electrically quiescent (Simons, 1981). While the trigger point itself is located in a portion of muscle fiber that demonstrates shortening of sarcomeres, or perhaps the empty sarcomeres, the sarcomeres are not in spasm. The sarcomere has functionally shortened, often resulting in lengthening of sarcomeres at the ends of the fibers. These muscles have been identified as fatiguing rapidly and functioning under a metabolic distress profile (Simons, 1988).

Upon examination with EMG biofeedback these muscles are often found to be inactive with functional movement. This inhibition of muscle activity is often related to gravity or function. Muscle testing frequently results in a normal grade of muscle strength, yet dynamic evaluation reveals that the muscle is not functionally active when compared to normal profiles.

This introduces the concept of inhibition which may be centrally mediated, and related to postural reflexes, overload phenomenon, or autonomic control. While the mechanism of the inhibition in muscles which are of normal strength cannot be delineated at the present time, its occurrence in patients with myofascial pain complaints more than three months post injury is impressive. These muscles appear to progress through various stages of dysfunction. Initial spasms lead eventually to overload and inhibition. Targeting the relationship of the trigger points and inhibition as related to postural dysfunction enables the therapist to direct treatment to the affected muscle groups, and introduce an effective therapeutic intervention. Postural dysfunction with asymmetries may be evident with muscle scanning, dynamic evaluations or direct stimulation of the trigger points.

The profound complaints of fatigue, inability to complete work tasks, and interference with function led the author to examine the relationship of trigger points to referred muscle spasms. This, when incorporated into the profile of postural dysfunction and pain, could account for the significant impairment in function experienced by these patients.

## REFERRED PAIN AND DYSFUNCTION

Original work done by this author, related to autonomic dysregulation and sympathetically maintained pain, incorporated an evaluation of four channels of thermal biofeedback at the distal points of the four extremities. Subsequent evaluation was often done with electrodermographic (EDR) assessments in the upper extremities. Patterns were found which were consistent with the theoretical and clinical hypotheses presented by Schwartz (1976, 1977) and Lacey (1958). The autonomic dysregulation evident in chronic pain patients was felt to be a contributing factor in muscle pain and dysfunction, if through no other means than the alterations in blood flow which occurred. Evaluations had been done on 52 patients with chronic neck and back pain, pre and post fusion patients, and those with reflex sympathetic dystrophy. Similarities between the RSD patients and the patients with back pain suggested the presence of a "subtle" sympathetic dystrophy occurring in approximately 63% of the back patients presenting for evaluation. Also evident among these patients was a frank "misperception" of the thermal information from the extremities. Often, the warmer foot was felt as markedly colder than the other foot.

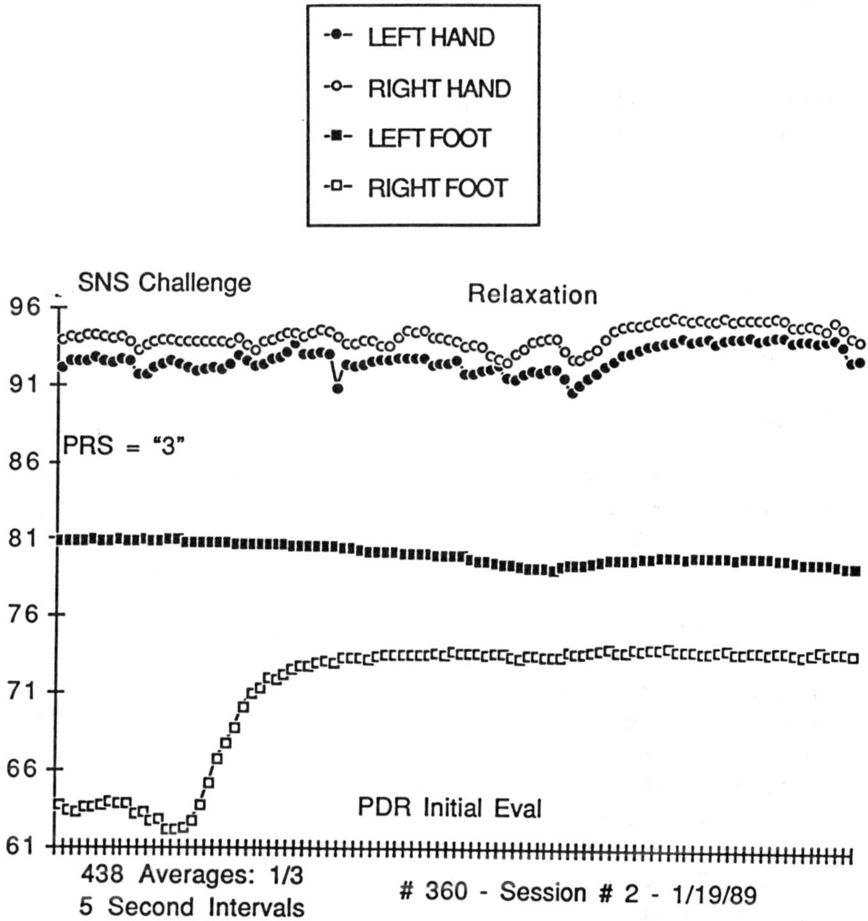


**Figure 1.** Sympathetically Maintained Pain Evaluation

This confusion, and extreme misperception, of the thermal differences was felt to be concurrent with many of the seemingly exaggerated complaints of pain reported by the same population. Re-regulation of the thermal asymmetries was concurrent with a loss of the profound misperception, and a reduction in the pain complaints.

The same criteria were applied in evaluating the profiles of 75 subsequent patients referred to the author with myofascial pain complaints. The comparison revealed that the first 52 patients who were referred for chronic pain management, or pre-fusion evaluation, had a higher level of autonomic dysregulation (or SSMP) than those patients referred primarily for complaints of myofascial pain.

**Table 2. PATIENTS REFERRED FOR TREATMENT WITH LOW BACK PAIN/DYSFUNCTION (Breakdown by Surgical Intervention)**

| | GROUP 1<br>N = 52 | GROUP 2<br>N = 75 |
|---|---|---|
| Conservative | 29 | 59 |
| Pre-fusion | 13 | 12 |
| Post-fusion | 10 | 4 |

Note: Patients in the second group reflect the tendency of the referral sources to send all pre-fusion patients prior to surgery as a result of the data and treatment results obtained in the first group.

Of the initial 52 patients, 60% (n=31) were positive for the criteria of subtle sympathetically maintained pain. Of the second group of 75 patients, 48% (n= 36) fit these criteria. While the second group contained a similar sample of chronic pain and pre-/post-fusion patients, the group also included a significant number of patients with soft tissue injuries, which would otherwise not have been referred to the author as early in treatment.

The prevalence of myofascial pain syndrome in the second group was 66%. The percentage of positive criteria for SSMP was lower in this group than in the initial group. This data suggests that the criteria for SSMP established with a model of autonomic dysregulation (measured by thermal biofeedback) is lower in the myofascial group than in the general chronic pain population. Yet the involvement of autonomic symptom- referral from stimulation of the trigger points is readily evident with many patients.

The inescapable fact that the various mechanisms of pain transmission, and its quantification, remain beyond our present knowledge base. This leads the practitioner once again to search for the link between pain and dysfunction. The ability to target treatment for dysfunction and produce changes in the pain experience continues to lend support to this direction of treatment. A quantified assessment of the dysfunction resulting from experienced pain leads to effective treatment.

The mechanism of the referred pain seen in myofascial pain syndrome may be more similar to the mechanism described in patients with amputations and phantom limb pain (Sherman, 1990), rather than the central pain theories of other chronic pain syndromes. For amputees, the source of the pain is clearly not where the pain is experienced, as the limb is absent. Unless one is to discredit the complaints of all these patients, the mechanism of pain and the source of the pain must be elsewhere. This analogy fits well with the problems inherent in the understanding of myofascial pain and its subsequent dysfunction.

Stimulation of a nerve can result in the experience of a sensation distant from the point of stimulation. One has only to hit their "funny bone" to realize their pain is first in the area which was hit, but then travels down the forearm and into the hand itself. The discomfort felt in the hand is from a distant source. In the amputee, the source of the pain may be at a number of locations along the nerve, but the brain has a "pattern" or conceptualization of where that pain is coming from. The brain remembers many things in "patterns", so that incoming information is judged in relation to how that information was processed in the past. Stimulation of the nerve, or alternations in its conduction by vascular or muscle tone changes, result in the information being relayed to the brain often in relation to where it "should have come from." Similar changes in vascularity and muscle activity have been described for those patients with myofascial pain syndrome.

The generation of the referred pain in myofascial pain syndrome, however, may well include more visceral-mediated pathways than dermatomal. The referred pain patterns in these patients often bears little or no resemblance to any one nerve root. The consistency of the pattern of referred pain across a large population of patients forces the practitioner to acknowledge other pathways for referred pain.

An analogy in the lower extremity to the "funny bone" might be the piriformis syndrome, in which compression occurs as the sciatic nerve travels through the piriformis muscle, and the sensations resulting from this are often felt in the thigh, calf, or foot. Patients will frequently be insistent that the problem is in the foot or at the level of the disc in the low back. The ability of the therapist to rapidly relieve all these symptoms by merely reducing the piriformis pressure is dramatic, and can facilitate the development of an understanding of the role postural dysfunction and muscle imbalances play in low back pain complaints.

Trigger points exist in many individuals without the profound dysfunction we see in many patients. This dysfunction may be mediated, in part, by the autonomic involvement independent of, or concurrent with, the referred mechanisms which are the hallmark of myofascial pain syndromes. The shortening of the muscle fibers and its fascia have been documented elsewhere (Simons, 1988). In treating patients using myofascial techniques with low back pain complaints, a therapist is able to reduce the functional shortening of an extremity by working on the full myofascial chain in an extremity. Patients with low back pain, upon successful myofascial work, often remark that their leg feels "longer," and that their foot feels flatter on the ground than it had prior to the treatment. The impact of this functional change in leg length on low back posture and muscle imbalances was investigated.

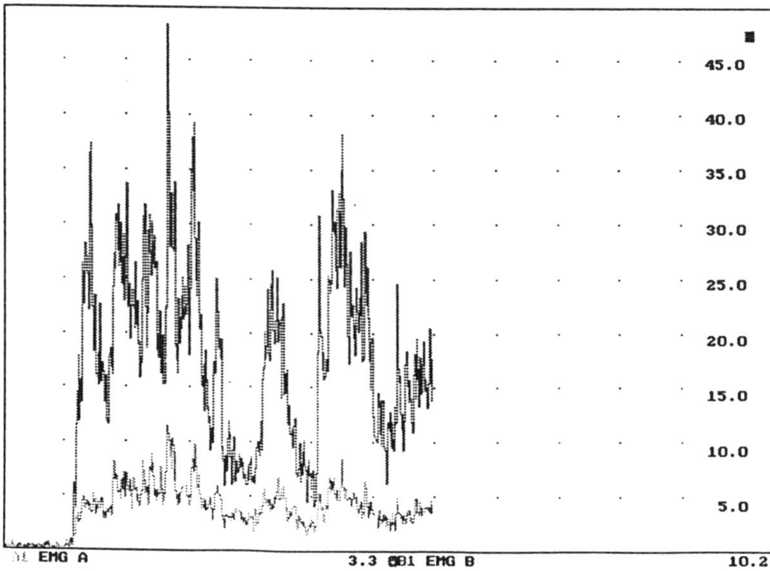


**Figure 2(a)** The right soleus muscle was evaluated prior to the beginning of the session and it was found to be referring a significant degree of muscle spasm to the right paraspinal muscles (a).

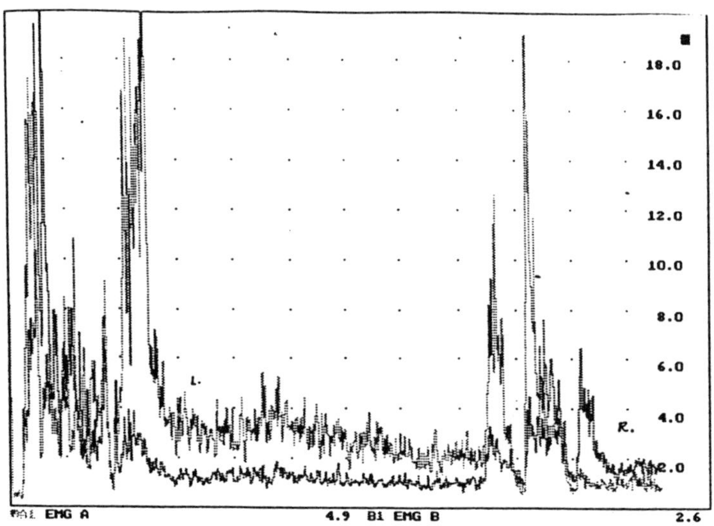


**Figure 2(b)** After the myofascial treatment, the referred muscle spasm was significantly reduced (b).

Trigger points were seen to have a profound impact on function, by the pain arising from the trigger point, or by its impact on the postural systems and its checks and balances. Pain effects these balances, as does the changes in functional length or fluidity of movement. Pain, then, leads to both impairment of mobility and a disorder of movement. In addition to addressing a patient's pain complaints, one must also construct the mechanism of the movement disorder so that intervention can be successful.

Using dual channel EMG biofeedback, dynamic low back evaluations were performed before and after treatment. It was found, with EMG monitoring of the low back paraspinals during the myofascial treatment, that "spasms", or at least a marked increase in muscle resting tone, were being referred to the low back from a number of points in the lower extremities. These points, in the tensor fascia lata for example, would result in ipsilateral, contralateral, or bilateral spasms in the paraspinal muscles during initial stimulation. After treatment of the trigger points, these spasms would often be reduced or absent. The patient would then frequently report that the leg pain had decreased, and often that there was also a significant reduction in the low back pain. This may have occurred with treatment directed only toward the leg muscles.

The trigger points which were most commonly found in these patients were then plotted, or mapped. It was found that there was a common pattern to the majority of patients with low back pain complaints, and the "chain" of trigger points could be identified.

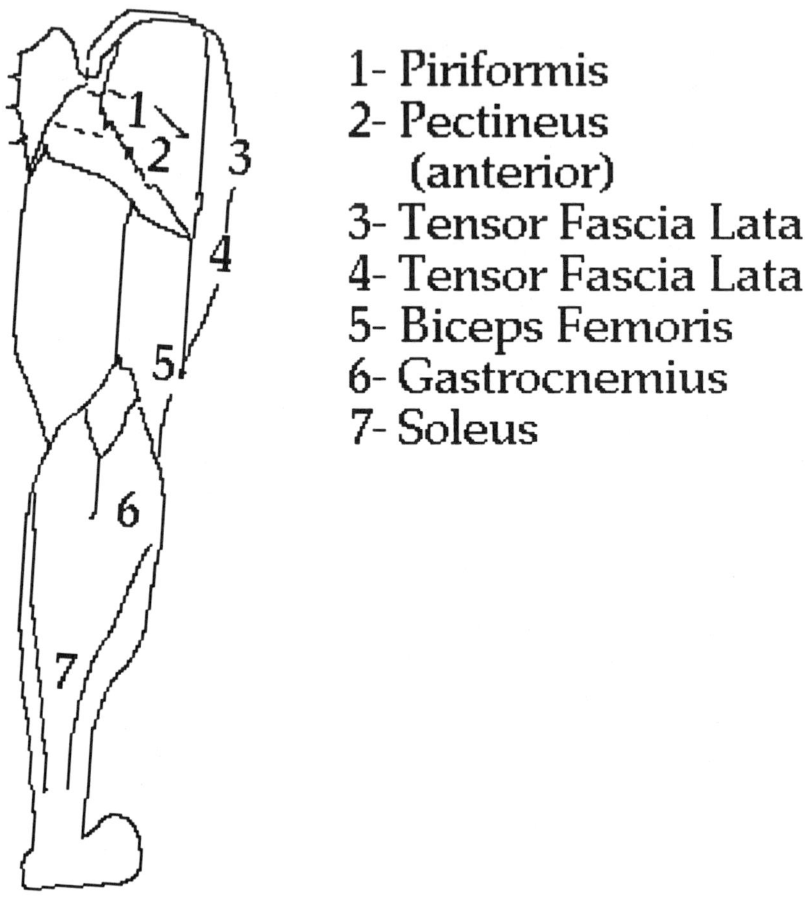


**Figure 3.** The Myofascial Chain - Lower Extremity

The impact of being able to "see" the spasm, or the interaction with their back complaints, was profound in many of these patients, and led to an immediate increase in level of compliance with treatment. The myofascial work involves a level of discomfort initially, often significant, and these patients were much more willing to work on the stretching and trigger points when they could so clearly see the interaction with their pain.

Assessment of over a hundred trigger point locations on a control group, and individuals experiencing no exacerbation of their chronic pain complaints, led to the determination that trigger points may be painful but do not contribute significantly to dysfunction. On the other hand, stimulation of trigger points which were rated as only mildly painful, or less painful than others, were often found to result in major referred muscle spasm in the low back in the patient population. These referred spasms were never significant in the control group. Treatment directed to these points resulted in reduction in low back pain complaints.

A similar pattern was found in the upper extremity. Although pain complaints were often reported in the cervical paraspinal or trapezius regions, this pain and its subsequent dysfunction were often found to be "set off" by trigger points in the contralateral upper extremity. The identification of a "chain" could also be made for the upper extremities, and very common repetitive activity could often be demonstrated as activating the spasm.

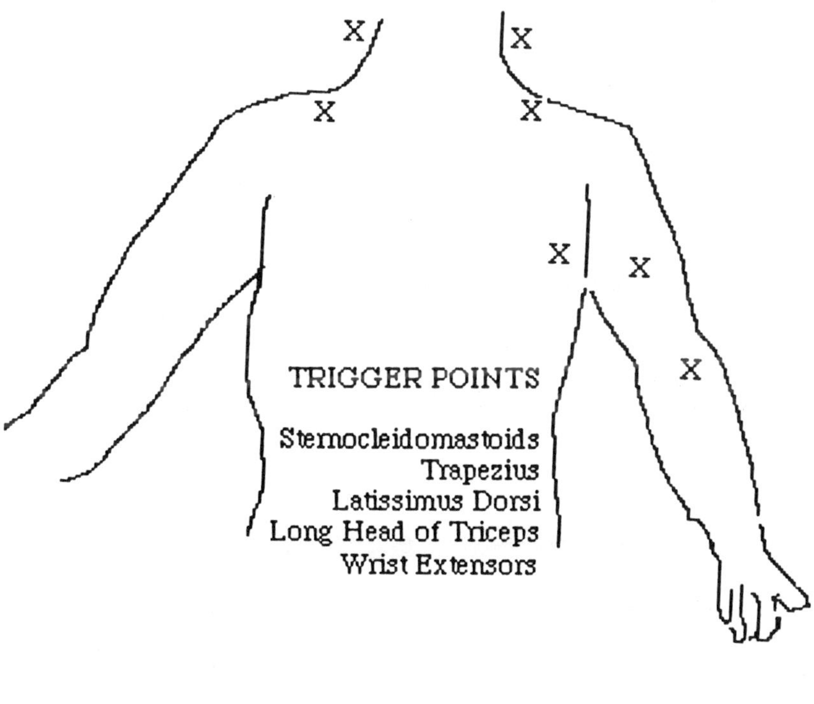


**Figure 4.** Myofascial Chain - Upper Extremity

With proper education into myofascial treatment and stretching, the neck or headache complaints could often be significantly reduced by working in the arm or contralateral upper quadrant.

The extent to which these patterns of referred pain do not follow the standard dermatome patterns is immediately evident. Autonomic hyperactivity is evident in many of these patients, who produce strong histamine responses with stimulation, histamine "wheels" and itching following stimulation. A characteristic of autonomically mediated pain, or central pain, has been suggested by Melzack in reference to the treatment of reflex sympathetic dystrophy and central pain. He demonstrates that the mechanism, which he describes as a Pattern Generating Mechanism, responds to hyper- or hypo-stimulation of the autonomic (sympathetic) nervous system. The most commonly used mechanisms are sympathetic blocks, sympathectomy and other interventions, which would effectively alter the autonomic activity so greatly that it would "get unstuck" and return to a state of re-regulation. A similar analogy has been reported in the literature on myofascial pain syndrome, whereby intense stimulation of the trigger point with injection, dry needling, or intense pressure often results in lengthy pain relief (Simons (1988); D.G. Simons & J.G. Travell (1983)). Multiple interventions of hyperstimulation are hypothesized to have a similar effect in myofascial pain patients as in the impact of hyperstimulation with RSD patients - alternating contrast baths is an example used with RSD patients.

The hyperstimulation applied to trigger points with deep friction can stimulate strong autonomic responses which are so severe that they may result in termination of treatment. The author has found that hypo-stimulation of the trigger points, with gentle constant pressure, may effect similar positive responses in terms of pain relief, without the adverse side effects.

## TREATMENT GOALS AND OUTCOME MEASURES

In addressing the interaction of myofascial pain syndrome and the resulting muscle imbalances, postural dysfunction, and limitations in activity, the following represent the primary treatment objectives:

1. Deactivation of trigger points to the latent stage.
2. Reduction in referred muscle activity and subsequent muscle imbalances.

3. Facilitation of proper muscle recruitment during activity to decrease postural dysfunction, decrease the fatigue factor, and decrease the frequency of exacerbations.
4. Promotion of a self-management program to decrease recurrence.

Addressing the postural dysfunction component requires the investigation of where the pain is coming from, rather than simply reduction of the pain complaint. Patients who have good inner awareness, or with whom this can be fostered, can provide many "clues" as to the sources of the dysfunction. Just as referred muscle hyperactivity may be contralateral to the trigger point, the pattern of referred activity may be from a low back muscle, such as the quadratus lumborum, to the cervical muscles. If the therapist remains limited to treating the part, the interactive forces of the whole upon that part are lost.

Assessment of trigger points cannot be limited by pain complaints. Visual clues, and information regarding functional activity may provide externally-mediated guidance.

The skilled hands of the therapist can often rapidly uncover trigger points which are less a point of pain than a dysfunction. While an entire "chain" may need to be treated to decrease dysfunction, rapid intervention at the primary sites of muscle dysfunction can facilitate reduction of postural asymmetry, functional ineptness, and the pain experience itself.

Not only must the physical components of myofascial pain complaints be addressed, but so must the interactive psychological components. Myofascial pain, just as many other forms of chronic pain, exists in a population greater than those individuals who choose to seek treatment. While the evidence of psychopathology may be no greater, there appear to be certain psychological profiles which are more resilient to the needs of adapting to chronic pain. In comparing those patients who do seek treatment, and those who do not but experience similar levels of chronic pain, some individuals are found to be more adaptive, feel like they have more skills or ability to manage their pain, and would appear to experience the pain as less threatening.

Thus, in those patients who do seek treatment, certain cognitive changes are also necessary in order to achieve a satisfactory outcome with self-management skills:

1. Pain is not seen as a "threat", but as something which they can challenge.

2. The ability to develop a sense of control over the pain experience.
3. Identification of inner tools with which to deal with the unpredictability of the pain and its interference with daily functioning.
4. Instructional information about exacerbations, to reduce fear of impending disaster.
5. Educational information regarding the nature of myofascial pain syndromes in relation to "disc" disease, and the patient's role in self-responsibility.
6. The transition into "Active Adaptation" with which one is willing to get on with their life, challenge their limitations, and no longer seek a "cure."

Biofeedback is an optimal tool for both the patient and the therapist in the treatment of this disorder. For the patient, it facilitates the development of a self-efficacy model toward long-term management. For both the therapist and the patient, it offers direction in treatment, both in the localization of the referred muscle spasm, and the development of appropriate muscle re-education. Biofeedback offers quantification of treatment outcome in response to the increasing demands for accountability.

Other outcome measures which are critical in this patient population, where soft tissue injury is not observable on the ordinary tests available, include changes in the level of function, the use of medications, and the long-term utilization of self-management skills rather than increased dependence on health care professionals.

## REFERENCES

Graff-Radford SB, Reeves JL, Jaeger B: Management of Chronic Headache and Neck Pain: Effectiveness of Altering Factors Perpetuating Myofascial Pain. Headache: April, 1987, pp. 186-190.

Lacey JI, Lacey BC: Verification and extension of the principle of autonomic response-stereotypy. Amer J Psychology. 71:50-73, 1958.

Melzack R. Neurophysiological Foundations of Pain, in The Psychology of Pain. RA Sternbach (Ed). 2nd Ed. Raven Press, NY, pp. 1-25, 1986.

Rosomoff H: In American Back Society Newsletter, 6:1, Winter 1990.

Sarno JE. Etiology of Neck and Back Pain. An autonomic myoneuralgia? J. Nervous Mental Disease, 169:55-59, 1981.

Schwartz GE: Self-regulation of response patterning. Biofeedback and Self-Regulation. 1:7-30, 1976.

Schwartz GE: Biofeedback and the self-management of disregulation disorders. Behavioral Self-Management Strategies, Techniques and Outcomes. Edited by Steward. Brunner, NY, 1977.

Sherman R: Phantom Limb Pain. Presentation at the Association for Applied Psychophysiology and Biofeedback, Symposium on New Frontiers in Chronic Pain Management, Washington, DC, March, 1990.

Simons DG: Myofascial Trigger Points: A Need for Understanding. Arch Phys Med Rehabil 62: 97-99, 1981.

Simons DG. Myofascial pain syndrome due to trigger points. In Rehabilitation Medicine, J Goodgold (Ed). The CV Mosby Company, St. Louis, MO, 1988, pp. 686-723.

Simons DG, Travell, JG: Myofascial Pain and Dysfunction. The Trigger Point Manual. Williams & Wilkins, Baltimore, 1983.

# CHAPTER 6

# *Use of Integrated Electromyography for the Assessment and Treatment of Musculoskeletal Pain: Guidelines for Physical Medicine Practitioners*

Glenn S. Kasman, M.S., P.T.

---

Surface integrated electromyography (IEMG) is increasingly recognized by physical medicine practitioners as a tool for the assessment and treatment of chronic musculoskeletal pain. Applications have been proposed for headache, neck pain, scapulothoracic dysfunction, low back pain, temporomandibular joint disorders, myofascial pain dysfunction syndromes, and certain extremity pain problems where inappropriate patterns of muscle activity are felt to be contributory.

For our purposes, we will refer to IEMG as the averaged or integrated sum of muscle action potentials as recorded by skin surface electrodes. Conceptually, this is in contrast to percutaneous, single unit EMG. Surface IEMG cannot be used to diagnose muscle and neuromuscular disease or neurologic impingement, as can "needle" EMG and companion studies. Within certain limits, however, IEMG can be used to assess the overall functional status of a given muscle. IEMG looks not at the activity of individual fibers or motor units, but at the ensemble of motor events that subserve useful activity.

Technical aspects of clinical IEMG recording and issues related to its validity and reliability are reviewed in Chapter 1 of this volume and in other sources (Krebs, 1987). Further consideration of limitations pertinent to clinical assessment will be made in this chapter where appropriate.

As IEMG is noninvasive and painless, its use tends to be readily accepted by patients for clinical procedures. Patients gain access to relatively instantaneous, performance-contingent muscle feedback by attending to the IEMG display. This exteroceptive biofeedback is intended to augment internal proprioceptive systems. The goal of IEMG biofeedback then, is to facilitate some improved motor outflow via central nervous system mechanisms involving sensori-motor integration and learning (Fischer-Williams, 1986). The reader is referred elsewhere (Wolf, 1987; Fogel, 1987; Schwartz, 1987; Basmajian, 1989) for broad discussion of IEMG biofeedback and physical therapies. Attention will be narrowed here to clinical experiences related to physical medicine and chronic musculoskeletal pain.

It is necessary to establish some conceptual scheme from which we may frame clinical IEMG procedures. What follows is a working, somewhat speculative model. Pain perception begins when nociceptive afferents are activated by the mechanical and chemical effects of tissue insult. This may involve a single traumatic incident, or repetitive stereotyped overuse. Resultant muscle spasm/splinting adds further mechanical stress and presumably produces a relative circulatory stasis, both of which contribute to more pain. With chronicity, there evolves a series of physiological and psychological events that bring about a downward clinical spiral. Habitual guarding and dysfunctional patterns of muscle recruitment figure in prominently. Chronically tense muscles and impaired blood flow are thought to perpetuate inflammatory responses and delay maturation of scar tissue. Range of motion is reduced and guarded, leading to a loss of myofascial extensibility. Tissue integrity becomes compromised and the subject becomes vulnerable to a cycle of injury and dysfunctional response. Normal muscle synergies are disrupted by hypertonicity at some sites, reflexive inhibition at others, and general disuse. Consequently, the normal balance of proprioceptive inputs from muscle and joint receptors is disrupted.

Proprioceptive feedback is an integral part of normal motor control. Neural structures are located in muscle bellies, musculo-tendon complexes, and joint capsules that are afferent to the CNS. These muscle spindles, golgi tendon organs, and other specialized encapsulated structures, as well as free nerve endings, make smooth regulated movement possible. Changes in the firing patterns of the afferent nerves signal changes in muscle length, velocity of contraction, intensity of contraction, joint loading, and limb position.

Signals may be routed segmentally at the spinal cord, or processed in exceedingly complex pathways through the brainstem, cerebellum, thalamus, and cerebral cortex. Reflexive and non-stereotyped motor responses are generated, more feedback is processed, and together with the other sensory and motor systems the subject responds to changes in its environment.

In the patient with pain and chronically aberrant muscle activation, perhaps this is the level where things really go wrong. Sensory and motor responses that are initially protective somehow become subverted and dysfunctional. Peripheral musculoskeletal receptors are thought to generate tonic activity, as well as phasic responses to acute conditions. When a muscle is held in an abnormally activated or shortened position, and tendons and joints are abnormally loaded, there is likely change in the tonic baseline feedback pattern, and likely change in the bias with which acute responses are signaled. For example, when a muscle is held in a static shortened position, the tonic activity of the spindles adjusts to that baseline length of the intrafusal and extrafusal fibers. Sporadic increases in muscle length, such as would occur with functional movements or novel exercises, would be vigorously responded to by the spindles, conferring a "hyperirritability" to the muscle's activation.

In addition to inputs from musculoskeletal proprioceptors, there is a wide range of somesthetic inputs from musculoskeletal nociceptors and from nociceptive and non-nociceptive cutaneous and visceral receptors. With repetitive noxious stimulation by mechanical or chemical inflammatory factors, there is recurrent activation of peripheral nociceptive pathways, perhaps with reduction of firing threshold. Potentiation phenomenon may be produced at synapses along central nociceptive pathways. Taken all together, the pattern of somesthetic inputs becomes changed, both with reference to the involved spinal segment and to the total suprasegmental pattern. These changes favor the perception of nociceptive signals relative to total background somesthetic activity, resulting in central nervous system responses that lead to dysfunctional muscle activity.

On the descending limb of the sensory-motor loop, there is influence by central nervous system nuclei associated with nociception, autonomic function, and cognitive processes. These effects are mediated by the brainstem reticular formation, diencephalic and limbic nuclei, cerebral cortex, and the hypothalamic-pituitary axis . Dysfunctional efferent programs might result in deviant controls of sensory information rising up along the neuraxis to further processing points, or by generation of aberrant outflow from central motoric pathways. Effects at alpha motor neurons would directly change patterns of extrafusal fiber firing and force generation. Aberrant gamma motor neuron activity would set an inappropriate "gamma bias" and contribute to dysfunctional reflexive control of extrafusal fibers.

Eventually, all the factors impinge at the spinal motor neurons, producing heightened, inhibited, and ill-timed motor unit activity.

It should be recognized that chronic muscle tension may lead to soft tissue problems at secondary sites. Resultant trigger points can produce local and referred pain. Dysfunctional muscle recruitment patterns may impose increased tensile and shear forces on non-contractile elements of the kinetic chain. These forces might then contribute to the degeneration of traumatized joint capsule, ligament, tendon, or intervertebral disc annular fibers. Also, chronic compressive, axial loading of articular cartilage or disc would inhibit normal fluid mechanics, and impair nutrient-metabolite exchange in those structures.

In a sense, then, one dysfunctional process reinforces others in complex, integrated loops of afferent and efferent neural activity and pathomechanics (Figure 1).

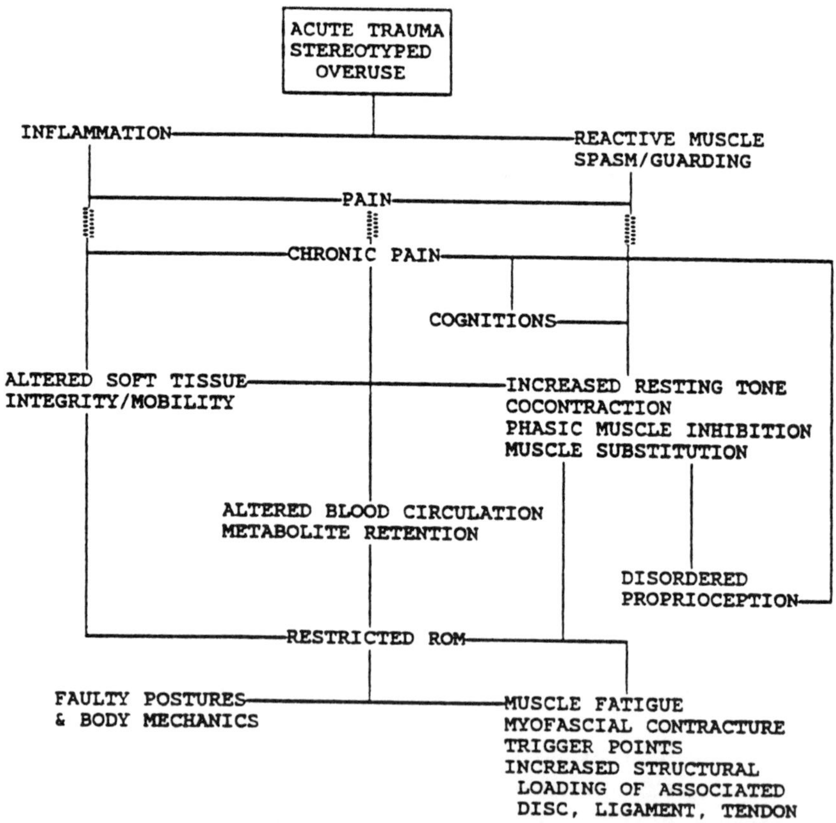


**Figure 1.** Inter-relationships of factors accompanying chronic pain and aberrant muscle activity.

Nociceptive events are processed within some cognitive context to yield the experience of pain. Of relevance here are the subject's arousal, attentional, affective states and expectational set. Anxiety, fear, depression, or anger can be especially problematic . A constellation of behavioral responses associated with chronic stress and increased sympathetic tone may be identified. Secondary psychosocial gains may complicate the situation enormously. The clinical picture tends to be one of slow progress, or more commonly, plateau or slow deterioration. It is recommended that a broad approach be taken to evaluation, so that the various physiological and behavioral issues can be teased out as clearly as is possible, and treatment planned accordingly.

## PURPOSE AND EFFECTS OF IEMG BIOFEEDBACK TRAINING

For most IEMG biofeedback patients, the main problem is one of excessive muscle tension. Muscle activity is usually elevated in an intermittent fashion, triggered by some mechanical or behavioral factor. Removal of the stimulus allows the tension to dissipate. In between, there may be an appreciation of pain but little discrimination of muscle tension per se. There seems to be an accommodation to the accompanying proprioceptive patterns, and small but significant changes in tension cease to be recognized. Tension discrimination for non-involved muscles tends to be quite normal.

Muscle tension-pain events, together with the effects of IEMG biofeedback training are schematized in Figure 2. At time (a), muscle tension begins to rise due to one or several factors. At time (c), a sufficient intensity and duration of muscle tension is produced such that there is onset of pain at time (e). Conceptually, the threshold "dose" of muscle tension leading to pain, and the latency to pain onset are patient and situation dependent. If the patient does not become aware of a change in muscle tension until time (d), that is, after the pain threshold is crossed, then self-directed intervention is not possible before pain begins. Pain, and muscle spasm or guarding, may cycle in a positive feedback manner, accelerating discomfort and rendering conscious muscle control ineffective. Clinical interventions at time (g) are designed to reduce muscle tension and pain, and restore normal relationships at points (h), (i), and (j).

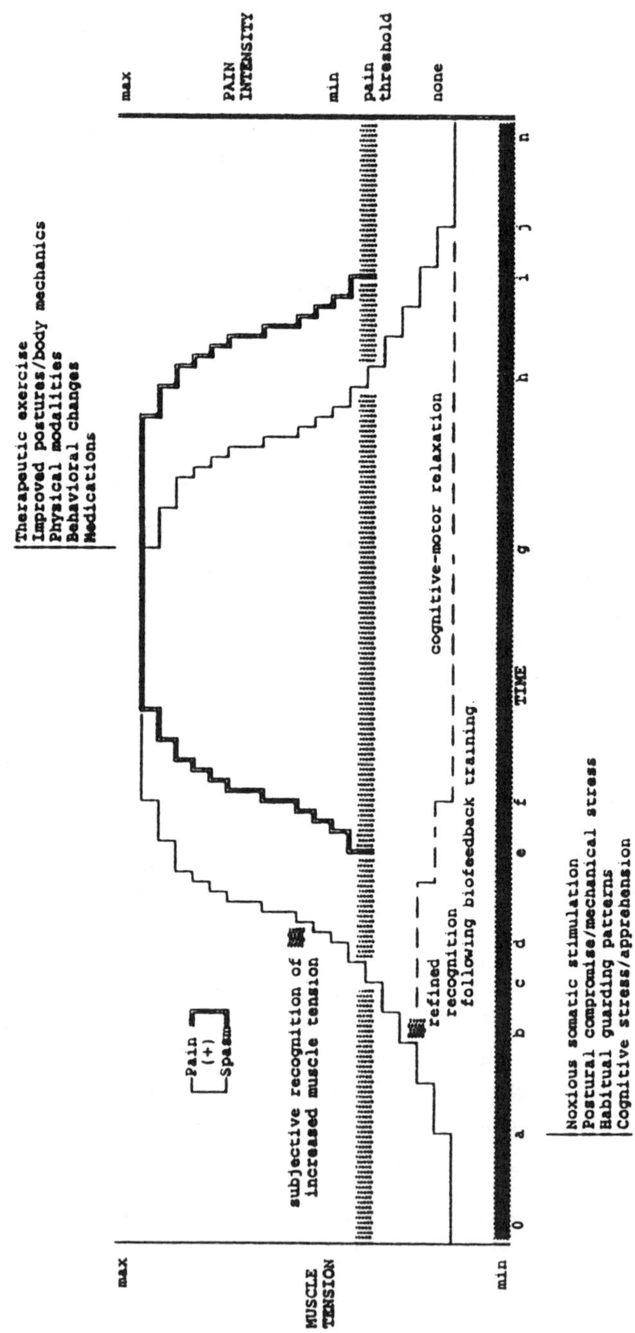


**Figure 2.** Purpose and effects of biofeedback training.

With biofeedback training the patient learns to recognize increasing tension at time (b) rather than time (d), that is, before the muscle tension-pain threshold is crossed, and before things cycle away from the patient's ability to control them. Cognitive strategies, postures, body mechanics, and therapeutic exercises are practiced with concurrent biofeedback. They are then applied to return muscle tension to baseline levels at time (f).

Interventions with IEMG biofeedback are designed to be preventive when feasible, and in all cases to reduce the physiologic consequences of increased tension. Patients learn skills that can be applied in functional situations, by themselves. They come to understand what things they do make them feel worse, and what they can do to make themselves feel better. In the process, they have opportunity to learn lifestyle skills that are of long-term, holistic benefit.

Some specific utilities of IEMG evaluation and IEMG-assisted biofeedback training are proposed as follows:

1. Assessments of muscle tension contribution to pain complaints.
2. Identification of specific aberrant patterns of muscle recruitment.
3. Assessment of contributory postural, dynamic mechanical, and psychophysiological factors.
4. Analysis of work station demands and potential modifications.
5. Objective documentation for medical and vocational staff, third party payers.
6. Facilitation of general or targeted muscle relaxation and kinesthetic awareness.
7. Facilitation of proper exercise technique and rationale.
8. Facilitation of postural, body mechanics instruction.
9. Provision of concrete, neutral, non-threatening targets for patient effort.
10. Facilitation of self-directed strategies to alleviate symptoms and advance functional status.
11. Adjunct for coordinated team planning and patient management.

## BIOFEEDBACK CANDIDATE SELECTION

Chronic pain patients referred to physical medicine for IEMG biofeedback evaluation have, by practical definition, failed to obtain a satisfactory result during the acute and subacute phases of injury. Some patients of this population will be found to have relatively straightforward problems that are amenable to biofeedback and other therapies. Others will have had multiple consults with a variety of health care practitioners, and will have consumed and exhausted a large range of conservative options. Some patients will have had invasive procedures performed, and others will go on to such. Not every patient that crosses the practitioner's threshold is an appropriate candidate for biofeedback training.

Ideally, candidate selection should begin before the biofeedback referral is generated. It is suggested that the biofeedback clinician review referral criteria with all associated health care providers. This should be done in general, and with case by case communications. Inappropriate subjects are those for whom there is no prior evidence that muscle activity is a problem. That is, there should be some patient report or clinical examination finding of chronic muscle spasm, splinting, guarding, or otherwise aberrant muscle activity pattern. Any referred patient should have a minimum "gadget tolerance", be reasonably cognitively intact and free of gross psychopathology and overt secondary gains. The latter two issues are not necessarily cause for exclusion from biofeedback, but rather biofeedback should be approached with caution and consideration of the likely cost-effectiveness of training attempts. Patients demonstrating marked pain behaviors within the context of protracted vocational disability may be best evaluated by members of an interdisciplinary pain management team. Given a structured, interdisciplinary setting, IEMG biofeedback is often an invaluable adjunct to the management of complicated chronic pain patients.

## PATIENT EVALUATION

Evaluation is a multi-dimensional process, tailored to the needs of each patient. We generally begin on a common track, and then diverge toward mechanical versus psychophysiological assessment procedures. Some patients will require exhaustive evaluation. This may be performed over the course of several sessions. The main components of the evaluation are:

1. Patient interview
2. Standard musculoskeletal exam
3. IEMG scan exam
4. IEMG psychophysiological stress profile
5. IEMG dynamic mechanical profiles

## PATIENT INTERVIEW

The subjective intake proceeds along standard lines. Medical, neurological, musculoskeletal, and psychosocial issues should all be carefully assessed. A thorough history will help guide the specifics of the IEMG examination that follows. Most important, the patient should reflect on what specifically makes them feel better or worse. This information is used to try to reproduce symptoms during IEMG monitoring.

## STANDARD MUSCULOSKELETAL EXAM

Included here are assessments of structural alignment, active joint range of motion, passive joint range of motion, accessory joint play, manual muscle testing, muscle length, muscle play and myofascial extensibility. Techniques should be used that isolate the involved peripheral joints or specific intervertebral levels, along with general spinal, extremity, and neurologic screening procedures.

Accurate assessments on these points are critical to the meaningful interpretation of IEMG. For example, an elevated shoulder girdle may have nothing to do with excessive muscle tension, but everything to do with a tight upper trapezius. It would likely be fruitless to try to correct this situation with relaxation techniques or volitional manipulation of IEMG activity in static postures; a therapeutic exercise program is needed. Biofeedback training might still be indicated for other tension problems, or it may be a useful adjunct to exercise, or it may be entirely unnecessary.

When elevated muscle tension is unequivocally present, it may be secondary to psychological factors, mechanical factors, or both. Bilaterally tight pectoral muscles could contribute to a forward head posture and increase the mechanical demands of anti-gravity postural control of the head, leading to elevated IEMG scan exam findings at the cervical paraspinals. This is a different type of problem than elevated cervical IEMG activity due to post-traumatic splinting or elevated cervical IEMG due to psychophysiologic stress responses, even in the presence of the same protracted cervical posture. It is improbable that the issues will be clearly sorted without an appreciation of the patient's functional anatomy and biomechanics.

Test procedures for joint range of motion can be found in standard texts of clinical orthopedic assessment (Hoppenfeld, 1976; Minor and Minor, 1985; Magee, 1987). Evjenth and Hamberg (1984) and Travell and Simons (1983) are useful sources for specific tests of muscle length.

Kendall and McReary (1983) and Daniels and Wothingham (1980) are excellent texts of manual muscle testing procedures. Assessment of joint and soft tissue accessory mobility is not always required, and tends to be a more complicated affair that should be left to those with formal professional training.

## IEMG SCAN EXAM

The IEMG scan exam is discussed in detail in Chapter 1 of this volume. Briefly, a hand-held probe is used to quickly sample muscle activity at standardized sites about the face, neck, and back. The scan is performed for left and right sides, with the patient sitting, and then standing. Scan results are compared to a normalized data base.

The scan exam lends itself to a far greater degree of precision and objective documentation than other, arbitrary assessment procedures. Gross patterns of activity are identified, as well as specific targets for further investigation. Evidence of asymmetry or postural influence may be revealed.

We regard the scan exam as the starting point for the IEMG evaluation. Upon further profiling, it may be decided that a positive scan finding is of major importance, or of little clinical significance. Also, there are cases where the scan exam appears benign, but aberrant IEMG patterns emerge with dynamic mechanical or psychological challenge.

We sometimes deviate from standardized scan placements to "fan" a distribution of focally elevated IEMG readings. Using two cylindrical probes held at a fixed interelectrode distance, one probe is positioned for example, at the center of a trigger point, and the other probe is used to sample points moving about in circular fashion. Alternatively, the probes may be "walked" along parallel to the fiber orientation of a given muscle. IEMG recordings with these procedures are not intended to be of precise diagnostic utility. Rather, they may be helpful in the identification of treatment sites for low frequency/high intensity or burst mode TENS programs, or for manual myofascial release and deep friction massage techniques.

## IEMG PSYCHOPHYSIOLOGICAL STRESS PROFILING

The IEMG psychophysiologic stress profile is recommended when the patient's subjective remarks and presentation suggest a behaviorally related component to their complaint. IEMG monitoring during the profile is continuous, with site selection based on the IEMG scan exam findings and to a lesser extent, the musculoskeletal evaluation. It is emphasized, though, that musculoskeletal factors must not be neglected. We have seen many patients in obvious psychological distress who have had muscle tension problems that were directly related to structural issues, or who posed some complex combination of behavioral and musculoskeletal problems. If available, it is helpful to assess sympathetic arousal with additional modalities for thermal, electrodermal, cardiovascular, peripheral plethysmographic, and respiratory monitoring.

The patient is monitored during:
1. baseline rest
2. mental arithmetic, verbal skills challenge, fine motor task
3. recovery rest
4. pain, stress, anxiety provoking imagery; guided conversation
5. recovery rest
6. volitional relaxation

The magnitude and time course of IEMG is examined at each step. Attention is paid to the patient's verbal remarks and non-verbal behaviors. These are used to define the emotional triggers of muscle tension, the patient's cognitive styles, and baseline relaxation skills. The practitioner is advised to watch for postural changes that might confound IEMG interpretation. Further details of test procedures can be found in Chapter 1 of this volume.

At times, the stress profile yields penetrating insight into the nature of a patient's problem. For example, we have seen patients complaining of neck pain status post motor vehicle accident with entirely benign radiologic, neurologic, and musculoskeletal exam findings, who have had extensive consultations and therapies, and who have showed unremarkable IEMG scan and dynamic movement profiles. Each of these patients, though, has verbalized considerable anxiety related to driving and has found their pain to be most exacerbated in that situation, seemingly unrelated to posture. Psychophysiologic stress profiles have been performed, the first few phases of which have also tended to be uneventful. However, when each of these patients has been guided through imagery related to driving, IEMG activity from their respective areas of complaint has become markedly elevated, producing asymmetries and other aberrant patterns that had not been previously observed. For some of these patients, the degree of IEMG elevation has seemed to be directly correlated to the intensity of the imagery relating to potential vehicular collision, increasing or decreasing in a remarkably consistent fashion. Most of these patients have verbalized recognition of tension sensations that, if sustained, would likely lead to pain. At the same time, they have had great difficulty returning their IEMG activity to baseline when instructed to try to do so. Since the patterns of elevated muscle tension are intermittent and behaviorally induced, standard clinical examinations that neglect an appropriate behavioral challenge are inconclusive.

In another case, behavioral challenge appeared to precipitate unilaterally increased IEMG activity at the scalenes, upper traps, and pectoral muscles of a patient with classic complaints of intermittent thoracic outlet obstruction. This was followed a few minutes later by onset of upper extremity pain, parasthesias, numbness, and measurably decreased hand temperature on the involved side. Previous radiologic and laboratory studies were unremarkable. Vascular and neurologic studies were somewhat equivocal, along with questionable findings on standard musculoskeletal maneuvers. The stress profile exam was the first time that symptoms were clearly reproduced in the clinic. It was thought that the patient's anatomy was probably of narrow dimension for the neurovascular bundle somewhere along its costoclavicular passage. Behaviorally induced muscle tension was felt to have somehow displaced the structural alignment of bony or fibrous structures such that the neurovascular bundle became intermittently compressed. Looking back on her history, the patient could correlate her symptomology to certain psychological issues. Based on these findings, treatment directions were established to deal with the relevant behavioral issues, as well as to address tight soft tissues and general deconditioning.

## IEMG DYNAMIC MECHANICAL PROFILES

Dynamic mechanical profiling is a most exciting part of the evaluation process. It is performed when there is suggestion of structural or neuromuscular factors that might predispose the patient to dysfunctional muscle activity. This is judged from the patient interview, musculoskeletal exam, and the IEMG scan exam.

Continual IEMG monitoring is performed at one or more sites. These may be left versus right, multiple muscle regions, or left versus right at multiple regions. Recordings from more than two to four sites can be very difficult to follow. Analog recordings with the ability to freeze or save images are required for all but the simplest set-ups.

The following are evaluated and compared for each channel of monitored activity, as appropriate to the particular profile:

1. Peak IEMG magnitudes obtained during tested activity.

2. IEMG timing patterns. This may simply be an assessment of prompt versus lagging change patterns on volitional movements or contract/relax commands. If desired, a more sophisticated approach can be taken (Robertson, et.al., 1982):

   a. rate of IEMG development, the differential or "upslope"
   b. "downslope"
   c. time to peak IEMG magnitude
   d. time to IEMG magnitude (X)
   e. time to peak IEMG differential
   f. IEMG at time (X)

3. Steady, controlled versus fluctuating IEMG during sustained contractions.

4. Overall shape of the recruitment curves. This will be detailed in due course but is basically an integrated impression of the above.

5. Defined range of joint motion of salient IEMG events, for example, the shoulder abduction arc during which pain is reported and discrete upper trapezius "spasm" is visualized, the knee angle of peak vastus medialis IEMG magnitude, the spinal flexion range of peak paraspinal IEMG asymmetry.

6. Time integrated IEMG values versus peak values (rms or otherwise). That is, the former represents the area under the curve of the latter, and conceptually would be analogous to work output versus peak torque. Indeed, several of the test variables are analogous to those torque parameters routinely examined with isokinetic devices (Davies, 1984).

7. Ratios of one IEMG channel to another on any of the above quantitative variables.

It may be helpful to repeat (+) tests with the patient positioned non weight bearing; examine if otherwise chronically hyperactive muscles relax when mechanical load is minimized.

A baseline activity pattern is established for the target muscle(s) and then variables are systematically introduced. By understanding the precise circumstances under which normal motor control breaks down, treatment resources can be expediently managed. There are many options for IEMG profiling:.

1. Extensive postural analysis
2. Active range of motion
3. Isometric contractions
4. Isotonic contractions at progressive intensities
5. Isotonic contractions at progressive velocities
6. Open versus closed kinetic chains
7. Tension recognition thresholds
8. Multiple muscle functions
9. Muscle synergy analysis
10. Standardized timing and coordination tasks
11. Functional activity analysis

1. Extensive Postural Analysis

The format suggested by Dr. Cram in chapter 1 is recommended.

Kendall and McReary (1983) discuss muscle length/strength balance as it relates to structural alignment. They define several postural types, how each deviates relative to the ideal postural plumb lines, and which muscles are likely to be tight in each case.

An important principle for IEMG interpretation emerges from consideration of postural alignment and the effects of gravity. Gravity will tend to induce a flexion moment at joint segments that have an axis of rotation in the transverse plane, posterior to the ideal sagittal plane plumb line. An extension moment will be induced by gravity at joints with a transverse plane axis of rotation that is anterior to the ideal plumb line. To maintain an erect posture against gravity, these moments are counteracted by passive soft tissue restraints and active muscle tension.

Patterns of IEMG over the neck and back may be predicted from a patient's postural alignment. In a given patient's neutral weight-bearing posture, flexion moments at the spine may be resisted by active tension of the spinal extensors, producing relatively increased IEMG at these levels. As an example, a patient with a "flat back", diminished lumbar lordosis inclined anterior to the plumb line would be expected to show relatively higher IEMG at all levels of the lumbar paraspinals.

Further, we tend to find relatively higher IEMG on the convex side of a spinal scoliosis. Take a more complicated example, of a subject with accentuated cervical lordosis and "forward head", accentuated thoracic kyphosis, and accentuated lumbar lordosis with anterior pelvic tilt, with convex left cervical-upper thoracic spine and elevated left shoulder girdle, and with convex right lower thoracic-lumbar spine and elevated left ilium. Considering the sagittal plane effects first, this patient would be expected to show relatively higher IEMG values at the cervical paraspinals and relatively lower values at the lumbar paraspinals. Thoracic paraspinal IEMG may be elevated, but more often is low. It is assumed here that the patient is restrained by passive ligamentous structures over the thoracic kyphosis (postural pain may result from chronic strain of these structures, as well as joint/ligament strain at the hyperextended cervical-thoracic junction of these patients, regardless of muscle tension issues). In the frontal plane, paraspinal IEMG may be higher on the left at the apex of upper spinal curve and higher on the right at the lower curve. IEMG asymmetry may or may not be observed at the elevated shoulder and ilium.

Postural asymmetry patterns can be very difficult to predict and interpret with respect to IEMG. This is the situation when spinal curvature has developed with skeletal maturation, or from an old injury to the spine or lower extremity. Ligamentous and myofascial tissues may then have accommodated to varying degrees as restraints of the curvature. The balance of active and passive compensations for the scoliosis will then be very patient specific.

Aside from observing plumb line relationships, it is meaningful to look at the overall dimensions of each body segment, and how each segment "fits" relative to the others. Areas that seem excessively narrowed or shortened compared to the opposite side contours may reflect tight myofascial elements. If there is less fascial slack in a kinetic system, then the joint segments will fail to attain their optimal mechanical alignment, and active muscle compensations will result. The practitioner examines how it is that gravity acts on the patient, such that some elevated or asymmetrical IEMG pattern is required to hold them upright.

Depending on the nature and duration of a postural deficit, it may or may not be appropriate to try to change the patient's alignment and concomitant IEMG patterns. Some IEMG patterns will be seen that are "abnormal" for a general population, but mechanically "normal" for a particular individual. We believe that there is a wide range of normal limits and that any and all deviations from the ideal are not necessarily clinically significant. Each case must be taken individually within the total context of the examination findings.

2. Active Range of Motion

IEMG activity is monitored as the patient moves the relevant joints through their range, to all cardinal planes. For the extremities, it may be helpful to include movement to functional diagonal planes; for example, proprioceptive neuromuscular facilitation (PNF) diagonals 1 and 2 (Sullivan, et.al., 1982). Normal and aberrant findings for the paraspinal muscles during spinal range of motion are described in detail in Chapters 1 and 2. A few further comments are offered here in regards to spinal movement and observed IEMG patterns.

During flexion/extension maneuvers, the transverse processes on the convex side of an passively maintained rotational scoliosis are displaced along a greater arc than those on the concave side. Paraspinal muscles on the convex side might then recruit to a greater extent to support such movements against gravity, and thus generate asymmetrical IEMG. The IEMG asymmetry would be greater if the subject favored weight bearing through the same leg as the convex side of the spine. Having the patient stand with each foot on independent scales can help clarify the situation.

A similar situation could result during flexion with unilateral facet restriction, with or without overt scoliosis. The paraspinals on the unrestricted side would fire more than on the side with reduced facet translation during flexion. Moving into extension, and depending on the nature of the restriction, the involved side paraspinals might fire more strongly moving against the restriction than the unrestricted side, conceivably resulting in a reduction or reversal of the asymmetry pattern seen with flexion.

Non-reciprocal IEMG asymmetries seen on side bending and rotational movements may be analyzed in like fashion. Each patient's IEMG pattern is evaluated relative to the mechanical constraints suggested by their structural alignment and functional anatomy.

In the presence of bilaterally reduced myofascial extensibility, shortened or less elastic dorsal ligaments, degenerative joint or disc disease, epidural scarring status post surgical procedure, or markedly tight hamstrings, flexion mobility may be restricted with symmetric but relatively high levels of paraspinal IEMG. Alternatively, range of motion may be relatively good, but the paraspinals may remain active throughout range, as opposed to the typical pattern of decreased IEMG beyond about 45 degrees of flexion.

3. Isometric Contractions

IEMG activity is observed during:

a. maximal isometric contractions
b. submaximal isometric contractions - these may be graded with a dynamometer, cuff weights, free weights, or elastic.

Formal manual muscle testing positions may help isolate the target muscle. More often than not, we find maximal volitional contraction IEMG to be roughly symmetrical in unilaterally involved patients with otherwise asymmetrical IEMG findings. IEMG does tend to be greater from the involved side at equivalent resistance submaximal levels. When asked to contract and sustain symmetric IEMG levels on the display, patients frequently state that it feels like it takes less effort to reach a given intensity on the involved side. At the same time, it often seems more difficult to discriminate and hold that particular IEMG level. The implications here are that the involved muscle tends to respond in a hyperactive fashion, with somewhat impaired fine motor discrimination and perhaps impaired proprioceptive feedback, without necessarily a change in peak recruitment ability.

4. Isotonic Contractions at Progressive Intensities (arbitrary velocity)

IEMG activity is monitored from the target muscle as it performs:

a. concentric contractions
b. eccentric contractions

These may be graded at progressive intensities with cuff weights, dumbbells, or elastic bands. If working with an extremity, a decision should be made if it is most appropriate to stabilize the proximal segments or leave them unstabilized. Whatever is closest to the patient's functional activities is usually the better choice.

There may be considerable differences in recruitment patterns between concentric and eccentric activity, especially when looking at the relative activities of muscles contributing to a synergy (Figure 3). Peak IEMG levels accompanying concentric paraspinal activity, that is, with extension to upright neutral from a flexed position, are generally greater than the eccentric counterpart, that is, flexion from an upright neutral position (see Chapters 1 and 2). When dealing with an interscapular site asymmetry and unilateral free weight lifting, we frequently see a large ipsilateral IEMG peak and relatively quiescent contralateral activity when the involved side moves, and a smaller ipsilateral IEMG peak with contralateral co-contraction when the uninvolved side moves. The magnitude of the reciprocal asymmetry tends to be different on the concentric and eccentric phases of the lift, as well as during a sustained isometric component.

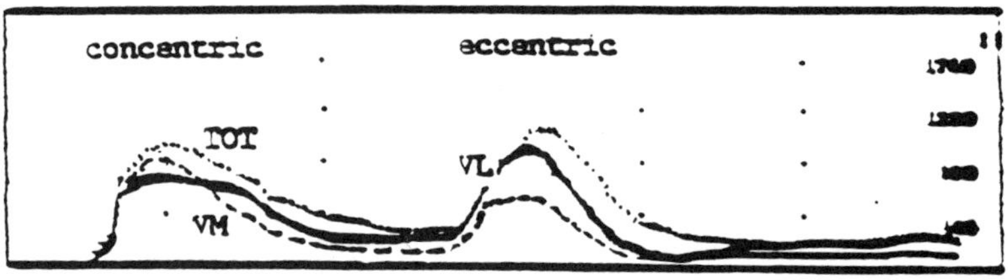


**Figure 3.** IEMG during concentric and eccentric quadriceps contractions in unilateral standing; VM = distal vastus medialis, VL = vastus lateralis, TOT = "total" quadriceps.

There are known differences in total tension generation and discussion as to the relative roles of passive tensile elements in concentric versus eccentric work (Davies, 1984). The relationship between IEMG and actual force generated may not be the same in the two situations (Komi et. al., 1987). Mechanically, one may be more painful than the other. It seems reasonable that concentric and eccentric neuromuscular recruitment patterns would be different in some cases, in response either to different mechanical demands or to pain-related reflexes.

Concerning pain-related reflexes, involved muscles may be seen to transiently increase or decrease their IEMG activity during a painful arc of the concentric or eccentric range. We interpret this as reflexive spasm and inhibition, respectively, and believe we have observed each in select cases of shoulder impingement and patellofemoral pain syndromes.

5. Isotonic Contractions at Progressive Velocities

IEMG activity is monitored as the velocity of joint movement is deliberately varied across repetitions. Trends may be observed simply with verbal commands to the patient to speed up or slow down. The exact velocity across repetitions, or at any given point in the range of motion of a single repetition cannot be controlled.

If desired, an isokinetic device may be used to hold velocity constant. This requires expensive, specialized equipment but has the advantage of simultaneous torque output readings.

We have seen cases where interscapular IEMG patterns were of quite symmetric timing patterns and peak magnitudes at slower velocities, but became markedly asymmetric at higher velocities. At higher velocities, the involved side IEMG tends to be of greater variance and peak magnitude, with slower downslope, sometimes faster upslope, and incomplete return to baseline level between successive repetitions.

6. Open Versus Closed Kinetic Chains

Target muscle IEMG activity is monitored as it performs concentric and eccentric contractions in an open versus closed kinetic chains. An open kinetic chain is one in which the distal segment is free to move, for example, knee extension in sitting. In a closed kinetic chain, the distal segment is fixed, for example, knee extension in standing. Mechanical and neuromuscular events may be rather different in the two situations. The closed kinetic chain position is generally more functional for the lower extremity. We have observed this to be a significant IEMG variable in musculoskeletal pain patients for the lumbar paraspinals, hip extensors, and quadriceps.

7. Tension Recognition Thresholds

These are used as a measure of a patient's ability to subjectively recognize changes in muscle tension from baseline. As the target muscle is monitored, the patient is asked to begin thinking about contracting the muscle, without actually doing so. The IEMG display is then turned away from the patient as she is directed to progressively increase tension in the muscle as slowly as possible. The patient says "now" when she is able to detect any active tension from the muscle, or when she can perceive a detectable increase in tension. At that point, the therapist records the number of generated microvolts.

The recognition threshold is most useful for comparison between left and right sides in unilaterally involved subjects. The threshold value is frequently higher from an involved hyperactive muscle than its contralateral homolog, even when the involved target is at that moment showing low baseline activity. The implication is that the patient's somesthetic system may have to some degree accommodated to increased tension levels, and no longer recognizes small tension changes.

8. Multiple Muscle Functions

Some muscles act in multiple, functionally different roles. A muscle may work toward tonic stabilization of a body part for weight-bearing posture, toward phasic movement of a body part, or toward phasic stabilization so that the prime movers can operate efficiently.

Consider the example of the mid trapezius. This muscle is active posturally at the upper back and shoulders, to different degrees with different postural types. It is also active during prime scapular retraction. Further, it helps to isometrically stabilize the scapula during rotation and certain functional upper extremity activities. We have observed cases where IEMG recordings were asymmetrically elevated and erratic during scapular stabilization for certain raised-arm work activities, but quite uneventful otherwise; or more commonly, much more symmetric otherwise. This may reflect differences in mechanical load, neurophysiologic synergies, or even the psychological valence associated with different types of functional activities.

9. Muscle Synergy Analysis

Activity is recorded from multiple muscles or muscle components that work together in producing a given movement or stabilization pattern. Recording activity from left and right paraspinals during spinal range of motion is a straightforward example. Here, the left and right paraspinals act as synergists during pure sagittal plane movement, and are expected to co-contract in a roughly symmetrical manner. During side-bending, however, the ipsilateral paraspinal initiates movement as the agonist, with activity from the contralateral side being antagonistic. Later in the side-bending range, both sides may co-contract to provide stabilization against the effects of gravity. The clinician, therefore, examines co-contraction and reciprocal recruitment/inhibition patterns, and the timing of component activities.

Figure 4 shows IEMG recordings from the quadriceps complex of a patient with chronic patellofemoral pain status post motor vehicle accident, refractory to multiple arthroscopic surgeries and several years of traditional physical therapy. Shown are IEMG activity patterns from distal vastus medialis, vastus lateralis, and "total" quadriceps channels with maximal volitional isometric contraction at full knee extension (% of maximal contraction IEMG from the "total" quadriceps leads has been shown to be linearly related to % of maximal contraction torque, in normal subjects, under isometric conditions at 10-15 degrees of knee flexion; Kasman, et. al., in progress). Note the diminished ratio of vastus medialis to vastus lateralis IEMG and vastus medialis to total quadriceps IEMG on the involved right relative to the uninvolved left. These observations correlate well with findings of right side quadriceps atrophy, especially at the distal vastus medialis, and known patellofemoral degenerative joint disease. We interpreted the findings as indicative of diminished medial patellar stabilization during quadriceps contraction, with resultant lateral tracking and degeneration of the retropatellar articular cartilage.

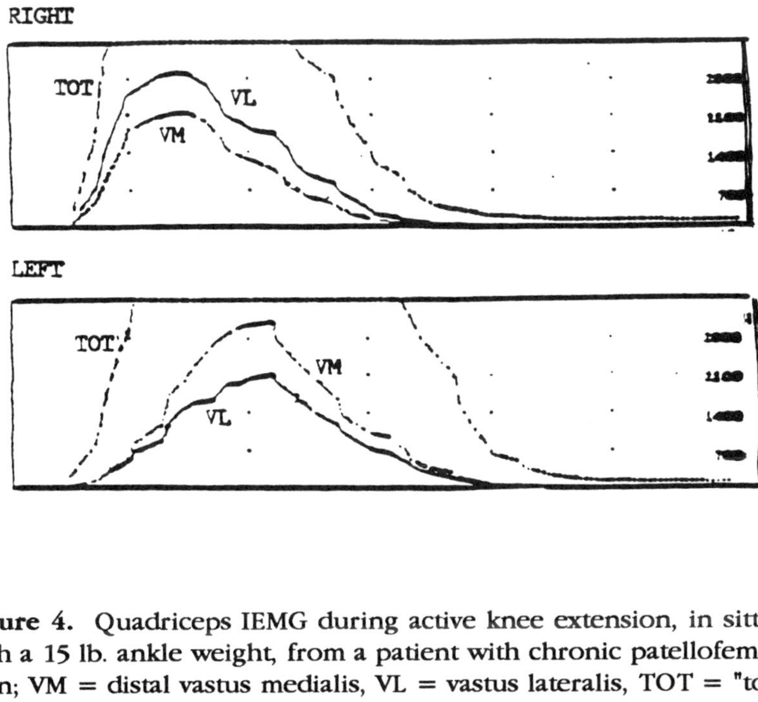


**Figure 4.** Quadriceps IEMG during active knee extension, in sitting, with a 15 lb. ankle weight, from a patient with chronic patellofemoral pain; VM = distal vastus medialis, VL = vastus lateralis, TOT = "total" quadriceps.

In Chapter 3, Dr. Taylor discusses muscle synergies with respect to the scapulothoracic complex. A significant proportion of IEMG biofeedback candidates report pain at lateral cervical, suprascapular, interscapular or shoulder sites. With multiple layers of muscle and complex mechanical relationships, the scapulothoracic complex is an especially intriguing, yet difficult area for study. Interpretation should be especially conservative as recordings may be contaminated by technical limitations peculiar to this type of complex set-up. Analysis may also be flawed by unaccounted neurophysiologic factors and dynamically variable relationships between IEMG and actual work output. These concerns notwithstanding, we do perform evaluations with recording sites selected from cervical, proximal suprascapular, distal suprascapular, pectoral, lateral deltoid, mid-axillary thoracic, infrascapular, and upper and lower interscapular regions. Figure 5 shows IEMG recordings from the left and right shoulder regions of a patient with profound shoulder pain and extremely complicated musculoskeletal and behavioral issues. It is apparent that there are substantial differences in the recruitment patterns between the involved left and uninvolved right sides.

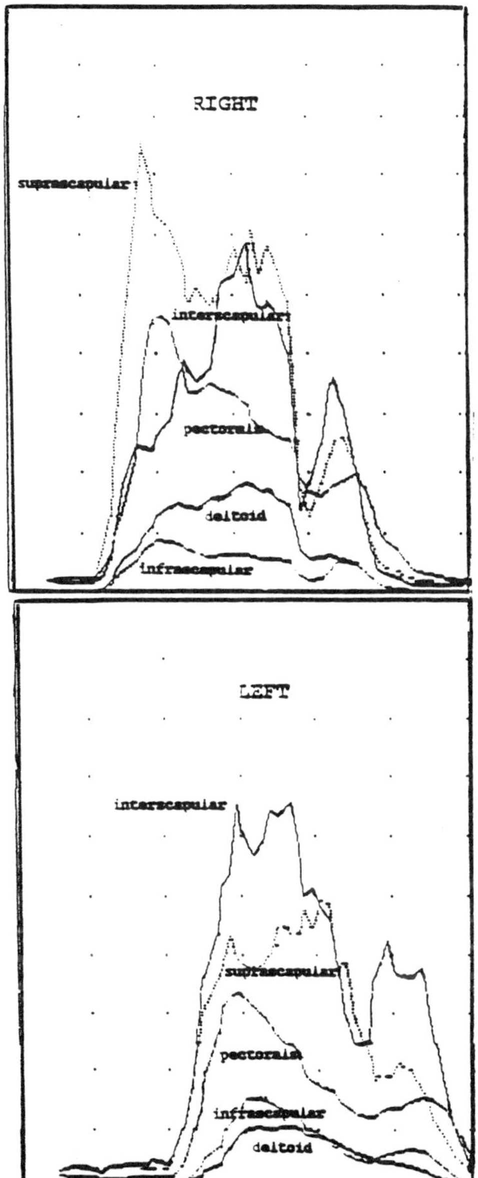


**Figure 5.** IEMG recordings from the scapulothoracic complex of a patient with left shoulder dysfunction and chronic pain. Note the respective timing patterns and magnitudes of IEMG signals, as labeled, on right versus left sides.

Another patient with multiple sclerosis complained of unilateral shoulder tightness, resting aching pain diffusely about the shoulder and especially at the lateral proximal arm, with superimposed sharp subacromial pain on abduction of the arm. On clinical exam, the humeral head appeared well-seated relative to the glenoid fossa, though posture was kyphotic and with bilaterally abducted, somewhat inferiorly rotated scapulae. Some glenohumeral capsular and pectoral muscle tightness was noted on the involved side. There was little palpable spasticity on isolated passive joint movements, but grossly aberrant scapulothoracic rhythm on the involved side during active abduction. That is, the scapula appeared to elevate rather than rotate cephalad. Resting IEMG was low and symmetric to the uninvolved side with the exception of the pectoral site, which showed modest but clearly increased activity. Comparing involved to uninvolved sides during active abduction, pectoral IEMG showed a faster upslope and much increased peak magnitude. Suprascapular and lateral deltoid sites displayed a similar pattern to the pectorals. Activity from the infrascapular site was of slow upslope and decreased magnitude. Interscapular and serratus anterior regions showed roughly symmetric timing, respectively, but markedly reduced IEMG magnitude. The findings were more pronounced with any combination of elbow flexion, forearm pronation, or wrist flexion.

Taking the IEMG findings within the context of comprehensive musculoskeletal and neurological exams, we concluded that there was clinically significant spastic hypertonicity in the pectorals, with insufficient stabilization and upward scapular rotation from the interscapulars and serratus anterior, potentially delayed and depressed activity from elements of the rotator cuff, and functional substitution from upper trapezius and an overwhelming deltoid. Mechanically, the humeral head would be powerfully loaded against the acromial arch with impingement of subacromial soft tissues. Positioning the upper extremity in progressive degrees of flexor synergy exacerbated the deficits. We felt that the patient's neurologic condition predisposed him to development of a musculoskeletal impingement syndrome. This patient did not respond to IEMG biofeedback, but pain was reduced with a program of manual joint and soft tissue mobilization, deep heat, and therapeutic exercises.

## 10. Performance on Standardized Timing and Coordination Tasks

The patient is asked to control the timing and magnitude of IEMG activity such that a standardized task is performed and scored on the display screen. Computer-assisted games may serve this purpose well.

A display signal from the target muscle may be used to follow a preconfigured template, or perhaps to allow manipulation of video game items.

There are many sensori-motor factors that could influence performance on this sort of task, not all of which are meaningful for the patient's pain problem. The tests are therefore used to evaluate relative ability on the involved versus uninvolved sides of a given patient. Numerical scores are used as indicators of fine motor discrimination and coordination abilities.

11. Functional Activities/Work Station Analysis

This can be among the most rewarding evaluation procedures for patient and therapist alike. Basically, the work or other activity environment is recreated as accurately as possible, and IEMG is evaluated along the lines discussed for the above procedures. Special attention should be paid to those issues that the patient has noted as exacerbating or alleviating of symptoms.

We have evaluated many patients as having entirely or relatively unremarkable exams to the point of getting them into a functional context. This has been the case for keyboard operators, telephone operators, assembly line workers, office professionals, mechanics, motor vehicle operators, homemakers, performing artists, and athletes. Occupational therapists may be of great help in creating job simulations.

It is not at all surprising that repetitive functional activities bring with them unique mechanical and psychological demands. The effects of these have a greater chance of being seen during actual performance of those activities.

## COMMENTS ON THE IEMG DYNAMIC MECHANICAL PROFILE

It is clear that IEMG analysis may be relatively simple, or quite complicated. The more sophisticated assessments will tax IEMG equipment, software, and clinician skills to their limits. We do not consider the IEMG dynamic mechanical profile itself, or any and all IEMG exam procedures by themselves to be diagnostic. From the IEMG findings, we develop a working hypothesis of what is happening at certain muscles and their associated tissues, and why those particular events are occurring. This working hypothesis must be taken in the context of all clinical examination findings. It would be highly irresponsible to diagnose some condition based solely on IEMG.

All of the IEMG procedures that have been heretofore described, and those that will follow, are predicated on the idea that IEMG accurately reflects muscle tension. Yet muscle tension is generated by events that follow the propagation of muscle action potentials, and the relationship between muscle action potentials and cross-bridging of actin and myosin filaments is likely not constant (Krebs, 1987). Further, we are recording some sum of muscle action potentials that happen to "reach" our electrodes at the surface of the skin. The biologic tissue that composes a patient's back or leg is a complex, nonuniformly conducting medium for electrical activity. Factors that variably influence the relationship between muscle work and IEMG include, at a minimum, sarcomere resting length (Lippold, 1952), contraction velocity, proportionate amounts and geometric distributions of type I and II fibers (Woods and Ritchie, 1983; Stalberg, et. al., 1989), angle of tendon insertion (Maton, et. al., 1987), fatigue (Gerdle et. al.,1968; Milner-Brown et. al.,1986; Grabiner et. al., 1988) and biochemical environment related to fatigue (Henriksson, et. al., 1974; Horita and Ishiko, 1987), and neurophysiologic mechanisms (Shakespeare, et. al., 1985; Fahrer, et. al., 1988; Krebs, 1983). In a clinical setting, these elements may go uncontrolled and confound our best empirical interpretations. A linear relationship between IEMG and torque can probably be consistently expected only for muscles of relatively uniform fiber type composition, under isometric conditions (Woods and Ritchie, 1983; Krebs, 1987).

With these limitations in mind, and with the understanding that there are few or no normative standards for the types of clinical assessments suggested in this section, it is recognized that it is easy to fall prey to faulty assumptions. Therefore, it is perhaps best to be conservative when making formal clinical judgements.

## PATIENT TREATMENT

Biofeedback training should be but one component of a comprehensive treatment approach. Patients should be instructed toward appropriate postures, body mechanics, isolated stretching and strengthening exercises, aerobic programs, and of course, relaxation techniques. Manual therapies, cold, or superficial or deep heating agents, TENS, etc., may be useful adjuncts for subacute and certain chronic problems (IEMG can be used to assess the efficacy of those treatments that are offered as muscle therapies). At a minimum, psychological intervention should be directed towards rationale interpretation of muscle tension-pain relationships, self-responsibility for change, pacing of physical activity, and autonomic psychophysiology. Specific coping strategies and cognitive restructuring may need be pursued with a psychotherapist familiar with physical medicine and rehabilitation.

In some cases, patterns of IEMG activity are found that are clearly aberrant, but are probably more reflective of, than causal to, the patient's primary problem. Mechanical soft tissue impingement or stretch imposed on inflamed joint capsule, as examples, can produce reflexive muscle spasm, pain-induced muscle inhibition, or quite reasonable apprehension and cortically-mediated protective guarding patterns. It is appropriate in those cases to defer biofeedback in favor of therapies that directly address the dysfunction.

The evaluation findings should guide the treatment plan. The focus for some patients will be relaxation training and cognitive intervention. Other patients will work mostly with postural, body mechanics, and exercise training. Others will concentrate on pure IEMG manipulation. Most treatment plans will include some combination of all elements. The goal is the development of normal motor control under the precise conditions that, in the evaluation, were accompanied by aberrant IEMG, and the transference of skills to the patient's functional environment.

Set the patient up for success:

1. Provide thorough postural support for patients treated with relaxation-based therapy, at least initially. This includes proper stabilization and support for proximal limb segments when working distally. Postural support may be deliberately withdrawn as the patient progresses.

2. Direct attention to the type of feedback that is favored by the patient. Different visual analog and digital graphics, audio inflections, etc., should be considered for those that best match the patient's learning style.

3. Discuss the theory, rationale, and goals of biofeedback training for each patient. Carefully explain what each type of feedback means. Do not assume that the patient understands the relationship between a computer display and their own biologic activity, or that they understand the implications of surface biofeedback electrodes relative to percutaneous EMG or electrical stimulation procedures that they might previously have experienced.

4. Start with easily achievable goals and thresholds.

5. Cue the patient to attend to internal bodily sensations associated with changes in IEMG, frequently and repeatedly. The patient must learn to substitute internal cues for external machine feedback if treatment is to carryover beyond the clinic visit.

6. Provide adequate rest periods. Allow for periodic changes of training venue and posture.

7. Strongly reinforce the patient's knowledge of performance and results. Acknowledge gains, encourage independent problem solving, emphasize patient self-responsibility.

8. Begin and end each session by soliciting questions or comments from the patient. Assess intra-session and inter-session carryover of training principles. Involve the patient in charting progress. Jointly monitor changes in IEMG values, musculoskeletal test results such as range of motion, and functional activity performance. Jointly review goals and plans. Training proceeds best by something of a team effort.

9. Plan a home program that will reinforce the supervised training sessions. Home programs typically involve structured practice sessions with audio tapes or handouts, abbreviated practice sessions throughout the day, physical exercises, and whatever else that is appropriate. Home diaries may be useful to identify problem areas, to chart progress, and to assess compliance. Diaries may include subjective discomfort scales, medication usage, major activities, and relaxation and exercise practice.

## IEMG TRAINING GUIDELINES

1. Relaxation-based IEMG downtraining
2. Tension recognition threshold training
3. Tension discrimination training
4. Generalizing from static training paradigms to movement
5. Left/right equilibration training
6. Postural retraining with IEMG feedback
7. Body mechanics instruction with IEMG feedback
8. Therapeutic exercise with IEMG feedback
9. Functional activity/work station modification with IEMG feedback
10. IEMG triggered electrical neuromuscular stimulation

1. Relaxation-Based Downtraining

The patient is instructed in standard relaxation methods while the effects are monitored with IEMG biofeedback. The goal may be generalized muscle tension reduction, or relaxation of specific target muscles. Programs usually begin with the former and progress to the latter. We try to select relaxation techniques that best match the patient's cognitive style (Davidson and Schwartz, 1976), that is, methods with somatic, visual-spatial, or logical-sequential basis. The effectiveness of any particular technique is judged by the patient's verbal and IEMG responses. The patient is encouraged to build a repertoire of relaxation skills and to mix and match them, until he arrives at a set of methods that is most effective for him as a unique individual.

In training relaxation at a specific muscle site, a threshold is set on the display below which IEMG activity is to be maintained. Progressive relaxation is "shaped" by slowly lowering the threshold, adjusting for about 90% success.

If appropriate, muscle relaxation and IEMG biofeedback may be integrated with training in systematic desensitization, stress inoculation,or visualization (McKay et. al., 1981). Psychotherapeutic interventions of this sort should be practiced by, or under the supervision of, a qualified mental health professional.

After several sessions, some patients appear to become classically conditioned to relax when coming into the biofeedback laboratory. That's all well and good, but of little significance if skills are not generalized to real life environments. Practice is therefore progressed to more complex, distracting settings. Postural support may be withdrawn, lights turned up, background noises imposed, etc.. If a portable IEMG unit is available, patients may practice in the office waiting room, physical therapy gym, or even better, at the home, workplace, shopping mall, or wherever else is relevant. Feedback cues may be gradually withdrawn to challenge the patient's independence.

Relaxation training should be reinforced with daily home practice. Written instructions or audio cassette tapes are standard aids. Custom audio tapes recorded during a live session seem most helpful.

2. Tension Recognition Threshold Training

Tension recognition threshold training is a direct outgrowth of the recognition threshold assessment procedure. A single channel of IEMG is displayed on the screen with the patient at rest. The patient is directed to observe the display screen and to produce as slowly a rising signal as possible with volitional muscle contraction. Internal somatic sensations are cued by the therapist. The patient attends to the quality, intensity, and distribution of tension-related sensations and passively observes how they change with increasing IEMG. As the patient becomes more familiar with these sensations, she is directed to focus attention at the initial IEMG rise from baseline. The goal is to progressively lower the IEMG value at which a change in sensation is first recognized.

3. Tension Discrimination Training

Tension discrimination training builds together with tension recognition threshold training to teach the patient to recognize subtle changes in muscle activity, before tension builds into the pain producing range. Training principles may or may not be applied first at a non-involved, easy to control site. The wrist extensors serve this purpose well.

The patient is directed to produce and maintain a particular IEMG value. While doing so, the patient is cued to attend to the associated internal somatic sensations. Training begins by discriminating wide values from baseline on a single recording channel. We typically start with 5-10 uv rms increments for 6-12 second trials. The internal sensations associated with baseline relaxation, and the tension recognition threshold, are used to anchor the lower end of the scale as a contrast for the sensations associated with higher activity levels.

The patient is progressed to smaller training intervals as she becomes able to find and hold target IEMG values. Discrimination increments are usually progressed to the 0.5-2.0 uv rms range for 5-7 second trials. Training moves from a large range of target values to an emphasis on the lower end of the scale.

Target values are first assigned with baseline relaxation preceding and following. Values are then assigned in ramped-up, ramped-down, and random sequences, in that order.

Once the patient is proficient with these tasks, visual feedback is withdrawn. The display screen is covered or turned away from the patient as she reproduces a given IEMG level. The patient responds with "now", or something equivalent, when she feels she is generating the correct IEMG value. Feedback is provided by the subsequent uncovering or turning back of the screen. As the patient progresses further, visual feedback is withdrawn entirely. Verbal feedback is then provided by the therapist along the lines of "higher, lower, correct, incorrect, incorrect - do you think it's higher or lower ?"

The purpose of all this is for the patient to internalize the microvolt scale as a means of reliably gauging muscle tension. Toward that end, the patient is instructed to briefly pause her activity, 6-8 times per day, and to judge tension in the target muscle in terms of perceived number of microvolts. Tension judgements may be made at prescribed times, or ad lib during common functional tasks. Some patients are instructed to write their perceived IEMG levels on a diary sheet along with pain ratings, activities, etc., a few times per day. After quantitatively judging tension, the patient takes another 30-60 seconds to practice an abbreviated relaxation technique. Carryover of skills may be assessed in advanced subjects by beginning each session with a blind judgement of IEMG level that is subsequently verified on the display screen.

IEMG levels from multiple muscle sites may be discriminated simultaneously in the most advanced patients. This is appropriate if multiple muscles appear to be involved, or for the purpose of distraction challenge to the original discrimination task. It may be helpful to begin multiple channel training with distinct visual threshold markers for each IEMG signal. Some computer assisted systems have the capability of generating distinctly colored, artificially generated templates for each channel that change value according to prescribed timing patterns. We have seen patients progress to the point of correctly discriminating randomly varying IEMG values for three different muscle sites, on independent timing patterns. These patients have progressed their proprioceptive-motor integration to "super-normal" levels. As might be imagined, these subjects tend to do well clinically and seem to maintain their gains over the long-term.

Another excellent way to challenge timing and coordination is with the use of video games. IEMG from the target muscle functions as a joystick control of the action. In a game setting, the target value varies very quickly, or continuously. The rapid changes and game graphics tend to preclude correlation of internal sensations with actual IEMG values, but this is not important in this setting. The games can be highly motivating tools toward the development of sophisticated motor control. Beginning patients, anxious patients, or those with extremely high performance expectations, however, may become frustrated. Patients should focus on simpler, albeit more mundane, tasks until competencies are firmly established.

Of course, the video games alluded to are produced from software designed for biofeedback applications and operated on hardware with appropriate patient electrical interfaces and isolation systems.

## 4. Generalizing From Static Training Paradigms to Movement

DeBacher (1989) describes steps for muscle relaxation training in the presence of spastic hypertonicity. We find a similar sequence to be useful for early skill generalization in the musculoskeletal pain patient. The patient has first learned to relax the target muscle while quietly at rest. He has also worked on lowering his threshold for tension recognition and has established some competency with IEMG magnitude discrimination.

The patient now learns to maintain relaxation during use of a non-involved extremity or distracting mental task. From there, relaxation responses are maintained with assumption of new static positions of increased muscle length. The patient next maintains relaxation during dynamic stretch of the target muscle via antagonist isotonic contraction. Examples of this would be maintenance of quadriceps relaxation during active knee flexion or right upper trapezius relaxation during active left side bending. Finally, the patient builds to prompt, appropriate patterns of reciprocal recruitment and inhibition. Thus, we would see alternate peaks of quadriceps and hamstring IEMG activity, or of left and right upper trapezius activity, with prompt upslopes and downslopes and complete return to inter-repetition baselines, during reciprocal knee extension/flexion and neck left/right side-bending, respectively.

Dynamic movements are initiated in small movement arcs in simple, cardinal planes. These are progressed to full range of motion and to diagonal and rotational combined planes that more closely resemble true functional activity. If appropriate, the patient may begin non-weight-bearing, well stabilized, concentrating only at the target kinetic segment. Appropriate control is then established with movement throughout the entire kinetic segment, progressing into weight-bearing and the closed kinetic chain. Velocity of movement and resistance intensity may be gradually increased for both concentric and eccentric contractions. Responses are practiced for all uni or multiarticular muscle functions in tonic postural stabilization, primary joint movement, and phasic stabilization.

It is important that the patient realize that he is not being taught to relax like a "wet noodle". This might achieve a goal of tension reduction, but it is hardly functional. Muscle contraction is necessary to effect skeletal movement. It is acceptable if, after training, 50 microvolts are required from the thoracic paraspinals to extend the spine from flexion with good body mechanics. The point is that the patient should learn to consistently perform the maneuver with 50 microvolts, and not 75 or 100 microvolts, and that he learn to promptly relax the paraspinals to their postural baseline with completion of the movement.

To reorganize proprioceptive and motor responses, the patient practices many repetitions at each step of the generalization process.

It may be helpful for the patient to try to match the IEMG signal from the target site to an acceptable tracing from a previous repetition or to a concurrent signal from an uninvolved contralateral site. Feedback may be progressively withdrawn from, for example, audio and visual cues, to audio feedback only, to verbal cues from the therapist only, to after-the-fact knowledge of results. Again, independent patient problem solving is to be encouraged as much as possible.

At this point in the patient's training, we sometimes find it helpful to adapt strategies from Yoga, Tai Chi, Aikido, and the like, as well as from the increasing wealth of western "mind-body" integration systems. Training protocols can become rather practitioner and patient specific.

5. Left/Right Equilibration Training

Equilibration training is simply a specialized application of the previous training strategies. First, the patient learns to produce symmetrical baseline IEMG with relaxation skills. This is combined early on with postural retraining, as appropriate.

Discrimination sequencing is then initiated, first on the uninvolved side alone, then on the involved side alone, then simultaneously on the left and right with symmetrical target IEMG values. From there, asymmetric values are assigned for left and right sides, with reciprocal target values assigned on consecutive trials. Finally, the patient matches left and right IEMG signals to randomly selected left and right target values.

Next, the patient moves on from isometric contractions to dynamic movements according to the guidelines described in section 4. The patient learns to produce symmetric IEMG for symmetric skeletal movements, building to progressively resisted concentric and then eccentric contractions. This leads to the production of symmetric or asymmetric recruitment patterns, as is appropriate to any type and plane of movement. Thus, a goal might be symmetric, conjoint paraspinal IEMG activity with spinal flexion, and symmetric reciprocal left/right IEMG activity with reciprocal rotation or sidebending. For interscapular muscle sites, we would hope to see symmetric IEMG activity while, for example, holding a piece of elastic apart with both hands, and reciprocally symmetric IEMG activity when a free weight is lifted by alternate upper extremities.

6. Postural Retraining With IEMG Feedback

The patient is guided to make specific adjustments, one at a time, to his sagittal and horizontal plane alignment, as effects on IEMG activity are observed. Again, we favor the outline described by Dr. Cram in Chapter 1. Mirrors, scales, and long cable leads can be helpful. It is instructive to point out the implications of reduced IEMG activity in terms of decreased energy expenditure and diminished mechanical stress. Thus, the patient learns to get the same thing, holding the head and trunk up against gravity, at a lower cost.

7. Body Mechanics Instruction With IEMG Feedback

The same principles used during postural training are applied to body mechanics instruction with IEMG feedback, the process is just more dynamic. Functionally related tasks are performed with correct and incorrect body mechanics and the effects on the IEMG display are monitored. Observation of reduced IEMG activity, while accomplishing the same task, can be highly reinforcing toward program compliance in the home and workplace.

Body mechanics guidelines and diagrams can be found in most contemporary patient publications for back and neck care. It is important to realize that adherence to the principles of efficient body mechanics will not necessarily result in decreased IEMG activity. These principles take more into consideration than the potential for musculotendinous strain. Some postures and lifting techniques are designed to protect the intervertebral discs and ligamentous structures. A few of these call for even more muscle activity than the "incorrect" methods. On the other hand, we find it generally useful to combine IEMG feedback with body mechanics instruction for patients with subacute disc and musculoligamentous injuries, who have no other need for biofeedback.

8. Therapeutic Exercise With IEMG Monitoring

Therapeutic exercises are prescribed for almost all of our IEMG biofeedback patients, by ourselves or by another referring therapist. It is often helpful to review the patient's exercise program with concurrent IEMG feedback.

Regarding passive and active-assisted stretches, the target muscle is brought to its endpoint by the prescribed technique. The physiologic end range of an irritable muscle may be substantially shorter than its anatomic end range due to spindle reflexes and guarding. Through the use of breathing, contract/relax, or antagonist inhibition techniques, the patient waits for the IEMG to decrease and stabilize. The muscle can then be stretched to a new endpoint and the process repeated. Dr.

Cram describes a specific method of combining breathing, stretching, and IEMG feedback in Chapter 1.

Strengthening exercises may be reviewed with IEMG feedback for appropriate muscle effort, symmetry, stabilization, and pacing. Muscles that are not mechanically required during the exercise should be of relatively quiescent IEMG activity. Patients may vary their mechanical and cognitive efforts so that they arrive at the most efficient IEMG patterns. If feasible, we use a portable IEMG device in the gym, working through the patient's regimen with machines, free weights, or whatever. Exercise prescription considerations for IEMG biofeedback patients will be discussed in a following section.

9. Functional Activity Performance With IEMG Feedback

Potent treatment ideas flow from the analysis of functional activities with IEMG monitoring. This phase of treatment is really continual with evaluation, though it may be approached at any time and repeatedly. For any given activity simulation, or actual performance, one element is varied at a time in an effort to reduce mechanical stress. Optimal parameters are determined by observing changes in IEMG activity, mechanical alignment, objective task performance, and of course, the patient's subjective impressions.

For many activities that are performed inefficiently, the stress induced with each repetition is relatively small. However, the cumulative effects that result from large numbers of repetitions are a source of great problems. It may take many repetitions or otherwise long time durations to observe the effects of activity modification. In other cases, modifications are suggested that can be immediately implemented and that return quick results.

Take the example of an accountant who presented with right side interscapular pain. This individual spent much of his work day operating a numerical keypad with his right upper extremity. The key pad was positioned at some distance from his body to accommodate other business related items. By changing the position of items on the desk, moving the keypad in closer at a slight tilt, and supporting the forearm on a chair rest, interscapular IEMG activity was shown to stabilize and decrease in magnitude. Symptoms were quickly reduced and the patient came to realize how mechanical stresses add up over the course of the work day, week, and so on. He learned that he had the power to make his symptoms better or worse, and saw the rationale for maintenance of appropriate postures, body mechanics, activity pacing and home exercises.

Work station modifications for the office may include the rearranging of desk items, switching chairs or otherwise adjusting seating parameters, use of a drafting table, use of a phone headset, adjusting the positioning of keyboards and video terminals, use of book stands, improved filing drawer locations, or any of a multitude of other things. Creativity will be rewarding for both patient and therapist.

This sort of reasoning may be extrapolated for any occupational, home, or recreational task. The patient changes his postures and body mechanics and/or the related equipment is modified in position or substance. It may be helpful to consult occupational or physical therapists familiar with industrial medicine. If the patient is receiving professional vocational assistance, the counselor can identify specific job demands and solicit employer cooperation. Occupational therapists can provide adaptive equipment and efficiency guidelines for home activities. Sports medicine physical therapists and athletic trainers can provide insights into improved athletic technique and equipment.

Two additional case histories are illustrative of how IEMG may be used to facilitate mechanical efficiency in unique functional contexts. Consider first the case of a mid thoracic paraplegic who complained of headache and severe upper thoracic back pain when seated in a wheelchair, during both activity and rest. It was felt that the pain presentation was due to several component factors that could be distinguished with the aid of IEMG. Static sitting posture was observed to be protracted and kyphotic, with scan findings of high IEMG at the upper cervical region and quiescent IEMG through the upper thoracic area. Upon performance of functional tasks, neck and scapulothoracic IEMG was noted to be erratic in magnitude, asymmetrical, and lagging in return to baseline with cessation of activity.

Looking at the increased resting upper cervical IEMG, the suboccipital musculature must mechanically work harder to hold the head up if it is protracted, and this may have contributed to occipital tension headache. Resting upper back pain with the thoracic spine positioned in kyphosis and with concomitant low IEMG (recording above the level of spinal lesion and assuming stable bony structures) probably reflected chronic passive stretch on deep soft tissues, that is, "hanging on ligaments".

This paraplegic was dependent on the muscles of the scapulothoracic complex as prime movers and stabilizers for most functional activities. The result was high energy expenditure and aberrant patterns of muscle recruitment through the neck and back with consequent tension myalgia, that is, upper back "burn-out".

One treatment goal was the achievement of a more retracted spinal alignment at rest, without excessive muscle tension at the upper back or neck. We also sought to unload the proximal arms and minimize muscle tension levels during activities of daily living. Modifications were attempted by raising and lowering the wheelchair back, adding passive lumbar supports and thoracic restraints, tilting the overall seating alignment, modifying the arm rests, raising the height and changing the position of selected items at the patient's work station, and providing simple adaptive equipment for dressing and personal care. By monitoring IEMG activity, we were able to quickly evaluate each modification and identify the most mechanically efficient combination.

Next take the case of a bicyclist who had recently increased her training schedule in preparation for a 200 mile race. She complained of pain at the cervical-thoracic junction that initially would come on at about 100 miles and would then quickly resolve with rest. Now, however, the pain was continuous at low level and exacerbated sooner with each ride. She described the problem in terms of the pain hampering her ability to concentrate and maintain speed over long distances. Upon thorough musculoskeletal evaluation, it was thought that the pain was due to accumulated intervertebral joint and ligament strain peculiar to the hyperextended cervical posture assumed on her racing bike. The patient was directed to return to the clinic after riding some distance, with her bicycle. Musculoskeletal and IEMG exams were repeated. In contrast to the initial evaluation, which had followed a rest day, cervical IEMG was extremely elevated and erratic. This was more the case when the patient assumed her riding posture on her bicycle. We felt that the aberrant IEMG activity observed in riding position was in part a response to the joint strain and pain, and in part a mechanical consequence of the forward head posture per se. The patient had recognized that there were possible variations of handlebar and seat alignment that would be acceptable for racing form, but was concerned that the race date was approaching and that it would take a long ride to evaluate each possibility.

The patient's acute issues were managed with standard physical agents and manual therapies. IEMG was monitored from the cervical and thoracic paraspinals as various combinations of handlebar and seat positions were tried on different bicycle frames. The patient settled on the combination that produced the lowest IEMG readings in static posture in the clinic. She went on to finish the race in good time, with much less discomfort than anticipated. She attributed the outcome to an improved riding alignment plus periodic, short breaks for postural relief and stretching.

## 10. IEMG Triggered Neuromuscular Electrical Stimulation

IEMG triggered neuromuscular electrical stimulation (IEMG-NMES) is an elegant technique that interfaces two traditional modalities together in what is hoped to be a synergistic combination. Fields (1987) provides an account of the use of IEMG-NMES for the rehabilitation of patients after cerebral vascular accident, and discusses the concept in general.

In the IEMG-NMES paradigm, an exogenous electrical stimulus is delivered to a target muscle once a prescribed IEMG threshold is exceeded. The electrical stimulus activates the muscle, via its nerve, to some greater intensity than the volitional contraction alone. Thus, endogenous neural activity results in motor unit recruitment that is discriminated, and then augmented by artificially induced motor recruitment. The augmented muscle contraction presumably increases the output of sensory receptors located in the muscle-tendon complex that are afferent to the central nervous system. This enhanced proprioception is somatotopically relevant and time-locked to the volitional muscle contraction. It seems reasonable, then, that the process would act to reinforce the central motor program that produced the initial, desired peripheral response.

Traditional biofeedback systems rely on the transduction of biologic activity to exogenous audio and visual cues. The IEMG-NMES paradigm is unique in that it directly facilitates endogenous proprioceptive feedback. (Figure 6). Because it is time-locked, somesthetic in nature, and somatotopically relevant in particular, the IEMG-NMES feedback loop perhaps has the advantage of a more direct sensori-motor linkage. That is, the spatial organization of the sensory and motor systems is preserved along the length of the neuraxis, and anatomically correspondent interrelations between the two can be found in the spinal cord, brainstem, and the cerebral cortex (Carpenter, 1978).

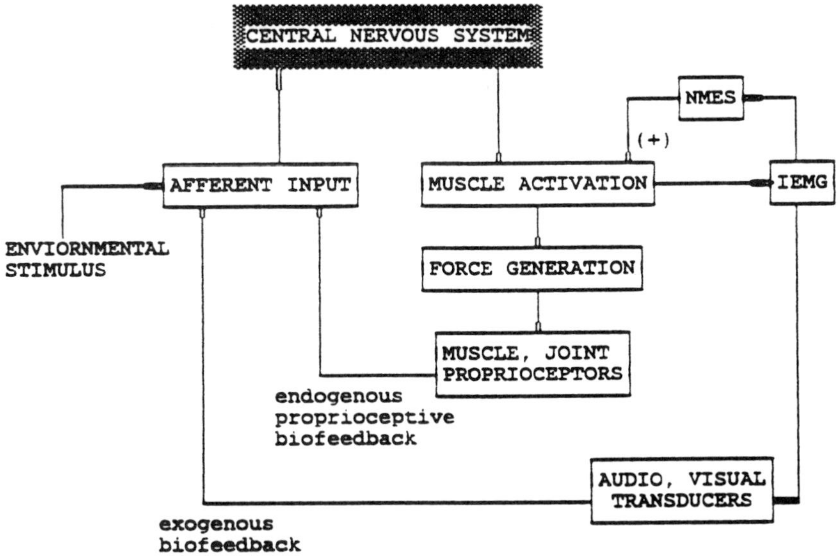


**Figure 6.** IEMG Triggered Neuromuscular Electrical Stimulation

With the IEMG-NMES system, augmented proprioceptive feedback is aggregative to traditional audio and visual cues. There is also the sensory feedback derived from electrical activation of cutaneous sensory structures. Heterosensory feedback of this type would access multiple levels of the neuraxis, maximizing the synaptic volume of opportunities for neuromotor plasticity and re-education.

Whereas IEMG-NMES is most commonly used for rehabilitation after neurologic insult (Field, 1987; Graup and Kohn, 1987; Hammond, et. al., 1987; and Kraft, 1988), we have found it to be a useful modality for the treatment of select patients with chronic musculoskeletal pain. Figure 7 shows IEMG activity from the left and right adductors of a 35 year old electrical engineer with chronic right adductor pain, atrophy, and weakness dating from recurrent strains several years prior. Note how isometrically resisted left adduction resulted in a large IEMG peak and resisted right adduction in a small IEMG peak, as would be expected. However, when left and right adductors were simultaneous resisted, IEMG activity was low on both sides. Anatomically, this implicated some central nervous system mechanism. It was additionally noted that the right adductors fired strongly when the patient's balance was suddenly challenged in standing or when he performed rapid closed kinetic chain activities with cognitive distraction. Previous x-rays, bone scan, CT scan, and percutaneous EMG studies were interpreted as normal. Results from an MMPI were evaluated as rather unremarkable on all dimensions. Several courses of therapies involving ice, ultrasound, multiple types of electrical stimulation, deep friction massage, myofascial release, and carefully graded stretching and strengthening exercises were without effect.

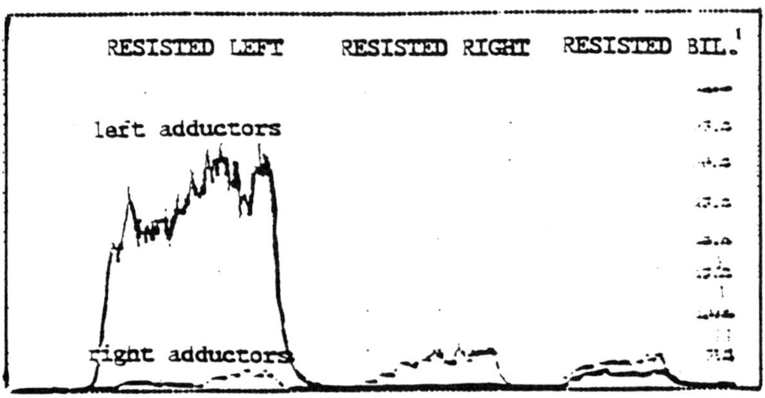


**Figure 7.** Left and right hip adductor IEMG activity during unilaterally resisted left adduction, unilaterally resisted right adduction, and bilaterally resisted adduction, in sitting, from a patient with chronic right adductor pain and weakness.

The patient was reinitiated on a gentle stretching program and started on a course of IEMG-NMES to the right adductors. As time went on, it became apparent that the patient engaged in a multitude of behaviors to avoid using the adductors during functional activities. This included a subtle change in gait, getting out of bed or out of a car in a certain way, and even turning away from urinals in a different way. This was done ostensibly to protect himself from further strain and pain exacerbation. We felt, however, that these sorts of things only contributed to a pattern of disuse weakness, disordered proprioception, and apprehension behaviors.

After six weeks of daily, home IEMG-NMES, the patient was able to generate smooth, tetanic isometric adductor contractions with IEMG peak magnitude of about 70% of the contralateral side. At this point, use of the IEMG-NMES system was tapered and independent strengthening and vigorous stretching exercises were accelerated. We chose to avoid further passive therapies and aggressively encouraged the cessation of compensatory functional behaviors. Over the following months, right adductor activity continued to normalize, pain abated, and the patient returned to normal functional activity.

Figure 8 shows IEMG recordings from the quadriceps complex of a patient with chronic patellofemoral pain, reviewed earlier with respect to Figure 4. The figure shows IEMG activity pre and post IEMG-NMES treatment at the distal vastus medialis, during isometric contractions at full knee extension. Note the relative activities of each channel before and after IEMG-NMES, and the relative ratios of one to each other. Clearly, there is more total activity, and more proportionate activity from the vastus medialis after treatment. Increased distal vastus medialis activity should, in theory (Wise, et. al., 1984), provide better medial stabilization and tracking of the patella in the femoral groove and reduce pain due to irritation of the retropatellar cartilage.

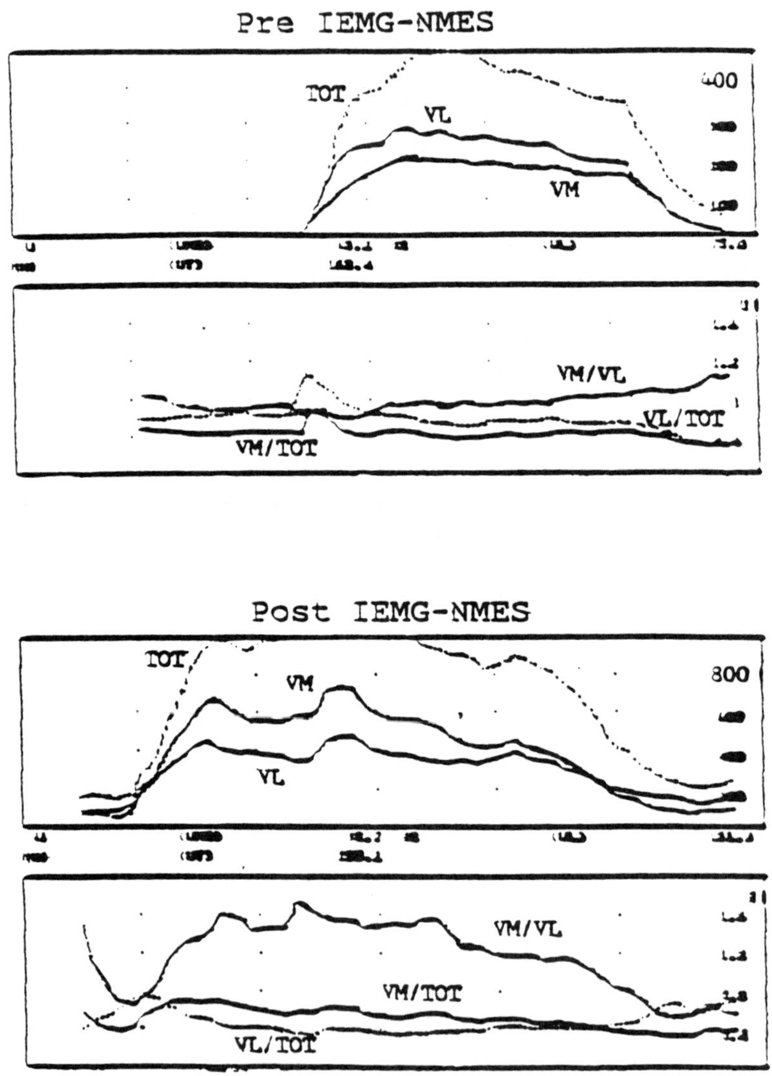


**Figure 8.** Pre and post quadriceps IEMG activity from a patient with chronic patellofemoral pain. VM = distal vastus medialis, VL = vastus lateralis, TOT = "total" quadriceps.

The patients just described had in common the fact that they had basically stopped using their injured body parts. For these particular patients, and others like them with pain at the hip, knee, and shoulder, the problem is one of muscle inhibition or aberrant muscle synergy. IEMG-NMES is most useful for neuromuscular re-education in such circumstances.

The reader is referred to texts of clinical electrophysiology for NMES protocols that may be adapted for use with IEMG (Wolf, 1981; Nelson and Currier, 1987).

As a peripheral comment, commercial IEMG-NMES systems are easily adapted for the treatment of spasmodic torticollis as proposed by Cleeland (1973).

## EXERCISE PROGRAM CONSIDERATIONS

Exercise prescription does not inherently have anything to do with IEMG, but it is so integral to biofeedback treatments for chronic musculoskeletal pain that some special considerations bear explanation.

Therapeutic exercises have the effect of improving blood flow, both acutely and by long term revascularization. Muscle biochemistry is shifted toward more efficient energy metabolism. Fluid mechanics inside synovial joints is improved. Histological changes are seen in the ground matrix and fiber structure of connective tissues that favor improved tissue nutrition and mechanical integrity. There occurs plastic increases in the length of myofascial, tendon, and ligament tissue. By virtue of the activation of low threshold and wide dynamic range mechanoreceptors, nociception is inhibited and hypertonicity is reduced. Along with the more appropriate activation of proprioceptor organs, there is an improved range and pattern of somesthetic inputs. Muscle synergies are restored to normal limits. Psychological confidence is facilitated. In brief, exercise is necessary to reverse the cycles of dysfunction that got the patient into chronic trouble.

Exercises for the chronic pain patient should be prescribed by qualified professionals. Standard principles of exercise prescription need be adapted for physiologic and behavioral factors that are unique to certain patient populations (Saunders, 1989; Kisner and Colby, 1985; Pollack, et. al., 1984).

Exercise principles for the IEMG biofeedback patient are outlined as follows:

1. Exercises should be specific.

If hypomobile tissues are not isolated, the imposed stretch will likely be taken up by other structures, or perhaps contribute to compensatory hypermobility at adjacent kinetic segments. Similarly, if strengthening is not isolated, exercise may merely encourage substitution patterns and abnormal synergies.

2. Consider the target muscle's functional role.
   a. tonic postural stabilization
   b. phasic movement
   c. phasic stabilization for other muscle actions

Plan exercises that coordinate muscle functions in synergy. In general, tonic muscles will be tight and tense. Phasic muscles may be either shortened, tense, with relatively normal peak strength, or alternatively they may be inhibited and weak.

3. Prescribe stretching exercises.
   a. low to moderate passive over-pressure
   b. progressively longer, sustained holds

The above mode is preferred to reset "spindle bias" and to allow elastic lengthening to give way to plastic change.

4. Prescribe strengthening/re-education exercises.
   a. Start with
      (i) high repetition (>30)
      (ii) low resistance
      (iii) slow speed

Use high repetition to encourage neuromuscular re-education. High resistance exercise is more likely to elicit pain, mechanically overwhelm healing tissues, or force muscle substitution synergies to predominate; low resistance is preferred most often. Slower speeds are preferred to mechanically load the tissue structure, muscle spindles, tendon organs, etc., at a lower rate, and allow some accommodation.

b. Progress by:
   (i) adding reps
   (ii) building multiple sets
   (iii) increased speed
   (iv) increased resistance

c. Work into:
   (i) full concentric, eccentric, isometric contraction patterns
   (ii) complete range of motion
   (iii) functional movement planes
   (iv) aerobic programs

## REFERENCES

Basmajian, J.V. (Ed.). (1989). Biofeedback: Principles and Practice for Clinicians, 2nd Ed. Williams and Wilkins.

Carpenter, M.B. (1978) Core Text of Neuroanatomy, 2nd Ed. Williams and Wilkins. Baltimore.

Cleeland, C.S. (1973) Behavioral Techniques in the Modification of Spasmodic Torticollis, Neurology, 23, 1241-1247.

Daniels, L. and C. Worthingham. (1980). W.B. Saunders Co. Philadelphia.

Davidson, R.J. and G.E. Schwartz. (1976). The Psychobiology of Relaxation and Related States: A Multi-Process Theory. In Motolsky, D. (Ed.): Behavior Control and the Modification of Physiological Activity. Prentice Hall. Englewood Cliffs. pp. 399-441.

Davies, G.J. (1984). A Compendium of Isokinetics.

DeBacher, G. (1989). Biofeedback in Spasticity Control. In Basmajian, J.V. (Ed.): Biofeedback: Principles and Practice for Clinicians, 3rd Ed. Williams and Wilkins. pp.141-152.

Evjenth, O. and J. Hamberg. (1984). Muscle Stretching in Manual Therapy. Alfta Rehab. Alfta, Sweeden.

Fahrer, H., Rentsch, H.U., Gerber, N.J., mBeyeler, Ch., Heiss, Ch. W., and B. Grunig. (1988). Knee Effusion and Reflex Inhibition of the Quadriceps

Fields, R.W. (1987). Electromyographically Triggered Electric Muscle Stimulation for Chronic Hemiplegia. Archives of Physical Medicine, 68, 407-414.

Fischer-Williams, M., Nigl, A.J. and D. L. Sovine.(1986). Human Sciences Press. New York. pp.38-85.

Fogel, E.R. (1987). Biofeedback-Assisted Musculoskeletal Therapy and Neuromuscular Re-Education. In Schwartz, M.S. (Ed.): Biofeedback: A Practitioner's Guide. The Guilford Press. New York. pp.377-411.

Gerdle, B., Johansson, C., and R. Lorentzon.(1988). Relationship Between Work and Electromyographic Activity During Repeated Leg Contractions in Orienteers, Eur J Appl Physiol, 58, 8-12.

Grabiner, M.D., Robertson, R.N., and K.R. Campbell. (1988). Effects of Fatigue on Activation Profiles and Relative Torque Contribution Of Elbow Flexor Synergists. Med and Sci in Sports and Exer, 20(1), 79-84.

Graupe, D. and Kohn, K.H.(1987) A Critical Review of EMG-Controlled Electrical Stimulation. Crit Rev Biomed Eng, 15(3), 187-210.

Hammond, M.C., Kraft, G.H., Fitts, S.S., and Nutter, P.B.(1987). Rehabilitation of the Upper Extremity in Chronic Stroke Patients. Arch Phys Med Rehabil, 68, 631.

Hendriksson, J. and F. Bonde-Peterson. (1974). Integrated Electromyography of Quadriceps Femoris Muscle at Different Exercise Intensities, J Appl Physiol, 36(2), 218-220.

Hoppenfeld, S. (1976). Physical Examination of the Spine and Extremities. Appleton-Century-Crofts. London.

Horita, T. and T. Ishiko. (1987). Relationships Between Lactate Accumulation and Surface EMG activities during Isokinetic Contractions in Man, Eur J Appl Physiol, 56, 18-23.

Kendall, F.P. and McCreary, E.K. (1983). Muscles Testing and Function, 3rd Ed. Williams and Wilkins, Baltimore.

Kisner, C. and Colby, L.A. (1985). Therapeutic Exercise: foundations and Techniques. F.A. Davis Co. Philadelphia.

Komi, P.V., Kaneko, M., and O. Aura. (1987). EMG Activity of the Leg Extensor Muscles with Special Reference to Mechanical Efficiency in Concentric and Eccentric Exercise. Int J Sports Med, 8,22-29.

Kraft, G.H. (1988). EMG-Triggered Muscle Stimulation. Arch Phys Med Rehabil, 69, 149.

Krebs, D.E. (1987). Biofeedback in Neuromuscular Re-Education and Gait Training. In Schwartz, M.S. (Ed.): Biofeedback: A Practitioner's Guide. The Guilford Press. New York.pp.343-376.

Krebs, D.E., Staples W.H., Cuttita, D., and Zickel, R.E. (1983). Knee Joint Angle: Its Relationship to Quadriceps Femoris Activity in Normal and Postarthrotomy Limbs. Arch Phys Med Rehabil, 64, 441-447.

Lippold, O.C.(1952). Relationship Between Integrated Action Potentials in Human Muscle and its Isometric Tension, J Physiol, 117, 492-499.

Magee, D. J. (1987). Orthopedic Physical Assessment.W.B. Saunders Co. Philadelphia.

Maton, B., Peres, G, and B. Landjerit. (1987). Relationships Between Individual Isometric Muscle Forces, EMG Activity and Joint Torque in Monkeys, Eur J Appl Physiol, 56, 487-494.

McKay, M., Davis, M., and P. Fanning.(1981). Thoughts and Feelings: The Art of Cognitive Stress Intervention>New Harbinger Publications. Richmond.

Milner-Brown, H.S., Mellenthin, M., and Miller, R.G. (1986). Quantifying Human Muscle Strength, Endurance, and Fatigue. Arch Phys Med Rehabil, 67, 530-535.

Minor, M.A.D. and Minor S.D. (1985) Patient Evaluation Methods for the Health Professional. Reston Publishing Co. Reston.

Nelson, R.M. and Currier, D.P. (eds.) (1987). Clinical Electrotherapy. Appleton Lange. Norwalk.

Pollack, M.L., Wilmore J.H., and S.M. Fox.(1984). Exercise in Health and Disease. W.B. Saunders Co.Philadelphia.

Robertson, R.N., Carlton, M.J., Carlton, L.J., and K.M. Newell.(19(1982). Height and Slope of IEMG as Predictors of Torque Parameters, J Motor Behav, 14,24-45.

Saunders, H.D. (1989). Evaluation, Treatment, and Prevention of Musculoskeletal Disorders. Anderberg-Lund. Minneapolis.

Schwartz, M.S., (Ed.).(1987). Biofeedback: A Practitioner's Guide. The Guilford Press. New York.

Shakespeare, D.T., Stokes, M., Sherman, K.P., and A. Young. (1985). Reflex Inhibition of the Quadriceps After Meniscectomy: Lack of Association With Pain, Clin Physiol, 5, 137-144

Stalberg, E., Borges, O., Ericsson, M., Essen-Gustavsson, B., Fawcett, P. R .W., Nordesjo, L. O. , Nordgren, B., and Uhlin, R. (1989). The Quadriceps Femoris Muscle in 20-70 Year Old Subjects: Relationship Between Knee Extension Torque, Electrophysiological Parameters, and Muscle Fiber Characteristics, Muscle and Nerve, 12, 382-389.

Sullivan, P.E., Markos, P.D., and M.A.D. Minor.(1982). An Integrated Approach To Therapeutic Exercise: Theory and Clinical Application. Reston Publishing Co. Reston.

Travell, J. and D.G. Simons. (1983). Myofascial Pain and Dysfunction: The Trigger Point Manual. Williams and Wilkins. Baltimore.

Wise, H.H., Fiebert, I.M., and J. Kates. (1984). EMG Biofeedback as Treatment for Patellofemoral Pain Syndrome, J Ortho and Sports Phys Ther, Sept/Oct, 95-103.

Wolf, S.L., (ED.)(1981). Electrotherapy. Churchill Livingstone. New York.

Wolf, S.L. (1987). Electromyographic Biofeedback: An Overview. In Nelson, R.M. and Currier, D.P. (eds.): Clinical Electrotherapy. Appleton and Lange. Norwalk. pp. 259-278.

Woods, J.J., and B. Bigland-Ritchie. (1983). Linear and Non-Linear Surface EMG/Force Relationships in Human Muscles, Amer J Phys Med, 62(6), 287-299.

# CHAPTER 7

# *Surface Electrode Electromyography Static and Dynamic Muscle Assessments*

## *An Overview*

Robert W. Tosh, D.C.

---

### INTRODUCTION

Chiropractors have for years relied on various and diverse methods for diagnosing a wide variety of subjective complaints. Surface electrode EMG is a very powerful tool for the chiropractor who wishes to better understand their patient's condition from a musculoskeletal stand point. With surface EMG the doctor can assess paraspinal muscular imbalances objectively, which results in a more specific treatment protocol. Another attribute of surface EMG is follow-up examinations performed at regular intervals during treatment which enable the doctor, and patient alike, to evaluate the treatment protocol for its effectiveness, and to show the need for further care (if that need exists). The third attribute to surface EMG in the chiropractor's office is that it helps to objectify the extremely subjective symptomatology of the soft tissue complaint.

The Chiropractic profession has come a long way in its diagnostic model in an attempt to objectify rather subjective conditions. Current methods include orthopedic testing, kinesiologic testing, and static and motion palpation. This list is by no means exhaustive, however, it represents a majority of basic diagnostic procedures used in chiropractic offices today.

Each of the above listed diagnostic techniques involve a great amount of input from the patient, is subject to interpretation, and/or is dependent upon the experience and expertise of the doctor. In an attempt to further objectify the analysis of the spine, many adjunctive modalities were introduced including NCM, DTG, postural analyzers, thermography, automated muscle testing, and computerized goniometry. Each of these have merit in assessing the overall neuromusculoskeletal systems, however, until recently there was no practical means to assess postural and functional components of the paraspinal musculature.

The use of surface electrode electromyography in the clinical setting has prompted a new age of objective evaluation of a patient's subjective complaints. Although the concept of interpreting activations of underlying musculature to assess a patient's condition is not new, it has only recently become affordable to the sole practitioner through the advent of sophisticated computer assisted electronics. Surface electrode EMG therefore brings to the practitioner an accurate objective method of assessing their patient's condition. The paraspinal activation levels are consistent with muscle tonicity of the underlying musculature, therefore, a properly performed examination will yield information (including, in most cases, a computerized statistical analysis) which the doctor interprets with respect to levels of hypertonicity, asymmetry, protective muscle guarding, splinting, and other related findings which will be discussed later in this chapter.

The examination is non-invasive, painless, and does not require much training on the part of the technician. It should be noted, however, that the practitioner must have a good understanding of muscle function of the areas of interest, as well as a conception of the ability and limits of the surface electrode EMG examination. Utilizing surface EMG in combination with X-ray, orthopedic, and other exam findings helps to complete a very effective diagnostic package. As with other examinations, the EMG exam should be interpreted with the proposed treatment protocol in mind.

For example, if a chronic lower back pain patient demonstrates high EMG activations on the side contralateral to the patient's complaint, residual protective muscle guarding is suspected. The practitioner must then make the determination as to how to modify the treatment protocol to lessen the degree of guarding. This will maintain appropriate paraspinal balance of the musculature which helps to ensure proper stability of the involved area post treatment. The interpretation of the surface EMG examination findings will be discussed in greater detail later in this chapter.

## ARTIFACT

Although the electronics associated with today's surface EMG units have many circuits which discriminate against artifact, there are some factors which the technician should control in the clinical setting. These potential artifacts, and how to identify and correct them should they occur, is an area which must be understood.

One such factor is the electrode-skin interface. If the two active electrodes become shorted together, a very low reading will occur (below one microvolt). This is due to the fact that the hardware utilizes a differential amplifier to assess the signal, and if the two electrodes are reading the same signal, the net result is near zero. Shorting of electrodes is known as "bridging". The most common cause of electrode bridging is the use of excessive amounts of conductive agent.

Another artifact which can occur is that of improper contact with the skin. This occurs when one of the three electrodes is not touching the skin, when an insufficient amount of conductive agent is used, or when the patient's skin has not been properly abraded. This is due, again, to the differential amplification system of the hardware. The amplifier in this case senses a great difference between the two electrodes without much in common, which results in a high, unstable reading on the computer. This can also occur if there is a break in any of the wires leading to or from the EMG interface.

A third type of artifact occurs when there is movement of the electrodes during a reading. This can cause multiple readings of the same activation which results in a high reading on the computer. A similar artifact occurs when the wires leading to the scanner are moving during a reading. The wires may act as antennae and pick up surrounding noise. The latter, however, may be greatly reduced or eliminated entirely if the preamplifier is housed within the scanner.

One other source of artifact is that of outside electrical influences such as a motor, overhead lights, the computer, or anything else which emits electromagnetic fields. A simple test to determine whether or not this is occurring is to connect one 10 kilaohm resistor between each active electrode and the reference electrode. In other words, one active wire connects to one end of a 10Ku resistor, with the other end connected to the reference lead. This, in turn, is connected to another 10Ku resistor, the other end of which is connected to the other active lead. This acts as a dummy patient load and should register as a very low, stable reading on the computer. With this setup in place, the technician moves the scanner around the room to ensure that the zero reading remains stable. This simple test should be preformed at regular intervals to analyze the equipment for reliability.

## CLINICAL PROCEDURE

The items which are needed to properly conduct a static muscle assessment are as follows: 1) computer with appropriate software for data collection and statistical analysis, 2) a scanner comprised of two active silver/silver chloride electrodes and a third reference electrode, and 3) an analogue to digital converter interface with differential signal amplification. In addition, supplies needed are alcohol wipes (semi-abrasive), and some type of conductive cream, liquid, or gel.

The patient's skin must first be prepped on any area which is to be scanned. This is accomplished by using a semi-abrasive alcohol wipe to vigorously abrade these areas. The abrade must be accomplished prior to scanning, regardless of the specific type of equipment being used. This will cleanse the skin of any material which could interfere with a proper skin-electrode interface such as oils, dirt, and make-up. Failure to comply with the skin abrade will result in abnormally high readings due to the fact that the electrodes may fail to pick up common signals, and therefore, such signals will not be rejected by the differential amplifier.

The procedure for the static scan begins with patient positioning. The scan is first done with the patient in the sitting posture, and then again in the standing posture. For the sitting posture the patient is instructed to sit comfortably with their hands in their lap and feet flat on the floor without having the back supported. The instructions here are deliberately limited in an effort to have the patient assume their "normal" posture. This procedure reveals a more accurate assessment of the individual's postural habits. The practitioner must therefore make available a chair without a back support, and which is adjustable in height to allow the patient's feet to be positioned flat on the floor. The patient is further instructed to look forward, and to stay in that comfortable position throughout the scanning procedure. Another aspect in patient positioning is that the computer monitor should not be in view of the patient. If it is, the patient will tend to slightly modify their posture to view the screen, thus possibly altering static posture readings paraspinally.

Once the patient is in the proper position, the reference electrode is affixed. The reference electrode should be placed away from the sites to be monitored, yet in a position which allows ease of movement of the scanner being used, taking into consideration the length of the wire connected to the scanner with respect to the sites to be monitored.

With the computer ready, the patient positioned, and the reference electrode attached to the patient, the actual scanning procedure begins. Basically, a site should be scanned with the active electrodes 2cm apart (usually affixed to the scanner), and placed parallel to the muscle fibers of the muscle of interest. For muscles less than 5cm in length, the electrodes should be placed at the central aspect of the muscle belly on either side of the neuromuscular junction. For muscles greater than 5cm in length, placement of the active electrodes should be at the midpoint between the neuromuscular junction and the distal tendon. Scanning the proper aspect of the underlying musculature reduces the sensing of unwanted activations from neighboring muscle groups.

With the patient in the seated posture, and the reference electrode attached to the patient, the technician begins the scan. The proper procedure is to moisten the scanning electrodes with a conductive medium (gel, cream, or solution) prior to touching down on the patient's skin over the desired site. Studies have shown that the electrodes should be on the site for the MINIMUM TIME NECESSARY TO ACHIEVE A STABLE READING. The average time necessary for a stable reading is 6-12 seconds. A stable reading is one in which each successive reading is within 10% of the previous reading for a period of 2 seconds.

Once the scanner touches down, there will be a very high reading on the computer for a moment as the electrode/skin interface is being established, and then the reading will drop abruptly to one of a stable reading. Once a stable reading is noted, it is recorded into the computer's memory either by pressing a button, or automatically, depending upon your software program. The operator then moves on to the next site to be scanned.

When all of the desired sites have been scanned in the sitting posture, the patient is instructed to assume a standing posture and the sites are reexamined. The instructions for a standing posture are simply to stand with the feet at shoulder width and arms down to the side, looking straight ahead (again, the patient should not be allowed to view the computer monitor).

The amount of pressure used on a scan site is of importance in surface EMG, in that too much pressure on the part of the technician may move the patient off their center of gravity (especially in the standing posture) which will adversely affect the reading. Therefore, the minimum amount of pressure needed to obtain appropriate contact with the skin is used. Evenly distributed pressure among both active electrodes is another consideration. Uneven pressure could cause additional contraction of the underlying muscle.

Yet another consideration is that of clothing. A patient undergoing a surface EMG should be gowned in such a way as to expose all areas of scanning interest. Failure to do so results in the necessity of moving the patient's clothing out of the way, which may cause the patient to compensate posturally, again, adversely affecting the readings. One last consideration along the same line is that of a patient with long hair. This poses the same type of problem as does the clothing and can be remedied simply by having the patient pin it up.

## INTERPRETATION

The basic concept of surface EMG scanning is that a muscle depolarizes yielding a generation of measurable bioelectrical impulses which are picked up by analog sensors. The signal is then passed through an analog to digital converter, which also acts as an amplifier and noise filter, and then is passed to a computer which analyzes the signal and displays appropriate information for the clinician to interpret. The underlying question is what do you do with this potentially valuable information once it is obtained. We will now explore some accepted methods of data collection and interpretation.

### What increases activation?

Since surface electrode EMG measures the effect of a wave of depolarization, we must understand when and why this occurs in order to make sense of the EMG readings. Activations of skeletal muscle is what is of interest to the EMG technician. Skeletal muscle will depolarize as the result of an impulse from an anterior horn cell of the central nervous system, or due to a reflex at the spinal cord level. This occurs when the subject makes a cognitive effort to VOLUNTARILY move a particular muscle group, or when there occurs a reflex muscle SPASM either in response to a noxious stimuli, or for ANTALGIC MUSCLE GUARDING. Other activations occur in response to proprioceptive changes of POSTURE, or to STRUCTURAL abnormalities such as SCOLIOSIS and VERTEBRAL SUBLUXATION. Another source of activation of various muscle groups may be in response to a NEUROLOGICAL DISORDER; however, such instances are beyond the scope of this book.

Voluntary

Increased activations due to voluntary movement are of interest only in that they may be caused by a patient's conscience attempt to avoid a painful posture (voluntary antalgia). Another instance of voluntary hyperactivation which should not be overlooked by the thorough clinician is due to patients with a secondary gain (malingering).

Spasm

A muscle group will go into spasm for several reasons. There may be a need by the body to protect an area of injury which is generally contralateral to the side of involvement, and is known as protective muscle guarding.

Similarly, there is a phenomenon known as muscle splinting which occurs for the same purpose as protective muscle guarding, and can be either ipsilateral or bilateral in nature. This is generally seen in the acute patient and is temporary in nature, usually lasting approximately 72 hours after onset. Another type of spasm is that which occurs when the muscle has suffered a direct insult, chemical imbalance, or neurological disturbance.

Muscle spasm is also a function of pain in that increased pain (noxious stimuli) will cause increased spasm which, in turn, causes increased pain and so on. This is known as a negative feedback pain loop. This pain loop should be halted to allow for proper rest and stabilization of the involved muscle groups.

Structural abnormalities

The primary structural abnormality of concern to the chiropractor is that of a scoliosis. Generally the convexity of the curvature will demonstrate higher surface EMG readings than that of the concavity due to postural accommodation. Functionally (with respect to the dynamic muscle assessment), the convexity will again demonstrate higher readings whenever that side is called upon to work.

## STATIC MUSCLE ASSESSMENT PROTOCOL

Once a proper site has been selected, the EMG technician touches down with the scanner on the site. The scanner should be held lightly, but firmly, allowing for a stable contact. Pressing too firmly could cause the underlying musculature to reflexively contract giving a falsely high reading, or may cause the patient to move off balance causing postural activations of the muscles.

The scanner is held in place until the reading on the computer (or digital readout) has stabilized. A stable reading is considered one in which each successive reading is within 10% of the reading before it for a period of two seconds. Once the reading has stabilized, its value is recorded.

Although surface EMG equipment is capable of scanning any attainable muscle group in the body, the area of interest is what is to be scanned. Most computer programs have a series of different sequences of scan sites which are preprogrammed, as well as the capability of user defined scans. The type of scan selected, depends upon the area of interest. In cases where area of involvement is not clearly defined, a general scan is indicated; however, in cases such as an evaluation of the anterior tibialis in foot drop, the practitioner would select an extremely specific scan site.

In most chiropractic cases, the area of interest is that of the paraspinal musculature, and of the face. The most commonly scanned areas of the face include the Frontalis, Temporalis, Masseter, and Sternocleidomastoid muscle groups. The electrode placements of these sites, and their generalized diagnostic significance will now be discussed.

The Frontalis is scanned with the electrodes approximately 1cm above the eyebrow, and centered over the pupil. This is the only site which is scanned perpendicular to the muscle fibers. In this way activations from a greater variety of surrounding muscle groups are recorded. An alternative placement would be a vertical positioning of the two active electrodes, centered over the pupil, with the electrodes centered over the muscle group. The Frontalis is the primary muscle of nonverbal expression. Therefore, increased activations here may indicate muscle tension headache due to stress, or muscle tension headache due to a synergistic action associated with the suboccipital muscles. The latter is generally seen as a unilateral involvement with concomitant increased cervical activations on the same side.

The Temporalis is scanned with the electrodes placed vertically, with the lower electrode just distal to the zygomatic bone and above the zygomatic arch of the temporal bone along the underlying muscle fibers. Increased unilateral activations of the Temporalis suggest mandibular involvement, as in a Temporomandibular Joint Syndrome, or bruxing. Patients with increased Temporalis activations should be instructed to preform relaxation exercises of this area in an attempt to avoid undo compression of the temporomandibular disc.

The Masseter is scanned with the electrodes orientated in line with the muscle fibers which run from the zygomatic bone to the angle of the mandible. With this site one must take into consideration the presence of facial hair. Beards will disrupt the proper electrode/skin interface which will cause falsely high readings. Therefore, this area should either be shaved or avoided if facial hair is present. Scanning adjacent areas will give an indication as to the activity of the Masseter, although such a deviation in procedure must, of course, be properly noted after the examination. Increased activations at this site suggest bruxing or Temporomandibular Joint Syndrome.

The Sternocleidomastoideus is scanned along the muscle fibers at the central portion of the muscle which runs from the mastoid process to the medial 1/3 of the clavicle. It is important that the patient be looking straight ahead while scanning this site as the SCM is a primary rotator of the skull. Increased unilateral activations at this site suggest latent or active torticollis, or post whip-lash type injuries. A bilaterally increased activation may be an indication of chronic muscle bracing.

The sites of the neck and back are determined by vertebral level. The spinous process is utilized as the level marker. Electrodes should be placed vertically, over the transverse/mammillary ridge, with the spinous process protuberance acting as the center marker between the electrodes. A scan of the full spine should include the following sites taken bilaterally: C-2, C-6, T-1, T-6, T-8, T-10, L-1, L-3, and L-5. It should be understood by the practitioner that the scanning electrodes in a paraspinal scan pick up activations from most of the underlying musculature, not just one specific muscle. This is desirable as it reveals a more accurate assessment of the overall activity of the paraspinal musculature, although studies suggest that it is not specific enough to detect individual vertebral rotation. This is probably due to the fact that the rotators and multifidi muscle groups (which are responsible in large part for rotation of the vertebra) are much too isolated to be picked up individually. Therefore, interpretation of the static muscle scan is directed to an overall functional impression of the paraspinal musculature.

## INTERPRETING THE STATIC MUSCLE ASSESSMENT

First, one must compare each of the activations against a normative data base to determine whether the activation is normal, or above normal (mild, moderate, or severe activation above normal). This evaluation is followed by the determination of paraspinal asymmetry. A site is considered to be asymmetrical when there is a difference of 1 standard deviation or greater. Finally, the sites are compared for postural components. An activation which is 50% greater than its contralateral counterpart is considered posturally aberrant.

An asymmetrical increased activation which is contralateral to the site of pain is suspected of protective muscle guarding, whereas, if it were on the side of pain, a simple muscle spasm may be present. Increased activations which occur when a muscle is placed on stretch, may represent a chronically shortened muscle. An irritable muscle will manifest itself as one which is highly active whether it is placed in stretch, or called upon to work.

In the case of a vertebral malposition, the paraspinal musculature about the level would be expected to be hyperactive in an attempt to compensate for the lack of mobility, and to attempt to stabilize the segment.

The static muscle assessment therefore helps in the determination of postural and/or asymmetrical variations of muscular activity in a static posture. To determine functional behavior of a muscle group, one must consider conducting a dynamic muscle assessment with the surface EMG equipment.

## DYNAMIC MUSCLE ASSESSMENT PROCEDURE

The area of interest for a dynamic muscle assessment is either the site of involvement, and/or a site which demonstrates aberrant activations with respect to a static muscle assessment.

The skin preparation is identical to that of a static assessment, in that a vigorous alcohol abrade of the electrode site is necessary for reasons previously stated. The electrodes which are used are either self-adhesive disposable silver/silver chloride, or permanent silver/silver chloride electrodes with an adhesive collar. Electrodes which are not pre-gelled must have a conductive agent applied to them prior to application to the patient. With this type of electrode, the silver/silver chloride electrode is recessed which allows for the addition of the agent. There must be enough conductive agent to fill the cup, which will allow contact between the electrode and the skin. A common mistake is for a technician to use too much agent which may cause bridging of the active electrodes, and may interfere with the adhesive. Another consideration to be made is the shape of the electrodes used. The electrode itself is usually round (1cm diameter is suggested), however, the adhesive backing may be rectangular. If this is the case, the electrode is positioned in a horizontal fashion. This will help reduce the effect of the patient's skin stretching beneath the adhesive which, therefore, allows for a better contact and less chance of an electrode becoming dislodged during a range of motion.

The dynamic muscle assessment which will be discussed here involves the assessment of bilateral paraspinal musculature for appropriate function during predetermined ranges of motion. The electrodes are placed such that the two actives are adjacent in a vertical orientation over the transverse ridge. The reference electrode is placed either directly below or above the actives, whichever is convenient. It should be noted that although the placement of the reference electrode is not critical, but the left and right side placements should be symmetrical to ensure consistency between examinations. The left and right side active electrode placements, on the other hand, must be identical to each other to ensure that the same muscle groups are being read on either side of the spine.

Once the electrodes are in place, the patient is positioned into the proper posture (sitting for a cervical placement, or standing for a thoracic or lumbar placement). It is important to carefully instruct the patient at this point as to what is to be expected in preforming the desired range of motion.

The computer will yield a bilateral graph which represents activations which occurred during the movement. It is important to be able to ascertain, via event markers on the graph, the exact time at which certain components of the motion began and ceased. In our clinic, each range of motion is marked for the following: when the motion began, when the full range had been achieved, when the patient began the return to the neutral posture, and when the motion was completed.

## INTERPRETING DYNAMIC MUSCLE ASSESSMENT

The interpretation of the dynamic range of motion study is broken down into symmetrical and synergistic motions.

### Symmetrical

Symmetrical paraspinal activations may be achieved via many motions such as flexion, extension, sitting, standing, kneeling, raising the shoulders, squatting, etc.. The most important aspect of a symmetrical motion is the pattern of activation. One would expect the right and left paraspinal muscles to behave almost identically. A bilateral excessive activation may indicate that the motion caused pain, or that it was difficult to preform due to anatomical abnormalities such as in a fixation or subluxation of the spine. An increased activation which is contralateral to the site of pain, is considered protective muscle guarding. An ipsilateral activation on the other hand would demonstrate muscle splinting due to pain on that side.

Two factors which must be taken into account in dynamic interpretation is handedness and scoliosis. The patient's side of handedness may demonstrate mild increases of activation over its counterpart due to overdevelopment of the muscles of that side, particularly on sites of the upper thoracic and cervical region. It is interesting to note that the opposite is true of the static muscle assessment. That is to say that the side opposite handedness will show mild increases of activation in the static assessment. This is probably due to a postural accommodation on the part of the weaker muscles. Unilateral increases of activation have also been attributed to muscles which lie on the convex aspect of a scoliosis. This is due in large part to chronic postural accommodation.

Synergistic

Synergistic ranges of motion for the paraspinal area include rotation and lateral bending. For an individual, the patterns of activation which are represented on the bilateral graph, should be similar to one another. For example, the graph of right lateral bending should appear similar to that of left lateral bending during the same session. Of course the right and left aspects of the representative graphs will be opposite, but the patterns should be quite similar. Any significant deviation of associated patterns suggests an aberrancy.

## GENERAL INTERPRETATION

When interpreting surface EMG, the practitioner must first consider what would be "normal" for their patient. There are several aberrancies which will alter range of motion findings. One such finding is that of a lateral spinal curvature. The practitioner considers how that curvature will affect posture, and then postulates how paraspinal muscular activity would be modified. For example, a left rotary lumbar curvature will cause the patient's center of gravity (for the lumbar region) to be moved slightly to the right. This will yield greater activations of the left paraspinal musculature in the lumbar region to maintain a standing posture. One would expect, therefore, greater activity on the left in this case whenever the muscles of the area are called upon to work.

In general, increased activations which occur on the side contralateral to pain, and which are hyperactive when they are called upon to work, may be considered activations attributable to protective muscle guarding. Muscle splinting is suspected when hyperactivations which are either ipsilateral or bilateral occur when the muscle is called upon to work. An irritable muscle group will cause hyperactivations during work or stretch and will be either contralateral or ipsilateral in nature. Another aspect of the irritable muscle group is that, generally, muscular rumination will occur. That is to say that the muscle may be relatively quiet on baseline, however, when challenged, it will become hyperactive and will not return to baseline, nor stabilize when rested. Another consideration is the muscle group which is hyperactive upon stretch, either contralateral or ipsilateral to the site of pain. This incident may be considered due to a chronically shortened muscle group.

The following chart (Table 1) represents the aforementioned considerations, however, the reader is reminded that it is to be used as a guide, for dynamic interpretation, and not as an absolute "diagnostic cook book".

It has been postulated, with respect to the vertebral subluxation, that higher surface EMG activations will be seen on the side opposite spinous rotation. Although more study is required in this area, it should be remembered that the activations of the small intrinsic muscles of the spine which are responsible for this activity are greatly overshadowed by the larger postural muscles. It is therefore suggested that interpretation be restricted to identifying the level of involvement as opposed to the level and vertebral orientation.

It should also be understood that a vertebral subluxation is more of a functional aberrancy than a structural one. It is for this reason that X-ray findings do not always correlate with surface EMG static muscle assessment findings. In other words, a functional subluxation which may not be apparent on X-ray, may indeed cause paraspinal muscle imbalances which are demonstrated with EMG findings.

**Table 1.** Guidelines For Interpretation of Dynamic Evaluation.

| *Muscle behavior type* | *Contralateral activation* | *Ipsilateral activation* | *Bilateral activation* | *Work* | *Stretch* | *Rumination* |
|---|---|---|---|---|---|---|
| Protective muscle guarding | X | | | X | | |
| Muscle splinting | | X | X | X | | |
| Irritable muscle | X | X | | X | X | X |
| Chronically shortened | X | X | | | X | |

## EXAMPLES OF DYNAMIC ASSESSMENTS

Figures 2a and 2b illustrate a fairly normal dynamic muscle assessment. This is a scan of the neck undergoing flexion, extension, left lateral bending, right lateral bending, left rotation, right rotation, patient stand, patient sit; and then flexion and extension are completed again for comparison. In flexion and extension we see similar activity bilaterally. Lateral flexion and rotation demonstrate ipsilateral increases in activation to pull the head to that side. Symmetrical motions are again achieved via standing, sitting, flexion, and extension. The important aspect here is that the left and right patterns of activation with respect to unilateral motions are similar. This will be discussed in greater detail later in this chapter.

Figure 1 illustrates the dynamic muscle assessment of a patient who complains of left upper neck pain, left upper back pain, and chronic headache. Orthopedic and X-ray findings were largely negative. The muscle assessment was taken with the electrode placement at the C-7 level paraspinally. The patient was directed through the following movements of the cervical spine: flexion, extension, left lateral bending, right lateral bending, left rotation, and right rotation. The solid line denotes the left side, and the dashed line represents the right side.

There is a hyperactivation noted when the patient returns from flexion on the right (working muscle). The right is again hyperactive in return from extension (working for increased stabilization). In both left and right lateral bending we see first ipsilateral activation followed by contralateral activation to right the head; however, in both cases, the right side is hyperactive while the muscle on the right is working. Left rotation demonstrates a strong co-contraction (working right muscle for increased stabilization). Right rotation shows a hyperactivation on the right.

This dynamic muscle assessment clearly demonstrates chronic protective muscle guarding on the right side. Treatment considerations in this case were directed to manual spinal manipulation (chiropractic spinal adjustment), and to strengthening the left side with rehabilitative exercises.

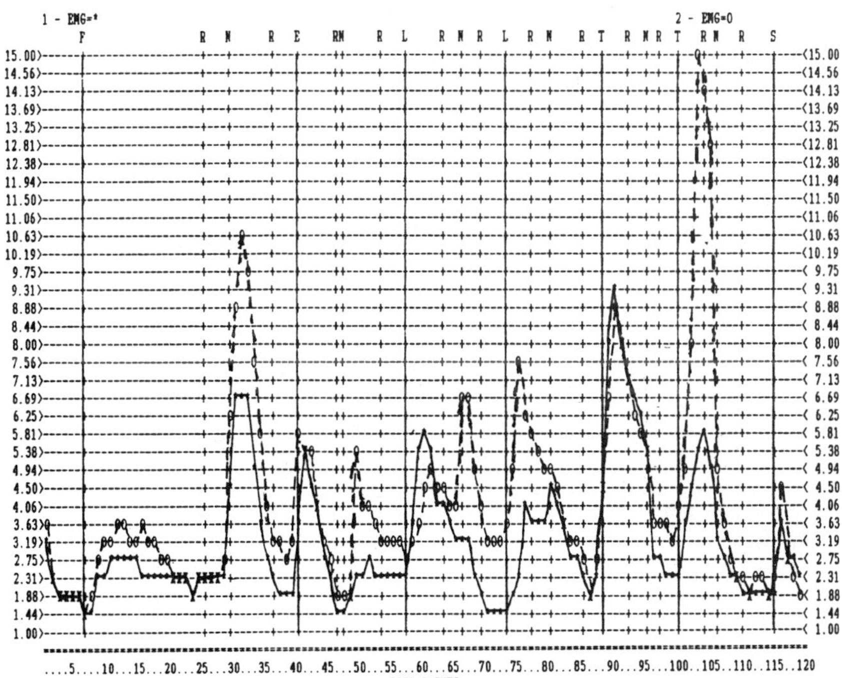


**Figure 1.** Dynamic Muscle Assessment: Abnormal Findings

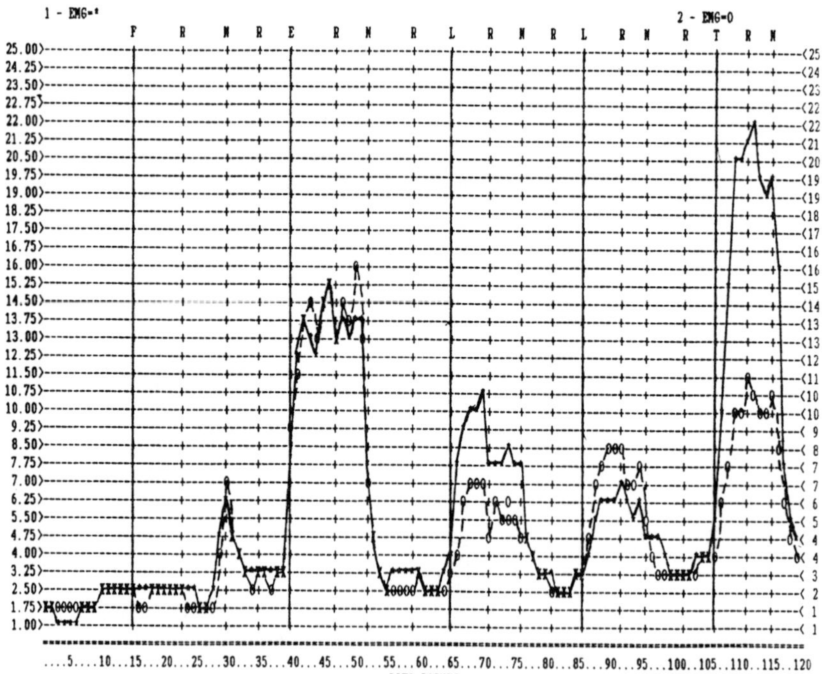


Figure 2a. Normal Dynamic Muscle Assessment

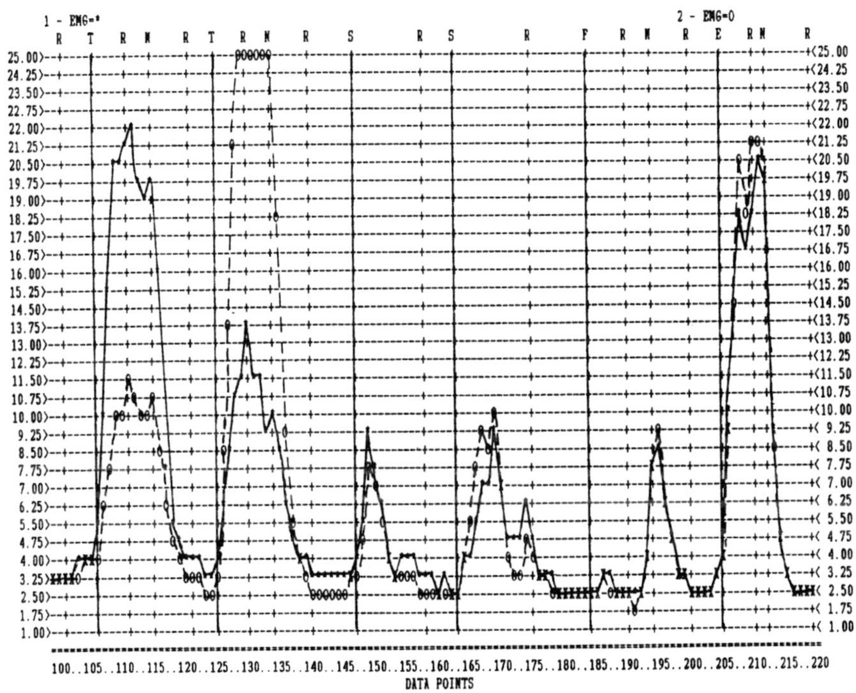


**Figure 2b.** Normal Dynamic Muscle Assessment (continued)

## STATIC ASSESSMENT OF NON-NEUTRAL POSTURES

Another type of "range of motion" study which is currently used is the static assessment of non-neutral postures. This is inferior to the dynamic study in that there is a great deal of muscular accommodation which takes place as a patient holds a particular posture for any length of time. For example, as a patient holds the position of left lateral bending, there will be a myriad of agonist and antagonist exchanges among the affected muscle groups as the muscles "take turns" at holding the posture. Also, the patient will have a tendency to sway slightly from position which will affect the reading obtained. More studies need to be made in order to quantify the above mentioned effects. It is not this authors intention to discredit alternative postural studies, but rather to caution the reader as to the potential artifacts which may be present. One thing remains clear, that the dynamic range of motion study is superior to the static in that the patient is allowed a natural movement; which is what the practitioner should be interested in. In the event that your equipment is not capable of bilateral scanning and monitoring, then the static alternative posture study will need to suffice.

## TREATMENT CONSIDERATIONS

The first consideration in the modification of treatment protocol based upon a surface EMG finding is that of the manipulative effort. Increased activations of EMG paraspinally often times indicate vertebral intersegmental dysfunction resulting in hyperactive musculature. The chiropractor, through palpation techniques of the area in question, determines how to correct the malposition resulting in decreased muscular activity, and increased stabilization.

Additionally, when a muscle is suspected of being chronically shortened, as may be the case in residual protective muscle guarding, a spray and stretch technique should be employed. T.E.N.S. therapy and electrical stimulation techniques are also indicated in cases where muscular rehabilitation and strengthening are required. Irritable muscle groups respond well to trigger point therapy, either manual and/or via electrical stimulation.

It is interesting to note that in the treatment of spinal-related conditions we have observed some patterns developing which seem to demonstrate results of the rehabilitative effort. That is to say that generally, the lower side of activation will increase first, and then the paraspinal muscle groups will begin to decrease in activations in a bilateral fashion as the condition improves. This phenomenon is merely an observation at this point, and requires further study in an attempt to determine the mechanism which propagates it.

The following chart (Table 2) summarizes the adjunctive treatment considerations which have been employed at our office with a good degree of success. It should be understood that the primary mode of treatment in all of our cases is spinal manipulation.

**Table 2.** Adjunctive Treatments For Diagnostic Findings

| *Muscle behavior type* | *Stretch high side* | *Strengthen low side* | *EMG biofeedback* | *Etiology treatment* |
|---|---|---|---|---|
| Protective muscle guarding | | 2 | | 1 |
| Muscle splinting | 1 | | | |
| Irritable muscle | | | 1 | |
| Chronically shortened muscle | 2 | 1 | | |

## CASE EXAMPLE #1

### History

The patient is a 38 year old female who suffered whip-lash type injuries as the result of a rear-end collision. Chief complaints are generalized neck pain and bilateral brachial paresthesia. Physical examination was largely non-contributory. Range of motion of the cervical spine was slightly restricted in all motions.

Initial surface EMG findings revealed the following:

**Table 3.** Initial Surface EMG Findings

| | Stand | | Sit | | |
|---|---|---|---|---|---|
| | L | R | L | R | Increased activation codes: |
| C-5 | 2.1 | 3.3 | 3.9(1) | 5.6(3) | (1)=Mild |
| Trap | 7.4(1) | 0.7 | 1.9 | 0.8 | (2)=Moderate |
| T-1 | 2.3 | 1.9 | 2.7 | 2.4 | (3)=Severe |
| T-6 | 2.8 | 2.1 | 5.8(1) | 4.5 | |
| T-10 | 5.0 | 5.9 | 6.0(1) | 6.5(1) | |
| L-3 | 1.1 | 2.0 | 1.4 | 2.5 | |

Mild to severe increases in activation are noted in the sitting posture for the cervical, T-6, and T-10 sites.

Treatment

This patient was treated with specific spinal adjustments and ultrasound therapy over a period of three months (Table 4).

**Table 4.** Post EMG examination

| | Stand | | Sit | | |
|---|---|---|---|---|---|
| | L | R | L | R | Increased activation codes: |
| C-5 | 1.2 | 1.7 | 1.9 | 2.4 | (1)=Mild |
| Trap | 1.7 | 2.7 | 1.6 | 1.2 | (2)=Moderate |
| T-1 | 1.9 | 3.1 | 2.5 | 3.0 | (3)=Severe |
| T-6 | 1.7 | 1.9 | 7.3(1) | 3.5 | |
| T-10 | 2.0 | 4.8 | 1.5 | 1.7 | |
| L-3 | 0.8 | 0.9 | 0.7 | 0.7 | |

A mild increase in activation is seen at the T-6 site on the left in the seated posture. This is consistent with handedness. The patient is asymptomatic and was discharged from care.

**CASE EXAMPLE #2**

History:

34 year old female suffering from a whip-lash type injury as the result of an MVA. Chief complaint consisted of overall neck pain, shoulder pain, and left arm pain.

Orthopedic findings included a left cervical antalgic posture, limited cervical range of motion, pain produced in all ranges of motion, but markedly with left lateral flexion. Minor's sign was present and pain was produced upon cervical compression and distraction. Upper extremity DTRs were within normal limits. Static muscle assessment findings are given below.

**Table 5.** Static Muscle Assessment Findings

INITIAL:

| | Standing | | Sitting | |
|---|---|---|---|---|
| | L | R | L | R |
| C-2 | 3.3 | 2.5 | 5.9(2) | 4.4(1) |
| T-1 | 3.5 | 2.6 | 4.1(1) | 2.5 |

Increased activation codes:
(1)=Mild
(2)=Moderate
(3)=Severe

2 WEEKS LATER:

| | Standing | | Sitting | |
|---|---|---|---|---|
| | L | R | L | R |
| C-2 | 3.3 | 2.5 | 2.9 | 3.2 |
| T-1 | 2.5 | 2.0 | 3.4 | 4.8(1) |

Mild to moderate increases in activation are noted in the following sites: Left C-2, right C-2 sitting and left T1 sitting initially, and right T1 sitting 2 weeks later.

Dynamic muscle assessment findings:

INITIAL:

The patterns of activation demonstrate overall excessive activations on the left, particularly with left lateral bending and left rotation. This pattern is consistent with an irritable muscle on the left (Figure 3).

2 WEEKS LATER:

Symmetrical motions demonstrate increased activations on the right. There was a greatly diminished ipsilateral activation with left rotation. The overall activations were much less than the initial EMG examination (Figure 4).

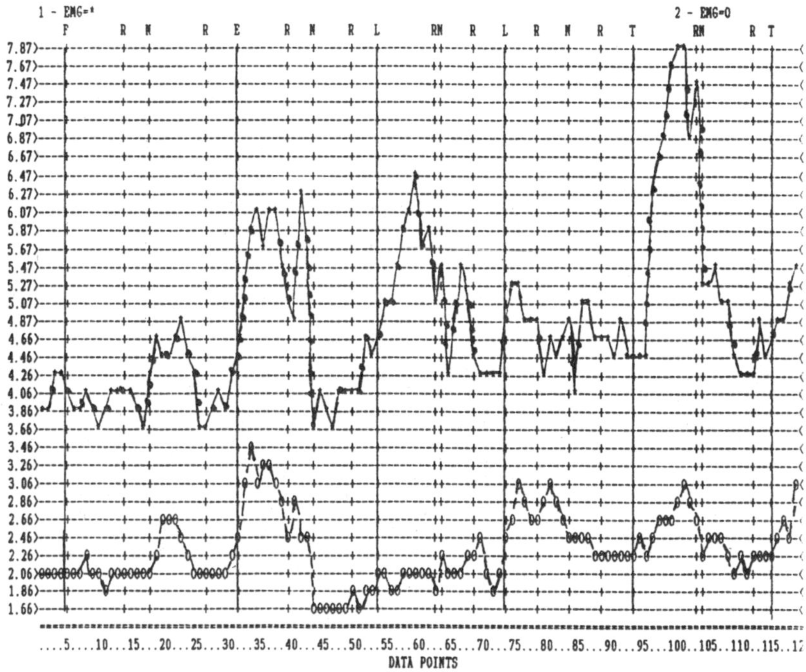


**Figure 3.** Irritable Muscle On Left

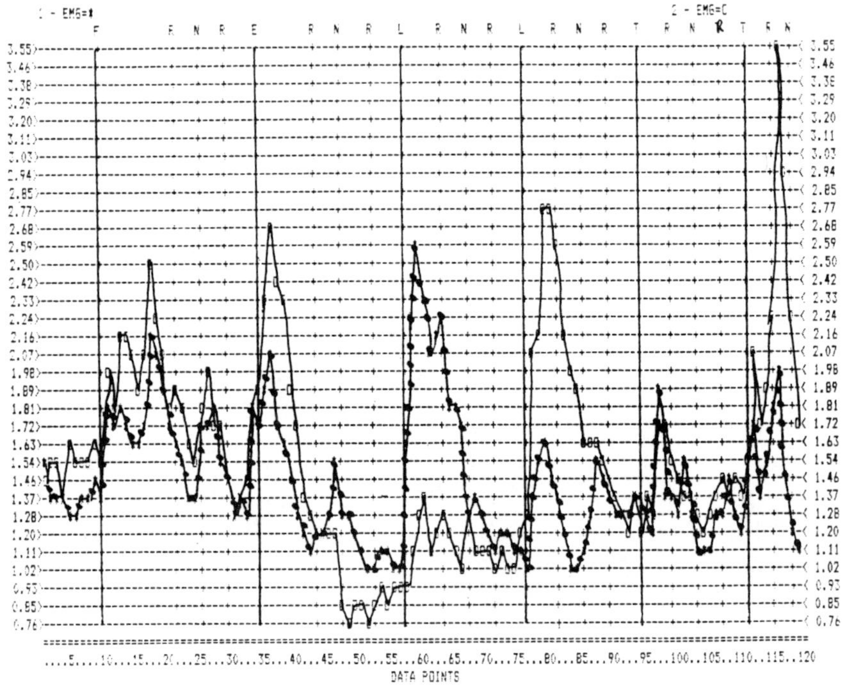


**Figure 4.** Asymmetrical Activation On Right

## Discussion

The EMG findings clearly indicate irritable muscles on the side of involvement initially. After two weeks of treatment the patient shows considerable improvement as represented by the decreased activations on the static muscle assessment and by the decreased activations and pattern improvement of the dynamic muscle assessment. The treatment utilized in this case consists of specific spinal manipulation and electrical stimulation.

## REFERENCES

Basmajian, John V.,M.D. and DeLuca, PhD. Muscles Alive: Their functions revealed by electromyography. Fifth ed., Williams & Wilkins, 1985.

Cram, Jeffrey R., Ph.D., Clinical EMG: Muscle scanning and diagnostic manual for surface recordings. Clinical Resources and J&J, 1986

Gray, Henry, F.R.S., et al., Gray's Anatomy. Twenty-ninth American edition, Lea & Febiger, 1973.

Guyton, Arthur C., M.D., Textbook of Medical Physiology, Sixth edition, W.B. Saunders Co., 1981.

Schwartz, Mark S., et al., Biofeedback: A practitioner's guide, The Guilford Press, 1987.

# CHAPTER 8

# *Myofascial Pain and the Temporomandibular Joint*

by

Leonard G. Hudzinski, Ph.D.
and
Gary Lawrence, B.A.

---

The term "temporomandibular disorder" refers to a constellation of conditions involving possible jaw dysfunction that range from muscular pain to displacement of the temporomandibular joint and its articular disc. Myofascial (myo or muscle) refers to tissue of the face. Myofascial pain associated with a temporomandibular joint (TMJ) disorder is generally referred to as a soft tissue injury or a muscle pain symptom of the connective temporomandibular joint. Facial pain is a common outcome of trauma such as results from acute over-stretching of muscle, a whiplash, muscle fatigue, oral surgery, or immobilization of muscle. The dysfunction can also occur in part or mainly as a result of emotional distress which frequently is a consequence of a muscle bracing oral habit, such as clenching and/or grinding.

Anatomically, the TMJs are found directly below and in front of the ears and work together with facial muscles to move the jaw. The TMJ is active in jaw movements responsible for talking and chewing. It is more complex than other joints because it moves up and down, back and forth and left and right.

The TMJ is moved by four different muscles (Table 1) of special interest to biofeedback research. Three of the muscles, the masseter muscles and the medial pterygoid move the jaw vertically. One muscle, the lateral pterygoid, moves the jaw laterally (See Figure 1). Activation of these muscles may be responsible for myofascial pain, neck pain and/or headache.

Numerous etiological factors involving intraoral, extraoral and psychophysiologic activity can participate in producing the complex TMJ myofascial pain symptoms reported by patients.

TABLE 1
MUSCLES ACTING ON
THE TEMPOROMANDIBULAR JOINT *

| | Muscle Function | | | | |
|---|---|---|---|---|---|
| Muscles | Opening | Closing | Protrusion | Retrustion | Later Shift |
| Muscles of Mastication | | | | | |
| Masseter | | + | + | ● | ● |
| Temporalis | | + | | + | ● (Ipsilateral) |
| Medial Pterygoid | | + | + | ● | ● |
| Lateral Pterygoid | + | | + | | + |

+ = major activity
● = minor activity
* From Hiatt, J.L., & Gartner, L.P. (1982). Muscles acting on the temporomandibular joint. Textbook of Head and Neck Anatomy (p. 223). New York, Appleton-Century-Crofts.

A number of theories have been proposed to explain the etiology of the TMJ disorder and their symptoms. Among the theories most widely held by dentists is the occlusal disharmony theory, which holds that poorly meeting teeth cause the mandible and its adjoining muscles to be responsible for forward bracing, muscle spasm and consequent pain (Thompson, 1951). More recently, psychophysiological theories have grown in popularity. These theories hold that in the absence of occlusal problems, psychological stress, habitual muscle contraction, or both can cause the muscles of the face, head and neck to be hyperactive (Laskin, 1969). These theories share the assumption that pain is the result of muscle hyperactivity.

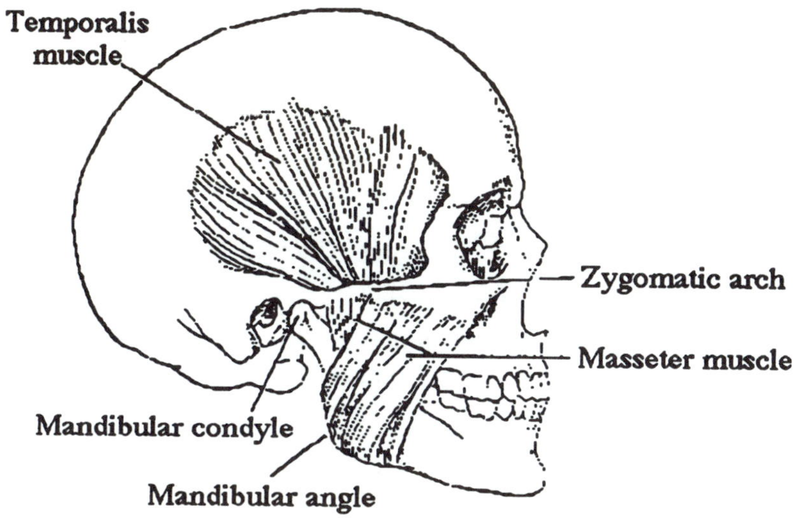


**FIGURE 1A.** TEMPORALIS AND MASSETER MUSCLES.

It is estimated that approximately 70 to 90% of patients with facial pain have reported either a primary or secondary muscle disorder; and that pain in these patients can be alleviated with biofeedback (Dohrman & Laskin, 1978). The majority of these patients are between 20 and 50 years of age, and most are women. Although symptoms are about equal in men and women, clinicians find that women are more apt to seek treatment (3:1).

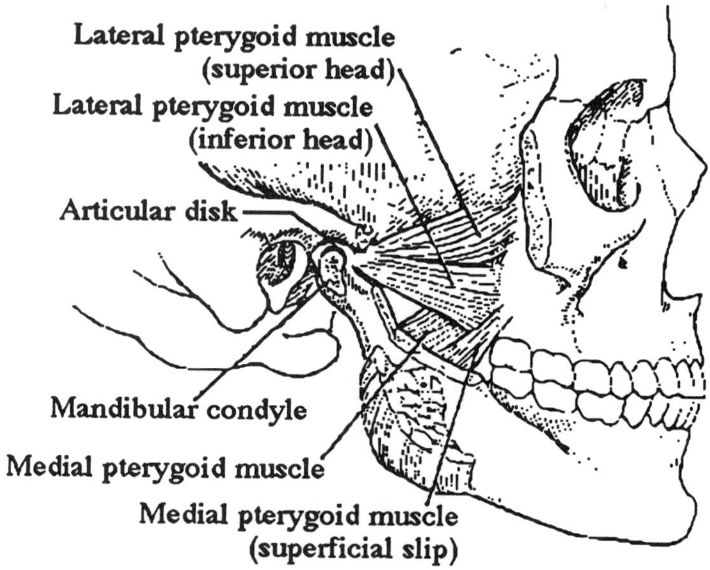


**FIGURE 1B.** DEEPER LAYER OF MUSCLES

The facial pain experienced most often by patients is described as a dull, deep ache that is usually constant and active for months, sometimes for years. It is often unilateral or one-sided. Often clicking or popping sounds are reported coming from the TMJ area. The pain is usually more intense on awakening. Mandibular movements are limited because patients have difficulty opening their mouth. The jaw often deviates to the effected and painful side of the face. The pain may radiate into the temple muscles and down into the sternocleidomastoid muscle and splenius capitis muscles as well as to the zygomatic process. Signs indicative of myofascial pain are: 1) Pain to the face, head and neck; 2) tenderness in the masseter muscle when palpated, often radiating to other areas; 3) inability to move the jaw properly; 4) popping or clicking sounds in the TMJ upon opening (however, only when present with points 1 and 2; and 5) accelerated tooth wear or bruxism.

Biofeedback is important in the treatment of TMJ dysfunction because muscle tightness and hyperactivity are reduced. Reducing muscle tightness and contraction generally reduces or eliminates the pain. When muscles surrounding the TMJ are contracted, they force adjoining facial structures into constricted positions, limiting range of motion, causing severe pain, and making coordinated movements difficult. Chronic contraction may result in muscle hypertrophy, reducing the joint 's range of motion; however, muscle contraction may be phasic. A spastic muscle may achieve normal tone at rest by contracting inappropriately during voluntary movement. After injury and/or a period of extreme pain, chronic immobility of the muscle can produce atrophy. Patients commonly experience muscle hypertrophy, muscle spasm, and muscle atrophy at the same time. Biofeedback treatment effectively relieves these muscle disorders and consequent pain.

A thorough history and examination of the patient are basic to biofeedback therapy. Collaboration of a biofeedback clinician and a dental TMJ specialist is encouraged. Remember that chronic facial pain symptoms are most effectively treated and relieved by a combination of physical therapy, medication, soft diet, muscle exercises, and biofeedback. TMJ surgery is generally the last measure used to correct a defect in the patient's TMJ; biofeedback has been found effective before and after surgery in reducing muscle tightness or spasm.

Since depression and secondary gain can be a consequence of chronic pain, psychological treatments, the use of anti-depressants, and/or psychotherapy may be warranted for patients who are unresponsive to conventional dental, biofeedback or behavioral therapy. Psychological treatments will be discussed in the Treatment section of this chapter.

## Evaluation

### CHIEF COMPLAINT:

After meeting the patient, elicit the history of the presenting problem. If there is a chief complaint, include a chronological history of the specific problem. Questions that should be asked are listed in Table 2. An index that is useful in documenting evaluation and treatment progress can be found in Figure 2.

**TABLE 2.** EXAMINATION

Location?
Onset?
Constant or intermittent symptoms?
Duration of episodes?
Course?
Pain (character, severity)?
Previous diagnoses?
Previous treatment and results?
Does chief complaint relate to anxieties?

**TABLE 3.** MUSCLE CONTRACTION RATING SCALE

A positive response to any of the following questions identified by a * indicates probable muscle tightness. Generally speaking, the greater the number of * items reported by the patient, the more severe the myofascial contraction disorder.

| | | | |
|---|---|---|---|
| * | ____Yes | ____No | Does it hurt when you open your mouth wide? |
| * | ____Yes | ____No | Does it hurt when you chew? |
| * | ____Yes | ____No | Do you have pain in or in front of the ear(s)? |
| * | ____Yes | ____No | Does your jaw feel tired after eating, after long conversations, or after laughing? |
| * | ____Yes | ____No | Do you have pain in the fact, neck, or temple region? |
| * | ____Yes | ____No | Do you experience chronic headaches? |
| * | ____Yes | ____No | Do you have pain or soreness in your gums or teeth? |
| * | ____Yes | ____No | Do you clench your teeth during the day? |
| * | ____Yes | ____No | Has anyone heard you grinding your teeth in your sleep? |
| * | ____Yes | ____No | Have you ever had chronic shoulder or back pain? |
| | ____Yes | ____No | Do you hear jaw clicking or popping sounds when you chew? |
| | ____Yes | ____No | Have you ever had nervous stomach, ulcers, or colitis? |
| | ____Yes | ____No | Have you ever had arthritis? |

## The Kinnie-Funt (K-F) Chief Complaint Visual Index for Head, Neck, and Facial Pain and TMJ Dysfunction

1. Please circle the number in front of the symptoms you regularly or occasionally have.

2. Indicate your main or chief complaints in order of their current importance.

(A). ______

(B). ______

(C). ______

3. Please draw areas of pain or distress on the picture below.

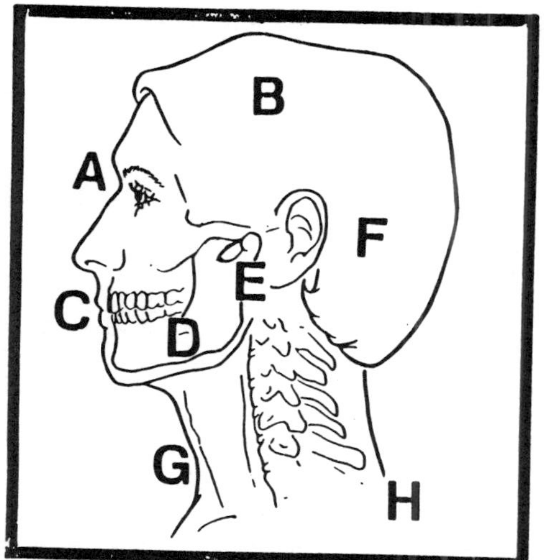




Name: ______

Age: ______ Date: ______

**A. *Eye Pain and Eye Orbital Problems:***
1. Eye (orbital) pain; above, below, behind.
2. Bloodshot eyes (hyperemia)
3. Blurring of vision
4. Bulging appearance (exophthalmia)
5. Pressure behind the eyes (retro-orbital pressure)
6. Light sensitivity (photo-phobia)
7. Watering of the eyes (lacrimation)
8. Drooping of the eye lid (ptosis)

**B. *Head Pain, Headache Problems, Facial Pain:***
1. Forehead (frontal)
2. Temples (temporal)
3. "Migraine" type headache
4. "Cluster" headache
5. Maxillary sinus headache (under the eyes)
6. Posterior back of head headaches with or without shooting pains (occipital headache)
7. Hair and or scalp painful to touch (parietal headache)

**C. *Mouth, Face, Cheek, and Chin Problems:***
1. Discomfort
2. Limited opening
3. Inability to open smoothly, evenly
4. Jaw deviates to one side when opening
5. Inability to "find bite"

**D. *Teeth and Gum Problems:***
1. Clenching, grinding at night (bruxism)
2. Looseness and or soreness of back teeth.
3. Tooth pain (toothache)

**E. *Jaw and Jaw Joint (TMJ) Problems:***
1. Clicking, popping jaw joints
2. Grating sounds (crepitus)
3. Jaw locking opened or closed
4. Pain in cheek muscles
5. Uncontrollable jaw, tongue movements

**F. *Pain, Ear Problems, and Postural Imbalances:***
1. Hissing, buzzing, ringing, or roaring sounds (tinitus)
2. Diminished hearing (subjective hearing loss)
3. Ear pain without infection (otalgia)
4. Clogged, stuffy, "itchy" ears, feeling of fullness
5. Balance problems, "vertigo" (disequilibrium)

**G. *Throat Problems:***
1. Swallowing difficulties
2. Tightness of throat
3. Sore throat without infection (coryza)
4. Voice fluctuations
5. Laryngitis
6. Frequent coughing or constant clearing of throat
7. Feeling of foreign object in throat
8. Tongue pain (glossalgia)
9. Salivation (intense)
10. Pain in the hard palate (posterior areas)

**H. *Neck and Shoulder Problems:***
1. Lack of mobility-reduced range of movement
2. Stiffness
3. Neck pain
4. Tired, sore, neck muscles
5. Shoulder aches
6. Back pain upper and lower
7. Arm and finger tingling, numbness and or pain

A visual clinical index correlated from the most frequently seen symptoms documented in head, neck and facial pain patients by Dr. Bruce H. Kinnie (Columbia, South Carolina) and Dr. Lawrence A. Funt (Bethesda, Maryland)

**Figure 2. THE KINNIE-FUNT VISUAL INDEX**

In the clinical biofeedback evaluation, the TMJ can be examined carefully, preferably with a stethoscope. Listen directly to the TMJ as the patient opens and closes the mouth. Abnormal sounds, including clicking, popping, and/or grinding noises, should be noted. The presence of these symptoms alone indicates disc TMJ dislocation requiring dental referral and attention. A muscle contraction rating scale is provided in Table 3.

## MYOFASCIAL EXAM:

A careful observation of the muscles of the face, jaw, head and neck for the presence of muscle tension or contraction is important. The patient's facial symmetry should be noted. The position of the head relative to the shoulders and of the shoulders relative to the thorax should be noted. Examine the degree of head flexion, extension, and rotation when determining areas of muscle tenderness through palpation. Listen not only to the patient's verbal responses but take into account all bodily responses indicating pain during the clinical examination, and record these findings.

Ask the patient to open the mouth as wide as possible. Measure the opening from the lower to the upper teeth using a small ruler. An opening less than 40mm is particularly suggestive of a muscle problem.

Have the patient move their jaw to the left, to the right, and forward in the same motion, keeping the teeth in light contact. Measure the range of movement. Bilateral movement of less than 8mm generally indicates dysfunction. Note measurement and any reports of pain made by the patient during the examination.

Electromyographic (EMG) biofeedback has as one objective accurately measuring electrical activity in contracted muscles and providing the patient with this information through visual or auditory means. Budzynski and Stoyva (1973) showed that individuals could decrease activity in the masseter muscles when utilizing EMG biofeedback. Over the past two decades, a number of investigators have reported the treatment of myofascial pain, including John Rugh (Katz & Rugh, 1986), a pioneer in dental/biofeedback therapy and research. In the recent past, Cram and Steger (1983) departed from traditional methods of muscle monitoring and began to utilize EMG scanning as a diagnostic tool for rapid and accurate measures of muscle areas including the face, head and neck. Schneider and Wilson (1985) have contributed to advances in myofascial pain monitoring and therapy using conventional one-channel EMG equipment. Hudzinski and Lawrence (1988) concluded that the traditionally used bifrontal placement of electrodes provided no meaningful diagnostic information about patients with facial and head pain, when compared with either the Cram or Schwartz-Mayo methods of electrode placement. They concluded that the frontalis muscle is not, for the most part, a reliable indicator of EMG activity in other skeletal tissue.

## MYOFASCIAL EMG ASSESSMENT:

The following procedures are designed to help establish the diagnosis of muscle contraction dysfunction and thereby direct the clinician to a treatment strategy. These procedures employ the use of a dual-channel EMG unit; however, a single-channel unit may be used by independently monitoring each side of the face.

It is recommended that the following procedures be undertaken 24 hours after temporary termination of tranquilizer and pain medication. If the patient is wearing a removable splint (oral appliance) it is recommended that it be removed during assessment and treatment.

In each of the following procedures, compare muscle function and be aware of spontaneous muscle activity at rest and during stretch reflex or spasm, as well as muscle recovery time. Throughout the assessment, observe whether or not the patient is able to relax the masseteric muscles.

## BASELINE:

Place the electrodes as shown in Fig. 3 as illustrated by Schneider and Wilson in their text (1985). See Table 4 for TMJ electrode placement normative data. Position the patient in a straight-backed chair with the hands in the lap and the feet flat on the floor. Instruct the patient to sit quietly with eyes open and to avoid moving. Ask the patient to relax the facial muscles, keeping the lips closed (the teeth need not touch); record average EMG activity levels for a period of approximately 3 to 5 min. Document the average EMG on the left side and the right side at the narrow and wide filter settings.

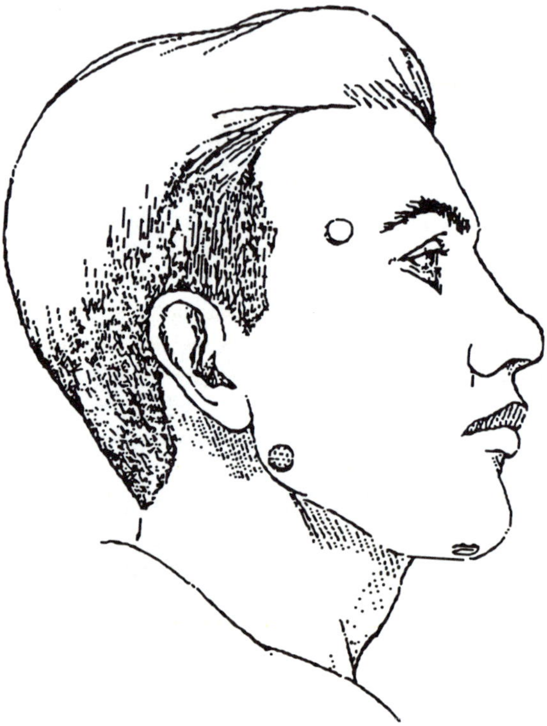

**Figure 3.** ELECTRODE PLACEMENT STRATEGY FOR TMJ DISORDERS (FROM SCHNEIDER AND WILSON, 1985)

**CLENCH:**

Measure muscle strength. Ask the patient to contract the masseters equally. Weakness will be reflected in EMG data. Observe muscle recovery patterns on the left and right sides of the face.

**STANDING:**

Assess the EMG level while the patient stands with the feet at shoulder width and hands at the sides, with the weight equally distributed. Instruct the patient to relax as much as possible. Record the EMG level for a period of approximately 3 to 5 minutes while the patient stands with lips closed (the teeth need not touch).

**TALKING (NON-STRESSOR)**

While the patient is seated, assess facial muscle activity during and immediately after the discussion about a non-emotional issue, such as activities that can be used to promote relaxation.

**TALKING (PERSONAL STRESSOR)**

Compare the non-stress data to muscle activity measured after 3 to 5 minutes of discussion of an emotional issue, such as the patient's facial pain and its impact on daily functioning. Monitor the patient's muscle levels and note differences between emotional and non-emotional material. Note bursts of activity and verbal stimuli accompanying EMG data.

**MENTAL MATH STRESSOR**

While the patient is seated, assess muscle activity during and immediately after a stressful task. Ask the patient to perform a serial seven subtraction test beginning with the number 100 and counting backwards by 7's. Have the patient stop after seven subtractions (93, 86, 79, 72, 65, 58, 51). Note whether the patient overreacts with elevated muscle activity. Equally important, observe whether the patient has a relatively normal response to stress but fails to resume pre-stress baseline levels after testing.

Assess the muscles of mastication for symptoms of pain or weakness. Stressing the main opening muscles of mastication is an invaluable test for determining whether symptoms arise from that area. If muscle problems are present, pain will be elicited during the examination. Testing the muscles for pain or weakness involves two procedures that can be employed while the patient is seated: 1) assessment of resistive opening to identify dysfunction in the lateral pterygoid muscles (opening muscles), and 2) assessment of resistive closing to assess problems associated with the masseter, temporal and medial pterygoid muscles (closing muscles).

To assess for symptoms in the muscles that open the jaw, have the patient open the mouth 1 to 2 cm and resist a force applied gradually against the chin. The patient's back should be supported by the clinician's hand in order to prevent backward movement during testing. To evaluate symptoms in closing muscles, instruct the patient to resist an attempt to forcibly open the mouth from an initial 1 to 2 cm opening. The clinician should apply opening force to the lower anterior teeth. When these methods are used, avoid hand contact with the TMJ or masseter to prevent confusion between artifact, symptoms and data that arise from muscle evaluation. Observe the EMG readings for bursts of activity and average EMG levels when force is applied.

## TREATMENT

Facial pain is an anxiety-producing experience for most patients. Therapy begins from the moment of the examination when the patient's worries and fears are addressed. It is important to explain to the patient that the pain is not unique and explain the relationship between stress, muscular hyperactivity and consequent pain.

Stress, muscle contraction and oral habits are the major factors that interact to cause most myofascial pain. Although stress is a major factor causing muscle contraction, the clinician may occasionally have to direct therapy toward breaking specific habits because these can maintain myofascial contraction and consequent pain. Biofeedback therapy focuses mainly on the muscles moving the TMJ, especially the masseter muscle, and emphasizes relaxation of all facial muscles. The literature supports the use of daytime (diurnal) feedback, as described in this chapter; however, nighttime (nocturnal) feedback can play an important role in the speed and quality of improvement. While the primary goal of diurnal feedback is to teach the patient to either increase or decrease facial muscle activity on demand, nocturnal biofeedback is usually used as a treatment adjunct to help patients suppress and control parafunctional oral habits such as clenching, grinding or gritting of teeth during sleep.

Begin treatment with a basic explanation of the relationships between muscle contraction, stress and perceived facial pain. In brief, stress may cause muscles to contract, and prolonged or frequent contraction of muscle causes and may sustain facial pain. The explanation is critical to treatment and it can be especially helpful for use in the patient's management of stress. Explaining how biofeedback and its equipment can reduce heightened muscle activity provides an understanding of pain and an improved sense of control, which in turn helps to relieve stress, anxiety and depression.

The clinician can anticipate a reduction in patients' stress and depression when a therapeutic relationship is cultivated based on verbal feedback and biofeedback. Elevated anxiety reported by patients may be a result of long-term pain and not a cause of pain. The same is true with depression. A depressed person commonly over-reacts to pain; however, it is just as likely that chronic pain leads to depression. Awareness of psychological issues affecting the patient is fundamental to successful therapy, and these issues can be identified in several aspects of myofascial pain. First, stress-related muscle hyperactivity and parafunctional oral habits can cause pain. Second, psychological states (i.e., depression) may interfere with a patient's response to therapy.

Finally, psychological factors have been used to explain why certain persons experience symptoms at an extreme, heightened level.

The biofeedback clinician is encouraged to use psychological consultation, particularly with patients experiencing or reporting moderate or severe depression. Persons who report or project helplessness, hopelessness or isolation may require psychological and/or psychiatric testing and consultation even though they may be making progress in biofeedback therapy.

In the first treatment session, acquaint the patient with the EMG equipment, review the myofascial evaluation data, and explain the treatment goals of relaxing the muscles that move the TM joint. The patient receives training in muscle awareness and receives biofeedback while being instructed to decrease muscle tightness, tension, or both. Training the patient to relax the muscles of mastication has been shown to be an effective treatment in reducing TMJ-myofascial pain.

After cleansing the skin and applying electrodes, as described in the evaluation section, treatment and initial training in facial muscle control are begun. Direct the patient's attention to the biofeedback equipment displays, and explain that EMG activity will increase with elevated muscle contraction and decrease with relaxation of the musculature.

Instruct the patient to lightly clench the teeth, and observe the increase in microvolt activity. Next ask that the subject not clench the teeth so hard, and again observe the level of contraction.

In initial treatment sessions, work toward the goal of facial relaxation and discrimination of muscle contraction. Ask the patient to reach and hold a specific level of muscle tension in the masseters ( i.e., 20% above the patient's average evaluation baseline level) and maintain that amount for a period of one minute.

Ask the patient to relax the facial muscle and to be aware of the sensations in the muscles of the face, head and neck. The aim in each session is to help develop awareness of muscle sensations and reduce contraction-tightness when there is a perceived increase in that activity. Ask the patient to decrease muscle tension while observing the EMG display for one minute. If the patient is successful, use verbal praise as positive reinforcement.

The patient is given a 30 second rest (artifact elimination period) in which swallowing, change of position, etc. are encouraged. Ten trials of controlling contraction in the myofascial muscles are given during each session (Table 5). Give verbal reinforcement only when the patient lowers muscle potential. If EMG activity is significantly lowered, reinforce the behavior correspondingly. If EMG activity is only slightly decreased, reward the patient comparably with mild verbal reinforcement and shaping techniques. (Shaping is the gradual development of a correct response by means of small initial attempts, the first tentative steps taken in the right direction toward a particular goal). Using the EMG threshold setting and reinforcement, a particular behavior may be shaped to a desired microvolt level.

Devote a portion of each treatment session to correcting the patient's dysponetic myofascial patterns as documented through the myofascial evaluation. Behaviors which evolve into inappropriate muscle bracing are often referred to as being dysponetic -- dys meaning faulty and ponos meaning energy or effort. According to Whatmore and Kohli (1983), they are behaviors that result from misdirected efforts. Whatmore observed in coining the term dysponesis, that whenever patients were involved in cognitive or physical activity they produced inappropriate bracing of muscle. He stressed that patients should become aware of their misdirected efforts, learn how to correct these behaviors, and learn to lower their overall tension levels. In a 6 year follow-up study, Whatmore and Kohli found that long-term clinical success correlated highly with the patient's skill in controlling minimal changes in muscle tension at the end of treatment.

**TABLE 5.** TEN 30-SECOND TREATMENT TRIALS

| Trials | Lowest μV | Average μV |
|---|---|---|
| 1 | ________ | ________ |
| 2 | ________ | ________ |
| 3 | ________ | ________ |
| 4 | ________ | ________ |
| 5 | ________ | ________ |
| 6 | ________ | ________ |
| 7 | ________ | ________ |
| 8 | ________ | ________ |
| 9 | ________ | ________ |
| 10 | ________ | ________ |

μV = microvolt

By definition, a dysponetic behavior is one that has a detrimental influence on a person. Since most of these behaviors are unnoticed, they usually are classified as being covert. These behaviors are similar in composition to many psychophysiologic disorders that are a consequence of the interactive process between the brain's sensory and neuromuscular systems.

Dysponetic myofascial patterns can include clenching, grinding, gritting, gnashing, and clicking of teeth. They may also include the contact of teeth during chewing of the cheeks or lips, or chewing of pencils, pipes, or even gum. Dysponetic patterns might also include less obvious activity, such as imbalanced breathing, or posture or contraction of muscles in the neck, shoulder, or thorax, all of which are diagnosable and treatable through EMG scanning methods or multi-measurement procedures.

Examples of dysponetic patterns may be helpful. Many patients with TMJ dysfunction experience gradual and increasing muscle asymmetries which result in pain. In an attempt to maintain a pain free state or to reduce pain, the patient may alter the position of the mandible very slightly. Although this may initially relieve pain, it often leads to soreness in other muscles of the face and jaw because of the strain and pulling of muscle, particularly on the opposite side of the face. Muscles not only become sore because they are contracted; they also shorten. This is usually accompanied by tight masseter, temporalis, and pterygoid muscles. These muscle imbalances are frequently responsible for soft tissue pain where muscle strain is experienced in areas of the neck, head, back, and even chest. In another example, chronic masseter contraction may lead a patient experiencing facial pain to clench the teeth, in order to create an analgesic effect that frequently occurs when pressure is applied to a painful area. This inappropriate clenching and bracing when increased in frequency which is repetitious, usually becomes unconscious and no longer is analgesic but instead causes and amplifies pain.

Group activities frequently cause social anxiety and dysponetic muscle bracing. Using biofeedback, patients with facial pain can learn how to keep the masseter, the temples, and the neck muscles relaxed while chewing, eating, or attending anxiety-provoking social gatherings.

Estimates of these dysfunctional oral bruxing behaviors vary widely. Estimates of their incidence among the general adult population range from as low as 5.1% to as high as 21%. Among dental patients, the reported incidence is 20%.

Clinicians may find some of the following observations, including those in Tables 6 and 7, useful in diagnosing and treating patients with dysponetic oral habits. Bruxers, whether they clench or gnash their teeth, tend to develop typical symptoms: 1) protruding masseter muscles; enlarged, over-developed masseter muscles commonly ripple and spasm and are visible to the clinical eye; 2) grinding, gritting, or visible wear of the teeth; 3) complaints of toothache pain. Patients who complain of pain similar to toothache may be engaged in bruxing. If a patient frequently clenches on a slightly elevated tooth or on a malocclusion, then expect symptoms resembling toothache; 4) chronic fatigue or ache in the jaw. This is reported most frequently on waking and is often a sign of nocturnal bruxing. These patients also may report tenderness or tension headache in the face, head, and neck muscles.

**TABLE 6.** HABITS TO AVOID

1. Yawning widely, testing the joint, or "playing with" the jaw
2. Chewing gum or tough, crunchy, or hard foods
3. Resting the chin and/or cheek in the palm of the hand when sitting or lying on one side
4. Eating, reading, or watching T.V. in bed
5. Clenching the teeth -- when tense, jogging, etc.
6. Licking the teeth with the tongue
7. Closing the lips to hide braces, splint, malocclusion, missing teeth, etc.
8. Nail biting
9. Lip biting or licking
10. Cheek biting or sucking

Although these signs and symptoms are commonly observed, they must be viewed with care because these very symptoms could be a result of other medical or dental conditions. Therefore, the clinician must consider these observations thoroughly, consulting with a dental specialist.

It is important to correct the bruxing behavior and/or dysponetic pattern. This is accomplished by placing surface electrodes on muscles where inappropriate function is reported or observed. With electrodes applied, the clinician should monitor oral activity, speech and movement patterns, the ways in which posture and breathing affect EMG data, and perhaps even how these patterns can be altered to reduce muscle contraction and tension. It may be necessary to work with a patient to change jaw posture, body posture, breathing, even factors in the environment such as how the patient's work posture affects muscle activity.

**TABLE 7. HELPFUL HINTS FOR TMJ PATIENTS**

1. Avoid protruding your jaw -- for example, when smoking, during conversation, etc.
2. Make every effort not to strain the ligaments of the jaw unnecessarily.
3. Do not bite food with your front teeth.
4. If you clench your teeth, try to remember to keep your lips together and your teeth slightly apart.
5. Avoid chewing gum.
6. Do not eat popcorn, crusts of bread, tough meat, or any food that will require prolonged chewing.
7. Cut all foods into small pieces; avoid opening the mouth wide.
8. Avoid pressure on your jaw during sleep. Sleep on your back if possible.

Before any set of behaviors can be altered, they must be recognized. There is no objective way to convey to a patient by words alone what is being observed. The patient learns mainly by direct experience. The electromyogram can be employed for teaching purposes to objectify unnoticed muscular efforts and bring them to a patient as an observable event. The EMG can be used so that the clinician, or both clinician and patient, can be observers. Once a person has learned to increase or decrease a behavior, it becomes increasingly controllable.

Once the patient has learned to recognize and control a given set of behaviors as part of biofeedback therapy, the issue of generalizing the learned behavior into daily activities will become a basic clinical goal. We will review generalization and transfer of skills separately in the coming paragraphs.

For clinicians with single-channel units, we encourage applying biotrode adhesive electrodes to anatomical areas of interest using alligator clip leads, and monitoring the electrode site by moving the clip after 5 to 10 seconds of data collection at each monitoring site.

Multiple-channel EMG units are particularly useful in diagnosing dysponetic patterns. In diagnosis, it is wise to monitor the patient's bracing efforts during stressful and nonstressful conversations. When a patient is not using bracing responses, the clinician should provide verbal feedback and encourage greater control. In some instances, the clinician might encourage the patient to observe feedback directly from instrumentation during a stressful task; at other times, the clinician may offer verbal feedback when nondesirable patterns are observed. After developing an awareness of bracing, the patient will progress to recognition and regulation of the specific muscle bracing activities involved. For instance, patients may learn that they brace certain parts of their musculature when they feel insecure. They can also learn that by not bracing at these times they will likely cope more effectively with the causes of their insecurity.

Patients should become aware of their maladaptive patterns and be skilled in their ability to cope with those patterns. Patients can learn to become aware of maladaptive bracing, and when it is detected, to bring the behavior into the normal, relaxed range. The myofascial pain sufferer is taught to be aware of slight increases in muscle contraction in the masseters, neck, temples, and face. Eventually, the patient can learn to relax away mild tension even in difficult circumstances.

The terminal treatment goal, usually introduced in concluding sessions, is to enable the patient to quickly and accurately control myofascial activity and to apply these experiences outside of the treatment setting. Without biofeedback, the patient is asked to tense the muscles to a specified mild to moderate microvolt level. When the patient believes that this level has been attained, feedback is provided and the actual microvolt level shown. This procedure of discrimination training is generally repeated six to ten times during a session (Table 8).

**TABLE 8.** CHART FOR EMG TESTING OF TMJ DISCRIMINATION

| Specified µV Level | Attained µV Level |
|---|---|
| Trial 1 _______ | Trial 1 _______ |
| Trial 2 _______ | Trial 2 _______ |
| Trial 3 _______ | Trial 3 _______ |
| Trial 4 _______ | Trial 4 _______ |
| Trial 5 _______ | Trial 5 _______ |
| Trial 6 _______ | Trial 6 _______ |
| Trial 7 _______ | Trial 7 _______ |
| Trial 8 _______ | Trial 8 _______ |
| Trial 9 _______ | Trial 9 _______ |
| Trial 10 _______ | Trial 10 _______ |

It is important for the clinician to help the patient generalize to environments outside the treatment setting. Toward that aim, other discrimination training procedures should be emphasized.

The patient should be asked to once again contract the facial muscles to a specified mild to moderate microvolt level. After maintaining this level for approximately 30 seconds, the patient is asked to reduce the muscle activity as rapidly as possible to the lowest level possible. The time taken to reduce facial tightness is discussed, and the patient is encouraged to learn more rapid release of muscle contractions. Each time the patient reduces the activity in a shorter period of time, a verbal reward for progress is given. This approach is repeated a minimum of six times (Table 9).

**TABLE 9.** EMG GENERALIZATION TABLE

| | Specified µV Level | Lowest µV Level | Time |
|---|---|---|---|
| 1. | ______ | ______ | ______ |
| 2. | ______ | ______ | ______ |
| 3. | ______ | ______ | ______ |
| 4. | ______ | ______ | ______ |
| 5. | ______ | ______ | ______ |
| 6. | ______ | ______ | ______ |

µV = microvolt

Generalization is critical in the transfer of results from the clinical office to the outside world. We believe generalization occurs with daily practice, and we encourage deep muscle relaxation for 15 to 30 minutes daily. We urge patients to use a variety of home relaxation program cassette tapes in conjunction with biofeedback training. Our program contains elements of progressive relaxation, systematic desensitization, and autogenic training. Patients thereby receive biofeedback training refined with strategies for relaxation. We have frequently recommended the Feldenkrais Functional Approach to TMJ Syndrome; a collection of six tapes, each of which systematically focuses on improving the functioning of one's lips, tongue, eyes, face, head, neck and entire body in relation to TMJ movement.

Over the years, we have found it useful to use systematic desensitization while engaged in biofeedback monitoring. For patients with significant anxiety or who experience variable improvement, we encourage visualization and ask them to describe hierarchies of their life events which they associate with increasing stress. If a patient experiences arousal at a particular level in the hierarchy, relaxation is encouraged and is accomplished with the aid of EMG biofeedback. With continued trials, the patient learns to recover from muscle contraction and stress in fewer and fewer seconds. Then the patient learns to recover without the aid of biofeedback.

Through the use of these procedures, patients will become aware of myofascial tension and contraction and can rapidly reduce muscle activity to a relaxed level. Most patients are seen once a week over 8 weeks and thereafter once monthly in a follow-up session, as often as needed.

## REFERENCES


Bell, W.E. (1969). Clinical diagnosis of the pain dysfunction syndrome. *Journal of the American Dental Association*, 79, 154-160.

Budzynski, T.H., & Stoyva, S (1973). An electromyographic feedback technique for teaching voluntary relaxation of the masseter muscle. *Journal of Dental Research*, 52, 116.

Cram, J.R., & Steger, J.C. (1983). EMG scanning in the diagnosis of chronic pain. *Biofeedback and Self-Regulation*, 8, 229-242.

Dohrmann, R.S., & Laskin, D.M. (1978). An evaluation of electromyographic biofeedback in the treatment of myofascial pain-dysfunction syndrome. *Journal of the American Dental Association*, 96, 656-662.

Glass, A.G., & Rao, S.M. (1977). Bruxism: a critical review. *Psychological Bulletin*, 84, 767-781.

Hiatt, J.L., & Gartner, L.P. (1982). Muscles acting on the temporomandibular joint. *Textbook of Head and Neck Anatomy*. New York, Appleton-Century-Crosts, p. 223.

Hudzinski, L.G., & Lawrence, G.S. (1988). Significance of EMG surface electrode placement models and headache findings. *Headache*, 28, 30-35.

Katz, J.O., & Rugh, J.D. (19,86). Psychophysiological aspects of oral disorders. *Annals of Behavioral Medicine*, 8, 3-11.

Laskin, D.M. (1969). Etiology of the pain-dysfunction syndrome. *Journal of the American Dental Association*, 79, 147-153.

Schneider, C.J., & Wilson, E.S. (1985). Special Consideration for EMG Biofeedback Training. Wheat Ridge, CO: *Foundation of Biofeedback Practice*, Biofeedback Society of America.

Thompson, J.R. (1951). Temporomandibular disorders: diagnosis and dental treatment. In B.J. Sorrat (Ed.), *The Temporomandibular Joint* (pp. 122-144). Springfield, IL: Charles Thomas.

Whatmore, G.B., & Kohli, D.R. (1983). Dysponesis: a neurophysiologic factor in functional disorders. In E. Peper, S. Ancoli, & M. Quinn, (Eds.), *Mind Body Integration* (pp. 380-394). New York, Plenum Press.

## ADDITIONAL READING

Agerberg, G., & Carlsson, G.E. (1972). Functional disorders of the masticatory system. *Acta Odontology Scandinavica* 30, 597-613.

Al-Hasson, H.K., Ismail, A.I., & Ash, M.M. (1986). Concerns of patents seeking treatment for TMJ dysfunction. *The Journal of Prosthetic Dentistry* 56, 217-220.

Baile, W.F., & Myers, D. (1986). Psychological and behavioral dynamics in chronic atypical facial pain. *Anesthesia Progress* 12, 252-257.

Brooke, R., Stenn, P., & Mothersill, K. (1977). The diagnosis and conservative treatment of myofascial pain dysfunction syndrome. *Journal of Oral Surgery* 44, 135-152.

Chaco, J. (1973). Electromyograph of the masseter muscles in Costen's syndrome. *Journal of Oral Medicine* 28, 45-46.

Clark, G.T., Beemsterboer, P.L., & Rugh, J.D. (1981). Nocturnal masseter muscle activity and symptoms of masticatory dysfunction. *Journal of Rehabilitation* 8, 279-285.

Eversold, L.R., & Machado, L. (1985). Temporomandibular joint internal derangements and associated neuromuscular disorders. *Journal of the American Dental Association* 110, 69-79.

Friction, J., Kroening, R., Haley, D., & Siegert, R. (1985). Myofascial pain syndrome of the head and neck: a review of clinical characteristics of 164 patients. *Oral Surgery, Oral Medicine, and Oral Pathology* 60, 615-623.

Gessel, A.H. (1975). Electromyographic biofeedback and tricyclic antidepressants in myofascial pain-dysfunction syndrome: psychological predictors of outcome. *Journal of the American Dental Association* 91, 1048-1057.

Haber, J.D., Moss, R.A., Kuczmierczk, A.R., & Garrett, J.C. (1983). Assessment and treatment of stress in myofascial pain-dysfunction syndrome: a model for analysis. *Journal of Oral Rehabilitation* 10, 187-194.

Heibert, A.N., Heloe, B., & Korgstad, B.S. (1987). The myofascial pain dysfunction: dental symptoms and psychological and muscular function: an overview. *Psychotherapy and Psychosomatics* 30, 81-87.

Hijzen, T.H., & Slangen, J.L. (1985). Myofascial pain-dysfunction: subjective signs and symptoms. *Journal of Prosthetic Dentistry* 54, 705-712.

Kleinknecht, R.A., Mahoneyu, E.R., Alexander, J.D., & Dworkin, S.F. (1986). Correspondence between subjective report of temporomandibular disorder symptoms and clinical findings. *Journal of the American Dental Association* 113, 257-261.

Lundeen, T.F., Sturdevant, J.R., & George, J.M. (1987). Stress as a factor in muscle and temporomandibular joint pain. *Journal of Oral Rehabilitation* 14, 447-456.

Malamed, B.G., & Mealiea, W.L., Jr. (1981). Behavioral intervention in pain related problems in dentistry. In: *The Comprehensive Handbook of Behavioral Medicine*, Vol. 2 (pp. 241-259). Culver City, CA, Spectrum International, Inc.

Pomp, A.M. (1974). Psychotherapy for the myofascial pain dysfunction syndrome: a study of factors coinciding with symptom remission. *Journal of the American Dental Association* 89, 629-638.

Ramfjord, S.P. (1961). Dysfunctional temporomandibular joint and muscle pain. *Journal of Prosthetic Dentistry* 11, 353-374.

Remick, R.A., Blasberg, B., Campos, P.E., & Miles, J.E. (1983). Psychiatric disorders associated with atypical facial pain. *Canadian Journal of Psychiatry* 28, 178-181.

Rugh, J.D., & Solberg, W.K. (1976). Psychological implications in temporomandibular pain and dysfunction. *Oral Science Review* 1, 3-10.

Salter, M., Brooke, R.I., Merskey, H., Fichter, G.F., & Kapusianyk, D.H. (1983). Is the temporo-mandibular pain and dysfunction syndrome a disorder of the mind? *Pain* 17, 151-166.

Schwartz, L. (1959). *Disorders of the Temporomandibular Joint*. Philadelphia: W.B. Sanders.

Scott, D.S., & Lundeen, T.F. (1980). Myofascial pain involving the masticatory muscles: an experimental model. *Pain* 8, 207-241.

Vaughan, R., Pall, M.L., & Haynes, S.N. (1977). Frontalis EMG response to stress in subjects with frequent muscle contraction headaches. *Headache* 16, 313-317.

Wilkes, P.L., Onmell, K.A., & Rohlin, M. (1980). Double contrast tomography of the temporomandibular joint: a new technique based on autopsy specimen examination. *Acta Radiology* 21, 777-784.

Wolfe, F., Cathery, M.A., Kleinkensel, S.M., et al. (1984). Psychological status in primary fibrositis and fibrositis associated with rheumatoid arthritis. *Journal of Rheumatology* 11, 500-506.

Yemm, R. (1976). Neurophysiological studies of temporomandibular joint dysfunction. *Oral Science Review 1*, 31-48.

# CHAPTER 9

# *EMG Diagnosis and Treatment of Pelvic Muscle Disorders*

John D. Perry, Ph.D.
and
Leslie B. Talcott, M.S., R.N.

---

**Pelvic Muscles in Perspective**

More recently than many of us would like to admit, phrenology was a popular movement in psychological science. But the hope that one could identify internal mental and emotional states by reading external "bumps" turned out to be more "wishful" than "thinking", and phrenology almost vanished until it was recently revived by biofeedback therapists. Like the man searching for his keys under the streetlight (because the light was better there), biofeedback therapists have used the frontalis muscle as an index of general muscle relaxation---even when not justified by logic.

The pelvic muscles are a case in point. Several studies have been published using frontalis EMG biofeedback to relieve spasm of the pubococcygeus and uterine muscles, located some three feet distant from the electrode site. No one seriously argued that pelvic muscle tension could be detected at the level of the frontalis; it was simply more convenient to place electrodes on the exposed frontalis than to confront the special problems of the most private of places in the human body.

Technology conspired with social mores to produce this modern-day phrenology. Prior to the introduction of high impedance input amplifiers in the 1980s, careful skin preparation and electrode placement by the technician were essential to obtaining valid and useful electromyographic recordings and feedback. At the same time, the identification and learning of internal psychophysiological cues requires a high level of patient comfort, relaxation, and personal security. Even if technicians could be coaxed into attaching EMG electrodes to their patients' private parts, the patients were left in an exposed and vulnerable condition that mediated against learning self-regulation, and thus severely compromised the therapy. Moreover, the technician-applied electrodes prevented at-home practice, further reducing effectiveness.

In the early 1970s, an intermediate solution was developed at the National Institute of Aging laboratories. A system of air pressure balloons, previously used successfully to study activity of the even-less-accessible gastro-intestinal track, were employed to assess and train activity of the anal sphincter itself. The new technique, called "ano-rectal manometry", achieved modest results in the treatment of fecal incontinence (Whitehead et al., 1985) and later urinary incontinence (Burgio et. al., 1985). Symptom-reduction rates (accidents per day) of 80 to 94 percent were obtained, (although the number of actual "cures" was small; typically 30-35 percent of patients).

Although ano-rectal manometers required no "surface preparation", the actual insertion process was an invasive medical procedure requiring physician supervision, and thus home practice was excluded. Moreover, the air-pressure sensors were "size-dependant" and provided only relative-scale data. The system could not differentiate between a "tiny" orifice and a "tight" one, and was therefore useful for training weak muscles, but unable to even identify tense ones. In addition, the insertion of the apparatus (several inches into the rectum) was extremely uncomfortable, and most patients would not tolerate more than three or four training sessions.

These limitations were overcome with the introduction of the "vaginal myograph", developed by the first author (Perry, 1978). The new sensor was easily self-inserted by patients, thus making disrobing and exposure of the private parts unnecessary. The vaginal EMG sensor was followed quickly by a smaller "anal" size for males. Each consisted of a short clear plastic rod with three silver EMG sensor pads glued to the sides of the barrel with 120 degree separation. A retaining bulb was then added to the internal end, to keep the sensor from popping out when the pelvic muscles were contracted, and another bulb was fixed to the exterior end, to keep the sensor from becoming lost inside the vagina or rectum. Together these bulbs served to automatically position the electrodes over the "orgasmic platform" of the vagina, or opposite the anal sphincter. The pads were then elongated to allow for individual differences in the depth of the muscle from the orifice.

By orienting the connecting cable, which exits at a right angle, it is possible for the patient to correctly self-insert the device, even at home, with a very high repeatability. When they were finally introduced into the medical marketplace, the devices were classified as "perineometers" (pronounced "para-knee-OM-eter") by the Food and Drug

Administration, since they performed the same functions as the manometric perineometer invented by Dr. Arnold Kegel in the 1940s.

The main advantage of the new sensors was their ability to make absolute-scale (electromyographic) data available for diagnostic and therapeutic purposes. The full impact of this began to emerge about the same time as Cram's announcements about norms for muscle scanning in general. Although mass data collection did not begin in earnest until the establishment (by the second author) of the Biofeedback Institute of Philadelphia in 1985, it was clear from the start that the pelvic muscles, monitored with vaginal and rectal EMG perineometers, were merely a special case of skeletal muscles in general. Typical non-patient populations averaged about 2.0 microvolts (RMS) EMG background or resting tension, although healthy persons average only 1.0 microvolt. Similarly, resting tensions of about 4.0 microvolts (corresponding to Cram's "mild" elevation) are typically found in patients with minor pain or frequent urinary tract infections. Tension levels of 5 to 6 microvolts or more are usually associated with distinct pain (and often anesthesia or lack of sensation).

Diagnostically and therapeutically, the new EMG perineometers permit a modern technological reintroduction of the pelvic muscle biofeedback therapy first introduced (without the name "biofeedback") by Dr. Kegel in 1948 (Perry & Talcott, 1989). Using an air-pressure vaginal cone and manometer, Kegel had pioneered an exercise approach to restoration of pelvic muscles that was highly successful and widely used in the 1950s. But like other self-help methods, it was discouraged by the medical profession, who advocated the more lucrative surgical and pharmaceutical treatments. Just as in childbirth and related fields, cost-containment and "healthy-living" issues have now come together to favor a return to non-invasive, non-chemical "natural" solutions, such as perineometry.

In 1988, at a major "Consensus Development Conference on Urinary Incontinence in Adults" sponsored by the National Institutes of Health, a prestigious government panel reviewed the evidence and recommended, for the first time, that "less invasive behavioral methods", including biofeedback, should now be employed **before** drugs and surgeries for treatment of many forms of incontinence. Thus after a 40-year fascination, the present trend is once again away from surgical procedures and pharmaceutical solutions, both of which have relatively poor track records and significant side effects.[1]

Considerable confusion surrounds the so-called Kegel Exercises, which are a special variation of voluntary exercise of the pelvic muscles as an attempt to restore urinary (and even fecal) control by strengthening the sphincter muscles. Because they have been described so frequently in women's self-help books and magazine articles, many people believe they know what they are and how to do them. But the exercises they have been taught bear little resemblance to those advocated by Arnold Kegel.

The exercises described in popular literature might better be called "Davies' exercises", after Joshua Davies (1938), M.D., who first advocated therapeutic exercise of the pelvic muscles in the medical literature. It was Dr. Davies who first suggested identifying the pelvic or "pubococcygeus" muscle by attempting to interrupt the urinary stream while seated on the toilet. Davies' exercises were never subjected to even clinical evaluation, although they became the model for subsequent recommendations that have been mislabeled as "Kegel's exercises".

Kegel's unique contribution was the addition of biofeedback to Davies' pelvic muscle exercises. He also developed a portable biofeedback device, the Kegel Perineometer, and a systematic educational and therapeutic framework for home practice with the device. In fact, Kegel anticipated in 1948 virtually all of the elements now recognized as defining the practice of Biofeedback (Schwartz, 1987; Perry & Talcott, 1989). The only exception was the use of "electronic" components, which were simply not available in 1948. Despite popular literature to the contrary, there is no such thing as "Kegel's exercises" without the use of (1) biofeedback and (2) take-home trainers.

Kegel even anticipated major tenants of behavior therapy, by furnishing his patients with matrix charts on which to record and plot their daily biofeedback practice. Kegel recognized in the early 1950s that patients were encouraged (rewarded) by the daily improvements in their perineometer scores, and thus stimulated to further progress.

The availability of inexpensive EMG perineometers has made possible a return to the biofeedback exercises first advocated by Kegel. The California Gynecologist reported a 92-93 percent success rate in curing incontinence in the late 1950s. Similar rates are now being reported for the EMG perineometry method.

### Diagnostic Procedures Using EMG Perineometry

Arnold Kegel is reported to have had superb "bedside manner"[2] which was a fortunate asset, because he lacked some scientific precision in reporting his clinical results. For example, in several different publications he reported slightly different "normal" figures for strength of muscle contractions, without indicating whether the changes represented an update of earlier estimates or simply a slightly different way of describing categories. More importantly, in only one of his publications does he mention the concept of endurance, or the length of time that contractions might be held. In a paper in the American Journal of Obstetrics and Gynecology (Kegel, 1948, p. 247) he noted that pelvic muscle contractions became both stronger **and longer** with practice. In a simulated strip chart diagram, he depicted contractions with an initial (pre-therapy) duration of about one second, progressively lengthening to eight to ten seconds near the end of therapy. But in his text he never elaborated on this issue, in spite of obviously having incorporating it into his clinical practice.

For nearly twenty years, from the late 1950s to the late 1970s, reports of research with the Kegel Perineometer virtually disappeared from professional literature. In 1979, Benjamin Graber and colleagues began to publish a series of papers in which they differentiated between short pelvic muscle contractions ("flicks") and longer, 10-second contractions ("holds"). In what was a major contribution to the field, they observed that the strength of short contractions did not correlate with subjective experience of orgasm, but the 10-second "holds" did. This finding has subsequently been replicated by others (Waletzky and Weiss, 1986; Eicher, 1984; Perry and Whipple, 1981). More recently, the correlation of holds (but not flicks) has been extended to urinary and fecal control (Perry, 1984).

Lack of specificity in measurement was an inherent limitation of the original Kegel perineometer. Quantification of manometer (air-pressure gauge) readings required the therapist to subjectively estimate the "peak" readings from observations of a rapidly fluctuating needle. Integrated averages over time were even more difficult to make. They required the therapist to observe 10 seconds of fluctuating readings and then mentally average the observations; obviously these could only be crude guesstimates at best.

The introduction of electronic monitoring of pelvic muscle activity opened the door to far more accurate methods of quantifying pelvic muscle contractions. The original research with the vaginal myograph was conducted using a Cyborg P-303 EMG amplifier and Q-700 data integrator, which provided convenient 10-second integrated RMS muscle values. A strip chart recorder provided a more accurate means of detecting "peak" contraction values (by actual measurement of deflections on the paper chart). The design of the Clinical Perineometer in 1981-82 (Farrall Instruments, Inc.) provided an opportunity for the first author to standardize data collection with the new vaginal myograph sensor, and thus eliminate some of the confusion which had surrounded the use of Kegel's device. Three different measurement standards were incorporated into the instrument to assess both phasic and tonic components of the pelvic musculature, and hopefully provide a basis for empirical adjudication of the question of the "ideal" measurement method for pelvic muscles.

(1) The "flick" test, or measure of maximum muscle contraction however brief, was incorporated in order to make comparison with data from Kegel perineometers possible. This test measures the strength of "phasic", or fast-twitch striated muscle fibers. A relatively new electronic innovation, the "peak sample and hold" integrated circuit, was incorporated to obtain this data.

(2) On the other hand, a timed endurance test was also included in the Clinical Perineometer. Following standard practice in physical therapy, the endurance test first measures the maximum contraction level, and then automatically counts the number of seconds for which the patient is able to sustain a contraction above 50% of that maximum. This provides an estimate of the strength of tonic, or slow-twitch striated muscle fibers, in the pelvic musculature.

(3) Theoretically between these two measurement types, a third test provides a kind of compromise measure. This 10-second "hold" average is a statistic (like so many others in modern science) that was not possible before the advent of contemporary electronics and computers. In retrospect, it has proven a valuable addition to our tools for evaluation of pelvic muscles.

Some lack of empirical data about the relationship of pelvic muscle contractions and significant subjective experience (such as urinary or fecal incontinence) may be attributable to psychometric mistakes of previous researchers. For example, some researchers have used the single "strongest" flick contraction in a series as their "data". But sound psychometric principles require that several samples be taken, with the highest and lowest values discarded, and the remainder averaged together to provide a "valid" estimate of the muscle's true strength.

A further refinement came with the monitoring of resting levels before, during, and after pelvic muscle contractions. Because air-pressure devices such as Kegel's are unable to measure muscle (resting) tension levels, Kegel never discussed the issue. But vaginal myography, which assesses electromyographic data on an absolute scale, is able to ascertain deviations of resting level from normal. It was quickly discovered in pelvic muscle training that almost all patients began to increase their baseline "resting" level as they strived to increase their contractile levels. To prevent patients from inadvertently learning to become tense, the practice was developed in which the average resting level is subtracted from the average contractile strength to yield a "net strength" figure. This "net" figure has proven to be the single most valuable datum concerning the patient's condition.

There are three broad categories of pelvic muscle patients that can benefit from this therapy. Initially, vaginal myography was developed as an adjunct to sexual therapy. Patients with weak atrophied muscles frequently complained of lack of orgasm or lack of orgasm during intercourse, and pelvic muscle reeducation has been very helpful in these cases. On the other hand, many patients suffer from painful muscle cramps or spasms; these include diagnoses of dysmenorrhea, PMS, and vaginismus. Such patients typically show either chronic or situational pelvic muscle tension, and a general lack of conscious control over their muscles. (Situational tension can best be assessed when it is active, such as on the first menstrual day, but it can also be elicited by having the patient fantasize the difficult situation.)

The sexual functions of perineometry were quickly overshadowed by applications in the field of urinary and fecal incontinence, especially but by no means exclusively, among the elderly. At the 1985 convention of the Biofeedback Society of America, President Neal Miller described the potential savings of "$13 Billion annually" to the national health care budget if biofeedback techniques were applied to the treatment of incontinence. As a result of Miller's encouragement, vaginal myography was quickly applied to the treatment of urinary and fecal incontinence in an aggressive way. Incontinence treatment programs were established, first at the University of Pennsylvania (in cooperation with Glouster County, New Jersey Dept. of Public Health) and then at Presbyterian-University of Pennsylvania Hospital in Philadelphia. Results from both projects proved the efficacy of the method, and a third clinic was established by the authors at Paoli Memorial Hospital in suburban Philadelphia (Perry and Hullett, 1988b).

The third application of EMG perineometry is associated with childbirth, both before and after[3]. Modern "natural" childbirth programs, such as Lamaze method, have strongly advocated the use of Kegel's exercise, without providing either evaluation of the need for this prescription nor assessment of its success. Likewise, most obstetric experts give at least lip service to the importance of pelvic muscle exercise post-partum, but again, seldom provide material support and follow-up for it. An encouraging sign is the great number of young physical therapists specializing in "ob/gyn" who have been trained in Perineometry. Such persons constitute the largest single profession trained by the Perineometer Research Institute in Philadelphia.

## Typical Diagnostic Session

The basic evaluation consists of observing and recording short contractions, ten-second contractions, and endurance. Conventional and computerized biofeedback instruments are available to automatically sequence through the several test parts, and record data for display, graphing, and analysis.

The assessment of pelvic muscle function itself can readily be accomplished in five to ten minutes. However, because of the nature of the measurement system, it is generally considered good practice to spend at least a half-hour or more obtaining patient history, educating the patient about the mechanisms of pelvic muscle function and subjective experience. Throughout this period, the establishment of rapport and alleviation of patient anxieties are unmistakable secondary goals.

The particular focus and content of the intake interview and history collection will depend upon the presenting complaint of the patient. The therapist needs to be aware that problems in the areas of continence and sexual function go hand in hand, and both are often (but not necessarily) triggered by childbirth experiences. It is neither appropriate nor prudent to assume that incontinent patients in their eighties are no longer sexually active. For example, one very proper 85 year old woman presented with urinary incontinence and made substantial progress in therapy until experiencing a sudden and dramatic reversal. It was later discovered that her mysterious relapse coincided with her husband's prostate surgery, which interrupted their previously vigorous daily sexual life.

On the other hand, a young female resident appeared for sexual therapy. Despite a year of talk-therapy, she was still primary anorgasmic. Upon evaluation, her pelvic muscles proved almost non-existent, and a program of pelvic muscle exercise with a home trainer was prescribed. As she was leaving the office, the author remarked "I'm surprised that with such weak muscles, you don't also have a problem with incontinence." "Oh", she said matter-of-factly, "I do." "How is it that you didn't mention that," we asked (since treatment for a physical problem is usually reimbursed at 80% or more, but anorgasmia is considered "psychological" and only reimbursed 50%.) "Well," she replied, "I'm a physician, and I **know** what they do to women with 'incontinence'. I'm not letting any surgeon near my private parts!" Apparently her urology rotation had not included the latest behavioral techniques.

Discussion of the interaction of incontinence, sexual, and childbirth experiences will often help a patient to put isolated events and recollections into proper perspective, and many help the therapist anticipate training problems. It will also enable the therapist to better explain the process of inserting the vaginal or rectal sensor, which should always be done by the patient if at all possible (since they will have to do it at home by themselves on a daily basis). If the patient is unable, due to very young or very old age, satisfactory results can be obtained by training the appropriate care-giver.

Ordinarily the patient is instructed in sensor insertion and sent to the lavatory to accomplish the task. The process of walking back to the therapy room actually helps to seat the sensor in the vaginal or rectal entrance, although sometimes it is dislodged. (That provides additional diagnostic information without the necessity of physical examination.) Whenever possible, evaluation and training are accomplished in a soft reclining chair (in typical biofeedback style). Patients whose ambulatory ability is compromised may require a hospital bed and pillows, however. In our own clinic, we use the egg-shaped "Alpha Chamber" (from Sensory Environment Engineers, Inc.) because it provides optical and acoustic isolation, and thus aids the patient's concentration.

When the patient is comfortable, the sensor lead (which should be hanging out of the patient's waistband or skirt) is connected to the EMG instrument and formal evaluation begun. The first step is obtaining a valid baseline (between 0.5 and 2.0 microvolts RMS) and eliciting a sample contraction (which helps the therapist select the most appropriate sensitivity setting for full-scale display). If the baseline is elevated (3 to 8 microvolts) the patient must be instructed in relaxation before proceeding. If it is extreme, a broken lead or expelled sensor may be at fault. Since many incontinence patients have reduced tactile sensitivity in the genital region (due to the anesthesia effects of wearing wet diapers) they are not necessarily reliable witnesses to the sensor's state of insertion. A simple electrode continuity test, using an ohmmeter, can establish whether the sensor is in fact making good contact with the vaginal or rectal skin, without forcing the patient to disrobe.

The diagnostic testing sequence is either automatically or semi-automatically provided by approved diagnostic instruments, which are required in order to purchase the patented PerryMeter brand sensors[4]. The standard diagnostic test consists of a series (1) baseline, (2) six short contractions, (3) five or six 10-second holds [with interspersed 10-second rest periods], and (4) the timed endurance test. Interpretation of the results is straightforward, and becomes easier with experience.

If the baseline is high, very irregular, or becomes elevated following contraction attempts, relaxation training is in order. If the short contractions are substantially stronger than the 10-second contractions, it is safe to assume that the patient has been doing some form of pelvic muscle exercise improperly. [We no longer prescribe short contractions at all, since many patients report "I didn't have time to do all my exercise on Wednesday, so I only did the short ones." Short contractions are of no therapeutic value whatsoever.]

If the 10-second contractions are weak, therapeutic exercise is obviously in order. Despite some researcher's attempts to set absolute standards, we have found that "strong" is best described in relation to a patient's over-all muscular condition. Eight microvolts (10-second average) would be very strong for an elderly woman confined to a wheelchair, but very weak for an athletic 30 year old. Healthy women in their 20s or 30s can usually generate 20 to 30 microvolts strength.

Graphic displays, such as are possible with computerized biofeedback instruments, are helpful in diagnosing 10-second contractions which are actually composed of a series of three or four 2 or 3 second (phasic) contractions. If there is great variation in the 10-second "hold" (i.e., deviation from an ideal "square wave" display) the patient will probably also show a short endurance test and need tonic exercise (longer practice contractions).

Patients with stress incontinence will often show only short endurance scores---five to ten seconds. Generally speaking, patients who can last 20 seconds or more [above 50% of maximum] will be able to last 30 or more, because they have good tonic muscle fibers. Usually this does not appear until after several weeks of at-home practice.

Before attempting to diagnose patients' conditions, the therapist should submit to the test himself or herself. In addition, one should practice evaluating other non-patients who are candid enough to talk about their incontinence and sexual experiences. At least one multiple-orgasmic woman should be evaluated, since they typically have higher scores than the population as a whole. In forming impressions of norms, recall that more than one third of women and nearly one fifth of men will become incontinent by the age of 60. Therefore, like Cholesterol levels in the American population, what is "typical" is not necessarily what is "good."

## The Process of Pelvic Muscle Therapy

Most urologists and gynecologists today give lip service to the "Kegel Exercises", but without any expectation that their patients will actually attempt them, or, if they do, that they will accomplish much good.

[One famous urologist said recently at a urodynamics training program that he always mentions them, but the reason is solely to protect himself against malpractice suits.] Unfortunately, what is understood by the "Kegel exercises" they describe bears little resemblance to the exercise program developed by Kegel. A one-time President of the Urodynamics Society instructed one of our male post-prostate surgery patients to "cross your legs and squeeze." That did no good of course, until he learned with EMG biofeedback how to do pelvic muscle contractions correctly, without the use of accessory muscles. As a result, he avoided a second operation for an artificial urinary sphincter, and a $10-15,000 surgical bill.

Popular versions of Kegel's exercises differ from the original in three crucial respects: (1) the use of biofeedback, (2) the use of home trainers, and (3) the amount of work involved. Kegel never claimed that patients could learn the exercises without visual biofeedback of internal muscle activity, but many well-meaning clinicians have assumed it possible---without scientific basis for that assumption. Moreover, Kegel issued his home training instrument to all patients, and required them to keep daily logs of their exercise and results. Finally, Kegel told them that "300 contractions a day" were necessary for therapy, and "100 a day" for maintenance thereafter for life. This is significantly more than the "10 contractions" repeated "several times a day" described by Masters and Johnson (1966), or "five contractions ten times a day maximum" cited in the popular book, "Ourselves, Growing Older" (1987). Those who describe these exercises as "simple" and "easy" do their patients a gross disservice. Since such minimal prescriptions do not cure their problems, patients assume that their condition must be especially grave, and they readily submit to the surgeon's confident invitation (Perry and Hullett, 1988a).

In three years clinical experience, these authors have concluded that it is primarily the **amount** of at-home exercise the patient accomplishes that determines how quickly (indeed, whether) they get better. Recently retired persons have the best success, since they have disposable free time in which to practice, and a realistic fear of being sent to a nursing home if they don't succeed. More than one such patient has reported that he figured "if two practice sessions a day was good, three would be better". Young married women often have the worse results, since so many of them are preoccupied with carrying out their duties as a mother (without regard for their own needs). Written, contemporaneous daily practice records often help patients develop the necessary discipline.

A "Patient Hot Line" has proven to be a valuable tool for compliance. A dedicated telephone line was installed with only a tape recorder available to "answer" the phone. Patients were required to telephone progress (exercise) results every other day. Since they were assured that they would **not** have to speak to the therapist directly, they frequently confessed to failures that would have been embarrassing to admit to a human being. Self-confrontation of their failure, in turn, often lead to a spontaneous remedy the next day. Patients were telephoned between sessions only when they failed to leave the scheduled recorded reports.

Technological advances have improved our handling of the compliance problem. The authors developed an experimental data-recording "Kegel Counter" under a Small Business Innovation Grant from the National Institute on Aging. The home trainer logs all practice sessions in computer memory, and has already been used to expose patients who falsified their practice records. Although the Kegel Counter is not yet in production, two somewhat similar instruments are commercially available already. The "Myo II" from Verimed, Inc., records statistics from all patient practice sessions for downloading at the weekly appointment. The "Bio-Prompt" from Physical Health Devices, Inc. not only records the actual exercise data for downloading, it also chirps to remind patients of prescribed exercise intervals throughout the day.

The typical course of therapy involves two to four weekly visits to the clinic, during which time the therapists seek to ensure that the patient fully understands the exercise requirements and is not utilizing extraneous accessory muscles. During this time, if the patient is practicing at home, a 50 or 100% increase in contractile strength will be recorded, and some reduction in frequency of urinary or fecal accidents will provide motivating encouragement to the patient. Then the interval between visits is extended to two or even three weeks. At check ups, the therapists begin to encourage and then demand generalization of the exercises to spontaneous times throughout the day. Finally, in our model, the patient is "graduated" from the program and the home trainer is returned. The sensor remains the patient's property, to use as a resistive device every week or two forever. (Insertion helps recall the program and the sensations of correct exercise.)

The typical patient in one project (Perry and Hullett, 1988b) required an average of 5.2 visits over a period of eight to ten weeks to achieve complete urinary (or fecal) control. The most difficult therapeutic intervals are usually the second and seventh weeks. The second week is sometimes hard for patients because the excitement has worn off, but significant results usually haven't been obtained yet. The seventh week is also a motivational milestone, because the patient is so pleased that what was once a thrice-daily disaster is now only an occasional weekly nuisance, and therefore, is ready to quit.

Fortunately for most patients, the experience of being dry (or orgasmic, if that was the goal) is inherently rewarding and highly motivating. Since most patients inappropriately associate shame with wetting their pants, they are extremely happy to be relieved of the burden. But not all patients want to get better. One woman dropped out of our program two days after having her first "dry" day in over sixteen years. Investigation revealed that she had been a recluse during this time, while her executive husband openly attended social functions with his mistress. Had the patient actually become dry and able to socialize, she would have had to confront her husband, and she was unwilling to take that risk.

Upon graduation from the authors' program, patients are given not one but two guarantees. We guarantee that (1) if they continue to practice Kegel exercises, they will continue to stay dry. Then, to underscore the point, we guarantee that (2) if they abruptly stop exercising, they will start to leak again in two or three months time. To soften the blow, and ensure continued compliance, patients are offered "free" follow-up evaluations at three, six, nine, or twelve months. Slightly less than half have taken advantage of this offer. Some have demonstrated continued muscular growth, while others remain at the level they were upon graduation, while a few have relapsed. If the follow-up visit indicates severe deterioration and/or return of symptoms, additional training sessions (at normal fees) are scheduled until the problem is controlled again.

## Limitations of Biofeedback Exercise Methods

The fundamental requirement for successful completion of a pelvic muscle restoration program is sufficient motivation to carry out the exercises prescribed, and sufficient strength to do so. A healthy fear of unnecessary surgery also helps, since many people are under the mistaken impression that surgical alternatives are "simple", "permanent", and "without risk". At least half a dozen patients have dropped out of our programs when informed that personal discipline would be required.

Mental presence, however, is only required on the part of the person who will supervise the therapy, which may or may not be the patient. We have obtained good results (cures) with several patients who suffered Alzheimer's disease, and did not appear to understand, but had an effective partner-caregiver.

Persons who have had significant amounts of prior pelvic surgeries take much longer than the average patient to obtain satisfactory results. Although we have encouraged the **rental** of home trainers as an expedient to avoid dependency and encourage generalization, we now make exceptions. One courageous lady grew tired of her colostomy and insisted her surgeon re-connect her rectum. He did so only reluctantly, arguing that after 10 years without it she would never be fecally continent again. But she regained control in only a couple of months. However, she discovered that she required weekly practice with the EMG home trainer to stay that way. Fortunately, her insurance company agreed, and she was allowed to purchase the device.

When "Continence Clinics" first began, we all assumed that proper pelvic muscle exercise would benefit only patient suffering from "stress" incontinence, which refers to leakage under conditions of physical exertion, such as lifting, coughing, or jumping. But many patients with "urge" incontinence began to benefit as well. Even patients with "uninhibited bladder contractions" (also called "hyperreflexive bladder") got better. Upon investigation, we learned that the distinctions between "stress" and "urge" incontinence were somewhat artificial[5], and many patients had elements of both. The common element in both types is weakness of the sphincter muscles (Perry and Hullett, 1988a). "Stress" incontinent patients exhibit "denial"; they are constantly surprised when they leak urine. "Urge" patients, on the other hand, exhibit "obsession"; they constantly empty their bladders to avoid having any fluid to leak out.

Some 10 to 15 percent of "urge" patients fail to become dry even when they obtain modest improvements in muscle strength. In all cases, we have provided them with a conventional biofeedback stress analysis, and in all cases found highly active electrodermal response (also called GSR). A brief program of biofeedback-assisted relaxation training has helped all such patients control their panic reactions to bladder sensations.

As more and more physical therapists have become involved in incontinence work, we have learned a greater appreciation for the talents and contributions of this profession. One particularly promising area is called "myofascial release" (Barns, 1984). Patients who have had multiple pelvic surgeries (in previous attempts to correct incontinence, and then to correct the correcting operation, etc.) are often left with massive amounts of scar tissue. Myofascial release is a form of deep, slow muscle massage that attempts to breakup or release adhesions of the muscles to the surrounding fascia. Patients with prior surgeries seem to benefit greatly from this therapy, although we have not done any systematic studies yet.

## References

Barns, J. (1984) Benefits of Myofascial Release, Craniosacral Therapy Explained. Physical Therapy Forum, 3:35.

Burgio, K., Whitehead, W., and Engle, B. (1985) Urinary Incontinence in the Elderly. Annals of Internal Medicine 104:507-515.

Davies, J.W. (1938) Urinary Stress Incontinence: The Anatomical Defect Found and a Rational Method for Its Treatment. Surgery, Gynecology, and Obstetrics. 67:273-280.

Doress, P.B. and Siegal, D. L. (1987) Ourselves, Growing Older. New York: Simon & Schuster.

Eicher, W. (1984) Die sexuelle Reaktion der Frau. Teil II: Geren eine Uberschatzung des Grafenberg-Spots. Sexualmedizin. 13:531-537.

Graber, B. and Kline-Graber, G. (1979) Female Orgasm: Role of the Pubococcygeus. Journal of Clinical Psychology. 40:34-39.

Kegel, Arnold (1948) Progressive Resistance Exercise in the Functional Restoration of the Perineal Muscles. American Journal of Obstetrics and Gynecology. 56:238-248.

Masters, W. and Johnson, V. (1966) Human Sexual Response. Boston: Little, Brown.

National Institutes of Health Consensus Development Panel (1988) Draft Report, Consensus Development Conference on Urinary Incontinence in Adults. Bethesda, Maryland.

Perry, J. (1978) The Vaginal Myograph: Measurement in the Second Stage of Sexual Arousal. Biofeedback Society of New England, Wakefield, Mass., November.

Perry, J. (1984) Handbook of EMG Perineometry. Grand Island, Nebraska: Farrall Instruments, Inc.

Perry, J. and Hullett, L. (1988a) Bastardization of Kegel's Exercises. Northeast Gerontological Society, New Brunswick, NJ.

Perry, J. and Hullett, L. (1988b) EMG Biofeedback in Treatment of Urinary and Fecal Incontinence. Biofeedback Society of America, Colorado Springs.

Perry, J. and Talcott, L. (1989) The Kegel Perineometer: Biofeedback Twenty Years Before Its Time. Association for Applied Psychophysiology and Biofeedback, San Diego, March.

Perry, J. and Whipple, B. (1981) Pelvic Muscle Strength of Female Ejaculators: Evidence in Support of a New Theory of Orgasm. Journal of Sex Research. 17:22-39.

Schwartz, M. et al. (1987) Biofeedback: A Practitioner's Guide. Guilford Press.

Waletsky, L. and Weiss, S. (1986) Orgasmic Status. Proceedings of the Banff Conference on Behavioral Therapy.

Whitehead, W., Burgio, K., and Engle, B. (1985) Biofeedback Treatment of Fecal Incontinence in Geriatric Patients. Journal of the American Geriatrics Society. 33:320-324.

## Notes

1. Surgical procedures typically produce dramatic initial results that last from one to five years, with urinary leakage symptoms gradually returning. Then the patient seeks out a new physician in the hope of obtaining better results. She claims the previous surgeon "did the job badly" or "used the wrong operation". A few years later, the cycle may be repeated. In the meantime, the patient has so much scar tissue (and, therefore, pain) that she stops seeking medical help, and all the surgeons think that (1) they've been successful and (2) only the other doctor's operations fail. In the case of drugs, many patients do find temporary relief of their symptoms, but the side effects (especially dry-mouth) are so severe that patients eventually reduce consumption or stop taking the medication altogether. As they become older, they succumb to the myth that incontinence in an "inevitable" consequence of aging anyway. The public introduction, in 1987, of "adult" diapers in drug stores and supermarkets has only reinforced this myth. Joseph Oslander, M.D., noted incontinence expert, has campaigned for "warning labels" on adult diapers sold over the counter. Such labels would have to state that incontinence was neither natural nor inevitable, and diaper users should consult a physician to find out about new treatment alternatives.

2. Personal Communication, from a long-time gynecologic patient of Arnold Kegel, 1982.

3. Many therapists are reluctant to introduce any foreign object into the vagina of a pregnant woman, not out of fear of harm so much as fear of being included in a shotgun lawsuit in the event of a birth defect. There is, however, an easy alternative. Such women can be trained rectally, instead of vaginally, for the same reason that ano-rectal manometry was used by Burgio et. al. to treat urinary incontinence. Most patients are unable to

differentiate the urethral, vaginal, and rectal components of the pelvic muscles, and thus exercise all of them as a group. Post-partum exercise should, of course, be restrained until tissues have healed and a physician approves the exercise.

4. The vaginal and rectal myographs, currently sold under the Trade name "PerryMeter" Single-User Perineometer sensors, are available only to owners of biofeedback instruments which collect and display data in the standard format described in this chapter. It is this format for which Medicare reimbursement is being sought. The format is available on a variety of commercial biofeedback instruments, which license the use of the "PerryMeter" trade name.

5. The distinction is extremely important to physicians, since surgery is always justifiable for "stress" incontinence, but only drugs are acceptable therapy for "urge". If the proposed therapy is neither surgery nor drugs, however, the significance of the distinction vanishes. In practice, all diagnoses are treated the same.

# CHAPTER 10

# *Emotional Self Regulation and Facial Expression Muscle Measurement and Training*

Robert Kall, M.Ed.

---

Our muscles not only support our frame and move us through our world, they also mediate our experiencing of it through pervasive psychoneurohumoral circuits and feedback loops throughout the body. EMG self awareness and control techniques can be used to train individuals to increase awareness and voluntary control of their emotional states, and emotional expression and experiencing abilities. This can aid individuals in the voluntary activation and intensification of salutary, positive feelings, attitudes and expectancies.

This chapter describes how conventional relaxation biofeedback, and zygomaticus biofeedback training paradigms can be readily integrated into an emotional self regulation model. This optimizes the individual's capacity and ability to make the most of positive experience opportunities and to maximize positive affect and attitude.

Biobehavioral patients (pain, stress, anxiety, somatic dysfunction, phobia, behaviorally exacerbated medical illness) tend to tighten muscles, constrict their peripheral vasculature and emotional response range, and narrow and rigidify their selective perceptual pattern of viewing their environment so that they often miss or less-than-optimally respond to positive opportunities. After they've learned relaxation and stress regulation techniques, they are still in need of emotional expression skills so they can make the most of opportunities for positive experience.

The ability to express emotion effectively, to be aware of feelings before one acts, and to perceive the emotional expressions of others has helped humans to survive the Darwinian evolutional selection process of the survival of the fittest. The Neanderthal who could sense and control his/her fear so he/she didn't scream, saved her/his family from detection. The warm and affectionate Cro Magnon man was more likely to connect with a mate and keep her with him so he could father several children -- thus perpetuating his genes. The Peking Man, with a sense of humor, could laugh his way out of an argument.

## Emotions: The Same Language All Over The World

Our feelings are connected to our faces. When we make faces, we activate a universal human response programmed into our bodies before birth. Charles Darwin wrote in his 1872 book, The Expression Of The Emotions In Man And Animals, (Darwin, 1872) that emotions are not learned, but rather, biologically determined. About 100 years later, research psychologists Paul Ekman and Carrol Izard independently travelled around the world to observe the faces people make to express different emotions (Ekman 1973; Izard 1971). In every culture, every country they studied, they found that people smiled to express happiness, scowled when feeling angry, and made the same faces to express fear, disgust, and other basic emotions.

This universal set of facial expressions strongly suggests that the most common emotional expressions are not learned, like the hundreds of spoken languages, but rather, are wired into our nervous systems. Learning does effect the way in which different cultures mask or use their facial and body emotional expressions. But the basic nervous system wiring remains the same, producing a smile in response to happiness and a disgust grimace in response to a foul odor. The Navajo or Japanese may mask the expression when compared to the southern Mediterranean, but even at a non-visible level, the muscle patterns remain the same. We smile or frown because the facial expressions are programmed, through our genes, into our being. The dog's bark and cat's meow, the shocked look of surprise, the sneer of disgust and the happy smile are woven into the spiral helix of the DNA that spells out the definition and specifications of each species.

Just as hair and skin color, height and nose shape are passed along from parent to child, many of the facial muscles and other body parts that express emotion are also hereditarily influenced, giving some people blinding smiles, with flashing gums, or distorted attempts at smiles which produce barely perceptible movements at the corners of the mouth.

## Emotional Illiteracy

Emotional expression starts at birth with crying and takes many weeks or months to begin blooming. Parents lovingly work to cultivate those first smiles, coos and laughs. Long before the child can develop language skills, he can cry, scream or smile anything out of Mama or

Papa. Once walking and talking begin, our culture shifts all the emphasis there, almost forgetting emotional communication skills entirely. Emotional expression training tapers off with few, if any, conscious efforts at the emotional equivalent of grammar training or vocabulary building, or further development of the language of emotions. We're left to fend for ourselves.

Over 100 years ago a debate about one of the most important aspects of human emotion began: What comes first? Does something which happens in our environment, like the screeching of tires, set off a racing heart beat so we feel the pounding in the chest and become aware of the emotion? Or do we hear the tires, feel the fear and become fearful, thus setting the heart racing? Pioneer psychologist, William James, took this position in the debate, (1899, James):

> An emotion of fear, for example, or surprise, is not a direct effect of these objects's presence on the mind, but an effect of that still easier effect, the bodily commotion which the object suddenly excites; so that, were this bodily commotion suppressed, we should not so much feel fear as call the situation fearful; we should not feel surprise..., but coldly recognize that the object was indeed astonishing. ...the mere giving way to tears, for example, or to the outward expression of an anger-fit, will result for the moment in making the inner grief or anger more acutely felt.

James' following observation spells out the heart of the pattern activation component of the Happiness Response Model of emotional self regulation.

> ...Action seems to follow feeling, but really action and feeling go together; and by regulating the action, which is under the more direct control of the will, we can indirectly regulate the feeling, which is not. ...Thus the sovereign voluntary path to cheerfulness, if our spontaneous cheerfulness be lost, is to sit up cheerfully, to look round cheerfully, and to act and speak as if cheerfulness were already there. If such conduct does not make you soon feel cheerful, nothing else on that occasion can.
>
> So to feel brave, **act as if we were** brave, use all our will to that end. ... and a courage-fit will very likely replace the fit of fear. ...To feel kindly toward a person to whom we have been inimical, the only way is more or less deliberately to smile, to make sympathetic inquiries, and to force ourselves to say genial things. ...One hearty laugh together will bring enemies into a

> closer communion of heart than hours spent on both sides in inward wrestling with the mental demon of uncharitable feeling. ...To wrestle with a bad feeling only pins our attention on it, and keeps it still fastened in the mind; whereas, if we act **as if** from some better feeling, the old bad feeling soon folds its tent ... and silently steals away.

The debate is still alive, but more researchers have, like James, now taken the position that both sides can be true.

### Making Faces Can Increase Body Muscle Strength

Several researchers have proved that grimacing actually increases hand strength. Making a face produces a direct effect on a seemingly unrelated part of the body. Just as grimacing intensifies grip strength, making strong facial expressions can intensify experience of other emotions.

Recent computerized assessment of multiple site facial EMG activity has demonstrated its superiority over observer visual assessment of subject's emotions.

During the past 10 years many studies have repeatedly shown high correlations between facial muscle activity and emotional state. Schwartz and Fair (Schwartz, Fair, et. al., 1976; Fair, 1988) reported that normals show stronger zygomatic response during positive affective imagery. Depressed patients exhibit stronger corrugator responses and weaker zygomatic responses. This seems analogous to the pattern physical therapists encounter when using biofeedback to rehabilitate weakened or atrophied muscles. One muscle (like the zygomaticus) is underactive. The antagonist muscle (the corrugator) is overactive and must be voluntarily inhibited and controlled.

### Smile Rehabilitation

We have been using Zygomaticus activation and Corrugator muscle inhibition EMG feedback for positive affect facilitation, intensification and "smile rehabilitation." Prospective data is being collected. When subjects are instructed to maximally activate the zygomaticus, readings range from 12 microvolts (100-200 hz bandpass) to 150 microvolts. Practice appears to dramatically increase contraction strength above initial levels. When subjects are induced to laugh or smile naturally, their EMG activity tends to be higher than during volitional efforts at maximal zygomatic contraction, or even maximal efforts to smile.

This suggests an inhibitory process at work, perhaps similar to what occurs during the early stages of thermal biofeedback training, when efforts to produce vasodilation often result in cooling of the fingers. Further zygomaticus activation and strength training, coupled with biofeedback monitored smiling and laughing to facilitate subject "connectedness" with the awareness of psychophysiological dimensions of positive emotion, seems to help patients develop the ability to equal and then exceed automatic or "involuntary" positive emotional response EMG activation.

Cacioppo (1988) mentions one Japanese study in which a group of human cadavers were dissected and two percent of them were found to be lacking their zygomaticus -- the primary smile muscle. Were they atrophied through lack of use or missing from birth?

Emotion researcher Sylvan Tomkins (1963) suggests that most people rarely express pure, uninhibited emotions. They transform, inhibit and modify feelings based on their acculturation. The feelings end up being blocked and are never really felt. Tomkins suggests that breathing and vocalization are the most strongly blocked. The facial muscles are used to prevent the feelings from being expressed. Instead of jumping and shouting joyously, we smile with drawn cheeks and pursed lips. We need to be able to control our emotional displays in some situations. The problem is, some people don't learn how to modulate the release of uninhibited emotional expression. They are either totally inhibiting or totally emoting. Raw emotional expression can be frightening, like an "alien force within" if it is only released on rare occasions -- during intoxication or under extreme circumstances. For feelings to be tamed, one must be capable of varying mixtures of voluntary and uninhibited control. One has to be able to modulate emotional letting go -- ten percent sometimes, ninety five percent other times. Practice can help.

If a positive experience opportunity presents itself, one must be able to quickly make the most of it by connecting with it as completely as possible yet appropriately, emotionally, mentally and physically. It takes training and practice to learn how to comfortably express and experience deep feelings.

I often ask seminar participants to smile at the very beginning of my presentation. A few don't smile at all. Some scowl. Some barely smile, and some let loose with strong smiles with minimal cause. I only allow about three seconds and then I say, "Stop. If you didn't smile yet, you lost your chance." A quick, strong positive experience reflex is necessary to get the most out of each minute.

We are conditioned to experience good feelings in response to smiles and warm, happy sounds: because most of the time, smiles, and happy or pleasure sounds and actions, are genuine parts of positive experiences which produce good feelings.

Paul Ekman (Ekman, et. al., 1983) told professional actors and actresses to make emotional faces, one muscle group at a time. This way, the instructions didn't cue the actors to the kind of face formed. They weren't told to make a happy or frightened face, but rather, to move their facial muscles in specified patterns (i.e., pull the brows together, pull the mouth back horizontally, raise the upper eyelids, etc.) Ekman found that different facial expressions produced different physiological response patterns. Just combining facial muscle activity patterns could produce predictable heart rate and hand temperature increases or decreases. Synthetic faces seemed to illuminate the whole body with emotion-appropriate patterns of physiological activities.

**Paralysis of Feeling**

If synthetic faces can turn on feelings, paralyzed faces can turn off feelings. Psychologist Ben Twerski (1987), co-author of a chapter in a book on paralysis, (Twerski & Twerski, 1986) discussed how facial muscle paralysis patients suffered more depression than other paralysis patients. In a conversation we speculated, that perhaps, their facial muscles weren't stimulating and exercising the conditioned feeling responses; the patterns that are turned on in the brain and which the brain activates when people make happy faces.

In one study, military veterans with spinal cord injuries reported decreased experience of emotion after their injury. The more extensive the spinal cord damage and resultant greater loss of body sensation, the greater was their loss of their ability to feel emotion. Patients described their feelings as cold, as mental rather than feeling and emotional.

"But I can't make a fake smile," so many people respond when asked to turn on a fake or synthetic smile. I tell them to do it anyway! The goal of emotional self regulation is to teach individuals to learn to find their own buttons for activating the patterns of good feelings built into their nervous system. The other side of the coin is the need to identify how they inhibit good feelings. I explain to patients, that when we create a synthetic smile, we usually experience feelings that flicker between the real; feeling-good and "I'm just faking this and I feel silly or stupid" feelings. The reason we can actually switch to feeling genuinely good just by creating a synthetic smile is response pattern activation. The activation of facial muscle patterns, usually genuinely associated with good feelings, actually facilitates the turning on of the real thing -- a feeling-good, conditioned response.

New patients often resist instructions to smile. They resist, saying: "I can't smile," "I don't want to smile," "It feels silly," or "strange" or "It doesn't feel real so I don't want to do it."

I ask one to smile and he says he can't. I pause, allowing the silence to grow pregnant, then nod my head with a whimsical smile and ask, "C,mon. You can't smile?" I knit my brow, perplexed. "Do you have a neurological deficit? No? Then you don't want to smile?" My patients appear annoyed, as though they were about to say, "Leave me alone. I don't want to smile." But they do want to smile. My little pre-schooler acts the same way when he's miffed. I joke with him and he flickers between laughing, smirking and frowning. The patients are stuck in old patterns of inhibition that prevent them from opening up to good feelings when they want to.

It's so common for people to be uncomfortable expressing their feelings. Smile muscle activation training functions as a form of systematic desensitization, enabling the trainee to become more comfortable with free, stronger expression of feelings.

Here's how I work with smile resisting patients: I suggest, "You do want to smile, don't you? Let me show you how." At that point, I demonstrate an exaggerated smile. Or I'll tell a joke or threaten tickling (if it's a group.) So far, this effort has never failed. Every patient smiles and usually laughs spontaneously. Part of the reason my little trick works is because there are two nerve pathways which control the smile and other facial expressions. The upper nerve pathway is connected to the gray matter motor cortex of our brain. When we decide we want to smile and think "smile," then the motor cortex activates our smile via the upper nerve pathway. Stroke patients with damage to their motor cortex lose the ability to voluntarily turn on a smile, though some smile rehabilitation physical therapy can help. But even after a stroke, patients with the lower smile pathway still intact will reflexively smile in response to something funny or ticklish, because the lower pathway is connected to the lower part of the brain, where emotions are mediated and where the stress and emotion mediating sympathetic nervous system is controlled. I get my patients to smile and laugh by bypassing their voluntarily or subconsciously disconnected upper smile pathway. I've tickled their lower smile pathway into activation by turning on synthetic smiles that activate the PE physiological response pattern, allowing them to feel the real thing.

The smile response pattern activation proves to my patient he can smile. When he sees my exaggerated smile an image of a smile is formed in his brain. This smile image helps to animate or rouse his own smile-conditioned response pattern homunculus and to facilitate the lower pathway smile response pattern. The image functions like a template or behavioral "mold," shaping and helping the release of his feeling-good response even though the patient had been stuck, inhibiting it a moment earlier.

Some patients are feeling so bad they say they don't want to or can't smile. Then, smile biofeedback can be useful. Patients use zygomaticus EMG feedback to tell them what strategies help to boost their smile muscle strength and boost the EMG amplitude.

They learn to voluntarily create a genuine smile that helps them let go of their positive experience inhibiting behaviors. The concrete muscle strength oriented feedback helps them to strengthen their smile reflex without initially expecting wonderful feelings. Over and over again we've found in our research that people, when asked to smile as strong as they can, work and strain to make a smile. Sometimes, in their beginning efforts to intensify their smiles, they screw their faces into distorted smiles, very artificial in appearance (just as other physical therapy patients suffer from unwanted co-contraction of antagonist muscles). We crack some jokes, kid around. Making any feeling good sound like laughter, humming, cheering, seems to strengthen the smile activity. When we coax them into laughing, the smile comes effortlessly with the help of the lower motor neuron pathway, and is 20% to 100% stronger than the strained effort. This approach actually teaches feeling relaxation -- the ability to stay relaxed and comfortable, while feeling deep emotions.

Kicking in the lower motor-neuron pathway is a very important step for the so many people suffering from alexithymia and other emotional dysregulation disorders. Learning to at first tolerate, then remain comfortable while experiencing strong feelings is often a sign of improvement for patients in many different forms of psychotherapy. The anxiety of feeling emotion is a common one that can be coped with very effectively. The feeling biofeedback and facial muscle exercises allow people to take small, safe steps. The patient's begin to take risks. The smile becomes more symmetrical, more natural and more robust. The goal of smile aided relaxation is to go beyond feeling nothing, to feeling deeply, to be able to comfortably enjoy strong, deep feelings. You can learn to turn on the full range of your emotions with comfort and joy.

**Smile Anatomy**

The most basic smile muscle is the zygomaticus, named after the zygomatic arch -- the cheek bone -- which it attaches to at one end. The other end of the zygomaticus attaches to the corner of your mouth.

**Activation of the Zygomaticus**

Try exercising your zygomaticus now. Pull the corner of your mouth towards your cheek. Palpate for the zygomaticus. Smile and tighten or contract the zygomaticus. Then relax it, so you can feel the muscle moving under your facial skin. Put your index finger near your cheek and your pinky near the corner of your mouth and contract your zygomaticus. Use all four of your fingers to feel along the muscle. Relax and then tighten the muscle so you can feel around your cheek to find the most activity. Try a little smile, then full faced, beaming grin. Feel your zygomaticus bulge as your smile intensifies. Palpate along your cheekbone then down towards your mouth until you find the most activity. Focus on the sensations in your skin as you smile.

Because of genetic variation, some people will have bigger zygomaticus muscles than others. Some zygomaticus muscles will be built up more around the front of their cheek, while in others the zygomaticus is further back on the cheek bone. Though there's little research, it makes sense that humanity's genetics has included smiles in its approach to biological diversity. There may be people who were born with easier, bigger, wider smiles. There may be people with more smile inhibition nerve paths. I instruct patients to get to know their zygomaticus, the most important and central of the smile muscles. The other smile muscles act as modifiers, adjective-like descriptors and smile flavoring components.

The eye muscles involved in smiling, the lower lateral orbicularis oculi pars palpabraeus, which crinkle up our lower eyelids and produce crow's feet, seem solidly connected with lighthearted, open-hearted good feelings and warmth. The eye smile muscles can work with the levator labii superioris -- upper toothy smile muscles -- to intensify the total smile. These lift the upper lip to show the teeth, and are connected to the orbicularis oculi pars palpabraeus eye smile muscles. Try making a little smile, then add your heart and turn the smile into a strong, genuine, high intensity smile, including your eyes. Darwin suggested that the levator muscles might be involved, and at the least, they were attached to the orbicularis oculi muscles, and they either aided in the lifting of the upper lip when the oculi muscles were involved or, acted as cables pulling up the upper lip through the activation of the eye smile muscles. Paul Fair, when a graduate student at Harvard, found that the lower outer portion is the only part of the ring of muscle encircling the eye that is significantly involved in smiling. I instruct clients to practice smiling with just the lower outer part of their orbicularis oculi, to be careful that they don't frown or just squeeze the entire eye muscle when they attempt to activate the key part of the muscle

Ekman reports that not everyone appears to be able to voluntarily control the lower orbicularis oculi muscle. His observation concurs with our own findings. Yet we've seen that with biofeedback aided training, people can usually learn to develop voluntary control of these muscles. It may be that learning to turn on smiling eyes is like learning to warm the hands. You can get good at the skill without being able to verbalize what you are doing.

It seems that smiling with the eyes makes it easier to connect with, and smile with and from the heart. The eye smile muscles become involved in more intense, intimate and open smiling. Barriers seem to melt when two people face each other eye to eye and smile a full faced smile with puffed lower eyelids and gleaming, squinted eyes. Darwin speculated squeezing the eyeball caused a change in its shape which produced the sparkle so often associated with smiling, cheerful eyes.

There may be some muscle activity involved besides the zygomaticus in raising the upper lip. The levator superioris muscle is definitely involved in the look of disdain, and may be involved in lifting the upper lip in some strong smiles. Ducchenne (1867) actually separately categorized the zygomaticus into two parts, one to raise the corners of the mouth and the other raising the upper lip. But many facial anatomy charts omit the zygomaticus minor.

Some people, particularly people who wore braces, are shy about showing their upper teeth. They inhibit or minimize the activity of the muscles that raise the upper lip. Experiment with adding lifting your upper lip to your smile. Pay attention to the different parts of your upper lip and how your feelings vary as you activate them.

Showing the upper teeth in a smile appears to always be accompanied by zygomaticus activity. But what happens if a person is shy or nervous and inhibits the major ingredient in his or her smile? If he inhibits the zygomatic "primary" smile activity but allows the "secondary smile activity" upper lip to be raised, or pulls his lips straight back rather than up towards the zygomatic arch, he may turn on a "nerdy" kind of grin like the Nerds in the Nerds movies, "Seymour" in The Little Shop of Horrors or Jerry Lewis in his earlier comedies. Or, the zygomatic activity may be replaced with a pursed lipped, forced smile produced by the buccinator and risorius or platysma.

## Emotional Expression Anatomy

It can be very useful to go beyond the smile ABC's to develop a smile vocabulary so you can clearly discuss with patients, and help them to think precisely about, what they do when they smile. The more they can understand the physiological activities and events that underlie the smile and positive experiences, the better they'll be able to control and enjoy them. The list of muscles, and the facial muscle anatomy chart next, are provided to help your patients connect with the pieces of their emotional expression tool kit. Even if one is already in perfect emotional health, the smile anatomy information can still be used to identify smile patterns and ways one can strengthen less active components.

Experiment with and try to identify each muscle's activity on your face. Tense or flex it a little, then a lot. Combine two or more smile muscles. What expressions, if any, do these produce on your face? What feelings do you experience?

## The Birth of Electrophysiology

The descriptions of the muscles are taken from my own observations, those of the many researchers I've interviewed and whose literature I've reviewed, and the work of Guillame Benjamin Amand Duchenne, (Duchenne, 1867) a 19th century french physiologist. In the 1830's through 1870's, Duchenne used a Faraday battery (similar to the kind Ben Franklin used for his famous key-in-a-kite experiment) to stimulate the muscles of cadavers and live subjects so he could observe the electrophysiology of motion. He observed the actions of the muscles to determine where they were attached and how they functioned. It's unlikely the same research could be performed now. He warns of the risk of burned out eyeballs in these experiments. The more anachronistic descriptions are Duchenne's.

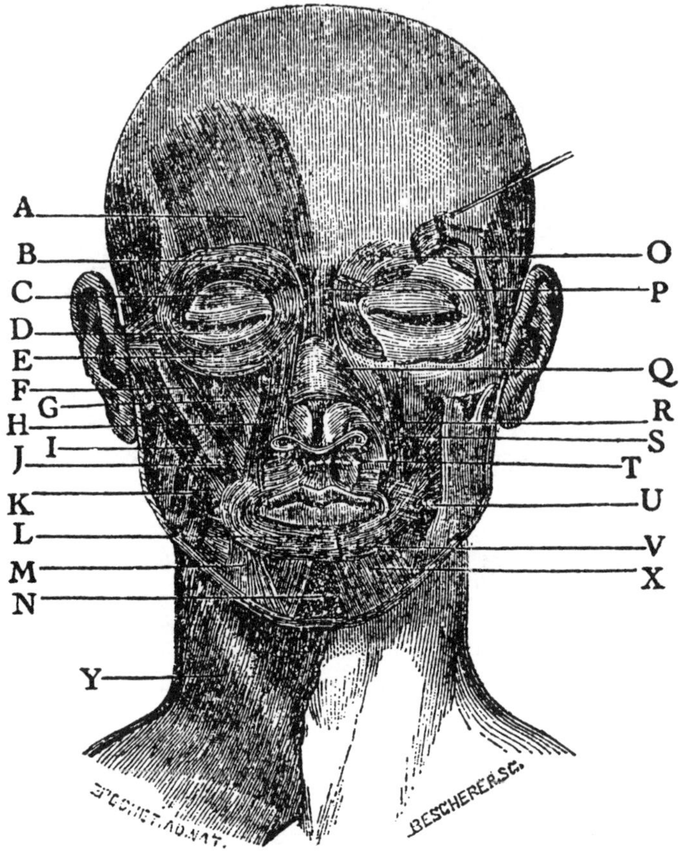


**Figure 1.** Anatomic Preparation of The Muscles of The Face

**Muscles of Positive Experience**

1) Zygomaticus -- lifts corners of mouth towards cheek bones; the primary smile muscle.
2) Orbicularis Oculi lateral inferior -- muscle of benevolence and frank joy; causes crow's feet, twinkling in the eye.
3) Nasalis -- lewdness and lasciviousness, flares nostril, creases central part of nose.
4) Pars alaris -- complementary with Nasalis for expression of passion, flares nostrils, lifts nose.
5) Mentalis -- shows lower teeth.
6) Platysma -- intensifier, opens lower lips, pulling them down and back.
7) Orbicularis oris -- modifies smile in kissing, tasting, sexual excitement, thrills.
8) Trapezius and cervical paraspinals --pull head back for laughter, sighs of contempt, breaths of sweetness.
9) Diaphragm -- belly laughing.
10) Arms and hands -- hilarious waving

**Aggression, Anger and Rage Muscles**

1) Corrugator-anger, anguish -- pulls eyebrows down.
2) Orbicularis Oculi Palpabraeus superior and inferior -- contempt, squints or narrows eyes.
3) Masseter -- clenched jaw of restrained anger.
4) Triangularis -- complementary of aggressive passions, pulls lower lip out to help form contemptuous face.
5) Procerus nasi -- aggression, frown, lowers, furrows forehead, creates the scowl.
6) Quadratus labii inferioris -- lowers lower lip.
7) Platysma- intensifier -- pulls lower lips down and back, helps show teeth.
8) Sternocleidomastoid -- pulls head toward chest or tightens neck.

### Fear and Sadness muscles

1) Frontalis -- muscle of attention.
2) Palpebralis superior and inferior -- muscle of contempt and complementary muscle of weeping.
3) Labii superioris - moderate weeping or grief, whimpering; this muscle lifts the upper lip; it seems to act as an intensifier for many feelings.
4) Orbicularis oculi -- the muscle encircling the eye.
5) Masseter -- may go slack during fear.
6) Triangularis -- muscle of sadness, pulls down corners of mouth.
7) Corrugator -- muscle of anguish, pulls eyebrows in and down towards nose.
8) Buccinator -- muscle of irony, a deeper muscle that pulls the mouth straight back.
9) Platysma -- a large sheet of muscle that attaches at the corners of the mouth, pulling down.

This is just a partial list. The dozens of facial anatomy maps I've examined illustrate the facial musculature differently, many omitting minor muscles or calling them by different names.

I suggest therapists have patients use a mirror, or touch your face and then their own with their fingers so they can feel how each muscle moves and on their own face, how it feels as they are flexed with different amounts of tension.

### Subliminal Smile Rehabilitation

Cancer patient counselor Yvonne White, takes an indirect approach when working with depressed or angry patients who don't want to smile. She assigns them exercises using muscles that are less obviously associated with the smile, like the platysma and mentalis muscles. She reports patients come back asking, "Are you trying to trick me into smiling? Because it's working."

Remember, facial muscles are but one category in the expression and experiencing of feelings. Posture, tone of voice, energy availability and previous activity all play important roles in what we feel and how we turn on our feelings.

## Progressive Smile Activation

This is a great exercise to do in a group, but is also useful to practice on a regular basis to build your smile reflex speed and strength.

Begin by turning on your zygomaticus. Pump a little zygomaticus iron for a few repetitions. Relax a few moments. Next, Start with activating your zygomaticus and add your lateral inferior (outer lower eye) orbicularis oculi and smile with eye eyes, making them twinkle. Pump this a few times. Next, start with the zygomaticus and eye-smile activation and add lower mouth, showing your teeth smiling. You can just add Mentalis and next Platysma, or take a short cut and do them together. Next, use your levator superioris to raise your upper lip and show your upper teeth. Make sure during all of this that you are not squeezing the rest of your eyes or corrugator or any muscles that have not been specified. Next, pull your head back, like you do when you really laugh. Next, breath from your abdomen. Next, move your arms and move your body from side to side. Add laughter or silly sounds and let yourself really loosen up enough to get silly and playful. You may want to experiment with tightening only one side of a pair of muscles. Try adding the nasalis and naris muscles -- flare your nostrils and scrunch up your nose. Remember, Duchenne suggested that was the muscle of lust and lasciviousness. See what feelings you experience when you use these muscles. Look at the activity in the mirror. What does it look like to you? Even better, do this exercise looking at someone else who does the exercise with you. This exercise was inspired by the work of Colorado Psychiatrist, Christian Hegaseth, author of The Laughing Place.

## Self Administered Smile Nerve Blocks

In case after case, depressed, angry and stressed patients have appeared to be creating their own upper smile pathway nerve block. Camille Palumbo, a counselor at Jefferson Medical Univ. calls this smile psychomotor retardation. Paul Fair (1988), an Atlanta psychologist, treated facial paralysis patients at Emory University. He found that the first thing he had to teach these patients was deep relaxation, so they could control the sometimes bizarre grimaces they would produce while attempting to smile. (I've often seen similar grimaces in normal, non-paralyzed people.) The facial paralysis patients would come in for two hour treatment sessions three times a week for several months. They were very motivated, and even small results helped them to feel much better about themselves and their appearances.

If you take just a little time to practice strengthening your smile reflex, you'll be helping yourself in several ways. First, you'll be better prepared to quickly smile into connecting with PE opportunities that present themselves to you. Second, you'll begin to develop a smiling face as your resting muscle tone for your face. Third, at a pre-conscious level, you'll begin processing and filtering your experiences with a better attitude that your smiling face sets up, reverberating throughout your body. The visuomotor behavior rehearsal training model (Suinn, 1984) can also be effectively used with smiling.

Patients can practice smile exercises almost anywhere. I usually combine them with my aerobic workout and my exercises on weight machines and Nautilus equipment, when I'm driving and when I shave in front of a mirror. I recommend they try to "pump smile iron" at least three times a day, doing sets of six muscle tensings. If you don't see immediate results, and most people do, persevere. Remember the flicker factor and relax to allow good feelings to pop in as a conditioned response.

**REFERENCES**


Black, D.W., 1984. Laughter. JAMA, Dec 7, 9252) 21, 2995-2995

Cacioppo, J., Dorfman, D., 1987. Waveform moment analysis in psychophysiological research, Psych Bulletin (102)3, 421-438

Cacioppo, J., & Tassinary, L.G., 1987. The relationship between EMG response and overt facial actions. Face Value; Facial Measurement Newsletter, (1) 3, 2-3

Cacioppo, 1988 Personal communication.

Darwin, C. 1872. The Expression of the Emotions in Man and Animals. New York: Philosophical Library, 1955

Duchenne, G.B.A., 1867. Translated into English and edited by Emanuel Kaplan, Philadelphia, J.B. Lippincott Company, 1949

Ekman, P. 1973. Cross cultural studies of facial expression. In P. Ekman (Ed). Darwin and facial expression: A century of research in review. New York: Academic Press.

Ekman, P., 1984. Expression and the nature of emotion, from Approaches To Emotion, K. Scherer & P. Ekman (eds.) Hillsdale, NJ: Lawrence Earlbaum

Ekman, P. and Friesen, W.V. 1975. Unmasking The Face, Englewood Cliffs, NJ: Prentice Hall 1978. The Facial action coding system. Palo Alto, CA: Consulting Psychologist's press 1987. Universals and cultural differences in the judgements of facial expression of emotion, J. of Personality & Social Psychology, (53), 4, 712-717

Ekman, P., Levenson, R.W., and Friesen, W.V., 1983. Autonomic nervous system activity distinguishes among emotions, Science, vol. 221, 9/16/83, 1208-1210

Ekman, P., Oster, H. 1979. Facial Expressions of Emotion, Am. Rev. Psychol, 30:527-54

Fair, P., 1988. private communication

Fridlund, A., 1987. Advances in analyzing the facial electromyogram. Face Value; Facial Measurement Newsletter, (1) 3, 4-5

Fair, 1976. Dissertation.

Izzard, C.E., 1969. The Face of Emotion, New York, Appleton Century Crofts

James, W., 1899. The gospel of relaxation; in Talks To Teachers On Psychology: And to students on some of life's ideals.

Lanzetta, J. and McHugo, G., 1986. The history and current status of the facial feedback hypothesis, paper presented at 26th Annual meeting SPR

McHugo, G. and Lanzetta, J., 1987. The multiple uses of Facial EMG. Face Value; Facial Measurement Newsletter, (1) 3, 1-2

Oster, H., 1978. Facial Expression and Affect Development, from M. Lewis & L.A. Rosenblum (eds.) The Development of Affect. New York, Plenum

Rickles, W.H., 1986. Self Psychology and Somatization: An integration with alexithymia. from Progress in Self Psychology, Vol. 2, Ed. by Arnold Goldberg, New York, Guilford Press

Schwartz, G.E., 1986. Emotion and Psychophysiological Organization: A Systems Approach, in PSYCHOPHYSIOLOGY, edited by Coles, M.G.H., Donchin, E., Porges, S.W., New York, Guilford Press

Schwartz, G.E., Fair, P.L., Salt, P., Mandel, M.R. and Klerman, G.L., 1976. Facial muscle patterning to affective imagery in depressed and non-depressed subjects. Science, 1976, 192, 489-491

Schwartz, G.E., Fair, P.L. Salt, P., Mandel, M.R. and Klerman, G.L., 1976. Facial expression and imagery in depression: an electromyographic study. Psychosomatic Medicine, 38(5):337-47

Suinn, R. M., 1984. Visual motor behavior rehearsal. The Basic Technique. Scandinavian J. of Behavior Therapy, 11, 131-142

Tomkins, S. S., 1963. Affect, Imagery, Consciousness Vol. 2 New York: Springer Publishing Co.

Twerski, A.J., Twerski, B., 1986. The emotional impact of Facial paralysis., from The Facial Nerve, May, M (ed), New York, Thieme Inc

Twerski, Benzion, 1987. personal communication

Zajonc, R.B., 1985. Emotion and Facial Efference: A Theory Reclaimed, Science, vol 228, 4/5/85, 15-21

# CHAPTER 11

# *Learning to Control the Tension that Underlies Your Pain -- The Real Challenge*

Edgar S. Wilson, M.D.

---

This paper has been developed and modified over the years to give chronic pain sufferers some guidelines and advice on how to control their pain problem. It is based on my experience with hundreds of patients and their journey of learning what pain has to teach -- how to listen to and love your own body. It may be reproduced in any form and is not copyrighted. It is dedicated to your own growth in self-awareness and self-regulation.

One very good method of increasing self-awareness is to keep a diary of your experiences. By tracking your experiences, you learn:

1. To observe your own attitudes about your pain.
2. To track the various difficulties and successes you experience.
3. To gradually identify the habits and lifestyle patterns that increase or decrease your experience of pain.
4. To modify those habits and patterns.

In your journal you might compare how you feel when you get up each morning with the way you feel as the day progresses. Add to this description all the things you tried to do in order to ease your pain. Then note whether or not they helped. Include the specific circumstances in which treatments worked or failed to work, the types of physical activities that increased or decreased your pain, and any foods or beverages that caused the problem to flare up or to improve.

It's particularly helpful to discover relationships between pain and stress you experience. Examples might be the type and amount of work you are doing, arguments that leave unresolved feelings, the number of cups of coffee you drink each morning, anticipation of major life changes, and so on. Next, make a complete list of the various stressors in your life:

1. Areas in which you feel frustrated, blocked, afraid to move, where you are forced to change or adapt.
2. Repeated situations where you are tense, angry or worried.
3. Intensely pleasurable events which leave you feeling off-balance and fatigued.

Then list what you are and are not doing about each of these factors and what you are willing to try to do about them. You might also list ways you can change your inner response to situations that you can't change directly. Write whatever makes good sense to you. Organize it as you will -- it is your journal -- but let it be a true reflection of your feelings and an assessment of your own approach to dealing constructively with chronic pain.

One caution: as you tally up the things you've been doing that seem to relate to pain, don't fall into the trap of feeling guilty and blaming yourself harshly. Excessive guilt and blame simply add another source of stress on top of everything else, and that's NOT part of the treatment.

## OVERCOMING BLOCKS TO SELF-REGULATION

There are several blocks or obstacles that one is likely to encounter in beginning a program of self-regulation. These include: dependence on drugs, a lack of personal support, and an underdeveloped sense of self to self-regulate. The latter problem is the most critical factor in any self-regulation program. Dr. Carol Schneider, an eminent psychologist, has devised a good self-development test.

Signs that you may have such a problem include:

1. Having very little self-confidence, little sense of self-worth or self-power. If we are lucky, our early experiences are full of pleasure in mastering new tasks. Our parents have praised our self-development, encouraged us to grow ever more able and learn more about how powerful our own self can be. If we don't feel we can master situations, it is very hard for us to self-regulate pain.
2. Having an extremely empty feeling when you are alone. If you have a strong belief when you are alone that "no one is there," then you have not developed a sufficient sense of your own availability to nurture, to care for, to protect, and to depend upon yourself. This leads to a profound dependency on others. If you feel others are primarily responsible for fixing your pain,

and that they should make it all right and "do it to you," then you will not be very able to "do it for yourself."

3. Lacking the ability to dialogue with yourself. By that we mean not being able to talk to yourself in a constructive, problem-solving and reassuring manner. The question: "Self, what are we going to do about this problem?" is never asked. If you have no access to or respect for that inner voice which talks to you about who you are, what you want and need, and how you should get it, you have a problem in self-development.
4. Very often feeling that life is out of your control -- even before your pain began. Life was chaotic with one crisis after another developing; work was never under control, relationships were out of control, your feelings of anger, sadness and anxiety were out of control because they could never be soothed through your own efforts. If you are going to have better control over your pain, you must have better control over yourself first.
5. Having a poor history of commitment. You never finished school; you are not committed to a career; you cannot keep going when there are obstacles to attaining a goal that you set for yourself. Then the goal of self-regulating pain becomes another goal you cannot reach.
6. You can never get close to another person without experiencing an intense fear that they will leave, reject, or abandon you, or that they will take you over, engulf your identity, and run your life. That intense fear makes you flee long-term committed relating, have repeated affairs, pick fights with those you love, or use a number of distancing techniques. If we comfortably love ourselves, we can comfortably love others. If your relationship commitments are always in a mess, then you first have to commit to continuing self development.

The inner task of self-development is an essential first step for many people as they seek control over their pain and over their lives as well. Without this step there is a lifelong dilemma of dependence upon doctors, drugs, and external treatments that fail to relieve the tension generated by a body that is essentially cut off from a well-functioning self. The task of developing the self takes time and commitment. It should be entered into with caution and concern for finding a helper and advocate who can guide you through the maze of self-discovery, but leave you free at the end to live your own life.

## THE ROLE OF PSYCHOTHERAPY

Before you can regulate yourself and begin the process of healing, you must have a self to regulate. If you begin to keep a journal and discover that you lack a sense of personhood with power over what is happening in your own life, or you feel utterly victimized by other people's ideas, feelings, and behaviors, then it is time to find a good psychotherapist who knows the problems confronting chronic pain patients. Hopefully, during the course of therapy you will receive the support and understanding necessary to gain insights, and develop new skills along with a deepening sense of self and personal power.

However, don't remain with a therapist who:

1. Tells you what to do and you resist and rebel.
2. Doesn't talk with you about or provide some structure within which your self can grow.
3. Blames your pain on you.
4. Doesn't like and respect your efforts.
5. You don't like or respect.

Liking and respecting each other will take you through the angry, upset times. So don't leave your therapy situation if your feelings suddenly change; make every effort to work through the problems between you and the therapist. But if your situation does not improve despite your efforts, consult another therapist. We have seen people whose lives, marriages, and recovery from pain have been compromised by psychotherapists who told them they needed their pain for a psychological reason and were therefore unwilling to give pain up. Keep looking if you get that message.

Chronic pain patients should be very selective in the type of psychotherapy they invest themselves in. Avoid therapies that are relatively unstructured and rely on the client to choose the direction and pace (such as client-centered or Rogerian counseling), or where the therapist tells you to expect multiple weekly sessions for several years (Freudian psychoanalytic therapy). Make sure your therapist is an effective agent for expecting and facilitating change and willing to help you initiate and carry through with as much self change as you need in order to bring your pain under the best possible control.

If you experience a lot of anger, it would help to learn assertion training skills and other cognitive and behavioral control strategies. If your problem has to do with early childhood abuse, find a therapist who is kind and empathetic enough to help you experience your inner child, to help it to weep and come to know the relative safety and power available to your adult of the "here and now." If inner feeling eludes you and you fear failure "again" with the more traditional forms of psychotherapy, don't give up. There are many other methods available that teach awareness, self-control, and responsibility in dealing with chronic pain. These can be used either alone or in combination with "talking" therapy. Biofeedback is the most widely utilized of such methods.

## DETOXIFICATION

The second major block to effective pain control is caused by dependence on drugs. Perhaps you suspect you are already dependent on such pain medications as codeine, Percodan, Demoral, morphine, Fiorinal, Valium, Ativan, Tranxene, or others. You may feel it is almost impossible to consider dropping these medications. In fact, you don't see how you could make it through the day and/or night without them -- they're the only friends you've had for a long time, and your attempts to wean yourself from them have failed. If so, then it's time to get "detoxed." Detoxification is not necessary for anti-inflammatory drugs or antidepressants, as these are used in long term control efforts. But do make sure you are taking these drugs in the proper dosages for you. Do not stop any medication suddenly.

Detoxification simply means withdrawal from drugs. Many people find the hospital is the safest and most effective place to go for detoxification. Attempts to withdraw on one's own at home are sometimes self-defeating and demoralizing, unless there is a lot of personal support. A good withdrawal program offers support and a variety of activities as your system adjusts to doing without medication. Hopefully, you will also learn techniques for dealing with stress in your everyday life when you are drug-free. The entire process of drug detoxification should prepare you to live more effectively in your own real-world situation without the constant need for medication, once you have conquered the physical aspects of addiction.

Detoxification is a necessary beginning in the self-regulation of chronic pain. This is because self-awareness is blocked or severely limited by the presence of drugs that dull the senses. It is very important to learn to recognize the tension and strain on muscles that are used to constantly avoid pain. Only when you are aware of the more subtle signals from your body are you able to reduce your tension or alter your posture to reduce pain. By blocking such awareness, particularly with the Valium-type drugs, you create a persistent barrier between the self and body awareness.

Sometimes it seems as though it is impossible to give up that last Ativan or Valium at bedtime; sleep never seems possible without it. This is particularly true if you have needed it for years. By cutting the dose down to a level as low as you find tolerable, it is still possible to learn to self-regulate your pain. Try not to take the medication except at bedtime. If you notice that you cannot resist escalating the dose during times of stress, then it may be time to detox completely. In any event, get your medication from only one doctor and be sure she or he knows your full medication history. Taking medications prescribed by several doctors increases the risks of drug interactions, dependency, and addiction. It can also cause a lot of strain in your ongoing relationship with your doctor.

When a potentially addictive or dependency-producing drug is prescribed, use it precisely as directed. Avoiding addiction is easy if one follows these simple rules:

1. Take the narcotic in adequate doses at regular intervals as long as the acute pain persists, not on an as-needed basis.
2. Taper off your use of the drug at a gradual pace, as your doctor prescribes. Suddenly stopping the drug may cause withdrawal symptoms.
3. Narcotics have no place in the routine management of long-term chronic muscle pain -- only in controlling acute pain such as that resulting from injuries or surgery. Terminal cancer pain is one of the few exceptions to this rule.

If you have occasional episodes of acute non-muscle pain (such as migraine), above your typical chronic level, there may be a place for a mild narcotic to manage that pain. But if such intense breakthroughs occur more than occasionally, you are probably not doing all you can to self-regulate your muscle problems. The easy availability of narcotics may reduce your motivation to take charge of what you can do to prevent acute episodes, and you should not let yourself get into that trap.

## USEFUL MEDICATIONS FOR CHRONIC MUSCLE PAIN

One thing physicians can do well, besides making effective diagnoses and ruling out diseases, is managing medications. This assumes they are aware of the problems of overmedicating and addiction in long-term care. Amitriptyline (Elavil is the best known brand) is probably the greatest mainstay of chronic pain therapy. It diminishes the resting muscle tension during sleep and provides a deeper sense of restfulness.

Unfortunately, even the low dosage of 10 mg. to 50 mg. at bedtime may produce dry mouth and drowsiness during the day and a weight gain of five to ten pounds, which is reversible when one goes off this medication.

Aspirin is a second mainstay in the drug therapy of chronic pain. If you use buffered aspirin with an antacid or with food to prevent stomach irritation, you can usually take six to eight tablets daily in divided doses until your tolerance level is reached. If your ears start to ring or you develop stomach discomfort, you are probably taking more aspirin than is optimal for you. Aspirin reduces the inflammatory phase of muscle pain and should be taken as needed during pain flare-ups rather than on a regular basis.

I have recently found Feldene to be very effective in controlling pain from inflammation. It works at the dosage of 20 mg. once a day and causes much less stomach irritation than aspirin. Unfortunately, it is expensive and not yet approved by the Food and Drug Administration for this purpose. Some patients have found any brand of ibuprofin, which is now available over the counter, to be helpful when 400-600 mg. are taken at a time. It is useful to experiment with anti-inflammatories until you find one that is helpful to you. Sometimes an anti-inflammatory does not help intense pain, or is helpful only in frequently repeated dosages. As your pain comes under better control, you will find anti-inflammatories helpful again and that taking them before you begin a physically stressful activity, such as marathon house cleaning or skiing, can help you enjoy more strenuous activities.

## BIOFEEDBACK AND THE SELF-REGULATION OF CHRONIC PAIN

We know that lowering the level of muscle tension is the most accessible and reliable way to control pain. Although as one becomes seriously involved in pain control there is an increasing awareness that it is very hard to feel muscle tension beneath the overwhelming awareness of pain. How can we successfully lower tension that we are not really aware of? The answer to this question can be found in the instrumentation used for biofeedback (BF). When muscles are tense, they produce more electricity. This can be recorded through electrodes placed over the skin, a painless experience. These electrical signals can be auditory feedback of the level of resting muscle tension in many areas of the body. In this way one actually becomes aware of minute changes in tension levels that are far below the threshold of sensory perception.

Electromyographic (EMG) feedback devices are commonly used in diagnosis as well as feedback for chronic muscle pain. Dr. Jeffrey Cram pioneered the use of muscle scanning and recording muscle tension levels throughout the body. This testing helps document not only the high resting tension levels over areas that are painful, but also shows the imbalance of tension in areas which are not painful but involved in perpetuating unbalanced tension throughout the body.

Stress profiling is a variety of procedures designed to show how you respond to stressful stimuli in a habitual and patterned way. These techniques help therapists devise treatment methods and establish goals that will help the pain patient. They also help document the existence of this disorder for insurance and legal purposes.

Biofeedback is not just an electronic technology, but an emerging field of health science as well. In the twenty years since its introduction, therapists have learned to use biofeedback tools to treat a wide variety of stress-related disorders. Not only is it possible to reduce resting muscle tension, but a variety of other physiologic processes as well. In addition, one can learn to alter the underlying physiologic "habits" that perpetuate hypertension, migraine, Raynaud's disease, and a host of other stress-related disorders -- not just in a laboratory, but during the day-to-day stresses of the real world.

Biofeedback therapy begins with learning to relate the physical changes recorded by the machines to the symptoms experienced by the patient. Once you can recognize your own bodily processes, you can learn to change them. The therapist helps to create an optimal learning environment by allowing you to experience the tension levels imposed by posture, performance anxiety, and other influences on resting muscle tension. Then it becomes a matter of learning to recognize internal cues which tell you when tension is building up and how to let go of it. Home practice using a variety of devices or aids can help you transplant the tension awareness and control from the laboratory into your real world, and increase your proficiency in controlling your tension and circulation. Symptom charting helps you to relate your real life experiences to the way your body feels.

Relaxation tapes have been of great help in home practice. Special tapes have been devised for deep relaxation, breathing, and progressive muscle relaxation. Each therapist has available a wide variety of tapes to choose from.

As biofeedback training progresses, there is a growing ability to achieve deeper levels of relaxation and symptom control. However, patients occasionally experience blocks or resistances to deep relaxation when they "let go." These blocks may have a lot to do with the person's fundamental dynamics or unshakable belief systems which tend to perpetuate high arousal states. These blocks may involve issues of control, fear of abuse, perfectionism, personal safety, or fear of losing one's mind. Overcoming the blocks to relaxation are important milestones of growth and awareness that should be worked through with the therapist. If it was easy to let go of tension and stress, one wouldn't need to have the insight and guidance of the biofeedback therapist. Often, however, psychotherapy is needed to unravel the knots that block the path to deep relaxation.

When therapy progresses further, one becomes both increasingly aware of tension in every day life and understands how to release it. At first this is frustrating because of the many waking hours involved in becoming aware of bodily processes that were previously taken for granted. Gradually we create new relaxation patterns which become habitual. Short relaxation breaks at red lights, in the office, or in the midst of stressful situations become habitual ways of counteracting the tendency to build up tension with the events of the day.

It is often helpful to utilize portable EMG units while driving to help desensitize oneself to the tension and anxiety associated with fear of repeated accidents. Pain patients often have excessively high EMG readings in the neck and shoulder areas at red lights. Systematic desensitization to fear and anxiety is one way of reducing a tendency towards bracing behavior.

Other techniques also help you learn to reduce and control your stress response habits. For example, learning to associate certain words with deep relaxation is helpful. "Raggedy Ann," "sleeping baby," and "melting like hot wax" are some of the inner descriptors our patients have used to recall their state of deep relaxation.

Visualizing calm and serene settings may also be used to generalize training effects. Pain patients really need to learn how to associate the rate and depth of breathing to a more relaxed state. Slow breathing that utilizes the abdominal muscles and diaphragm rather than the thoracic and neck muscles helps remind the nervous system to regain a relaxed state. Patients are encouraged to find their own private relaxation space through active imagery, and to return to this space frequently during the day to bring their tension back under control.

Training goals are flexible and geared to meet the needs of each patient. A good biofeedback trainer individualizes training methods and goals to meet the needs of the patient. She or he devises ways of breaking through old habits that get in the way of personal control of tension. Good biofeedback training integrates all of the other treatments being employed for pain, and provides meaningful stress management counseling as well as standard training techniques and methods. Learning to problem-solve, pace daily activities wisely, and acquire better assertiveness skills may be part of the training program.

In the twenty years since biofeedback was introduced, a new type of therapist has emerged who takes full advantage of the emerging technology and incorporates it into the ongoing art and science of counseling and changing behavior. The result has been a more cost-effective and efficient method of self-regulation without the long-term side effects of medication and surgery. In summary, selecting a good biofeedback therapist involves evaluating the following criteria.

The good BF therapist:

1. Listens to your report of your pain and your lifestyle and individualizes treatment to your needs.
2. Provides a wide variety of relaxation techniques and actively helps you learn to use the ones that work for you; methods of deep relaxation as well as short techniques that can be used throughout the day.
3. Uses EMG placements on a variety of muscle groups and in different postures.
4. Is supportive, non-blaming and works toward self-regulation.
5. Helps you problem-solve, sort and make sense out of all the possible physical and emotional stressors.
6. Assists in understanding and coordinating all treatments to allow for a more cost effective, individualized program of pain control.

## THINGS YOU CAN DO FOR YOURSELF

As a chronic pain patient, one of the most important things you can do for yourself is to arrange your day-to-day living environment so it will not create or contribute to the continuation of muscle spasms.

Since sleep is such a problem for most pain patients, it should be a high priority to create sleeping conditions that minimize problems. It is best, for example, to sleep in a relatively warm bedroom since the neck and shoulder muscles tend to contract in a cold environment. When you pull your head and shoulders down under the covers as far as you can to cope with cold room temperatures, you may provide your body with warmth, but you are also likely to trigger muscle spasms. If you can't warm up the room, it is wise to sleep in a warm garment with a high but loose neck or a turtle neck. This will keep your neck warm and make it unnecessary to lie beneath the covers with hunched shoulders.

Our patients almost invariably tell us that waterbeds and both soft and very firm mattresses make their pain worse. Some people have found relief by using a firm mattress on top of which they put a two-inch foam pad under the mattress cover. This allows some resiliency and enough firmness to give the body the support it needs.

Finally, window coverings should block out the early morning light. Many of our patients are so light-sensitive that they awaken at the break of dawn, no matter when they went to sleep. Once the room is almost totally darkened, it is also important to minimize the amount of noise you can hear. The slightest noise can be as disturbing as light. If you can't eliminate all sources of noise, then wax or foam ear plugs seem to be a workable, relatively comfortable solution.

Reducing noise may also mean tackling the difficult problem of snoring in your spouse. There are effective methods to reduce snoring; the best known being to keep the snorer off of his or her back during sleep. Recently, ear-nose-and-throat specialists have devised new methods for surgically firming up the soft palate to keep it from fluttering in the wind as one sleeps. If snoring continues to be a serious disturbance, you and your bed-mate might want to check with one of these specialists regarding the latest developments in treatment. A much less extreme method of controlling snoring involves using pillows to prevent the snorer from rolling over onto his or her back while sleeping.

**Postural Support Systems**

Pain patients know that chairs, sofas, and office furniture were not designed for human shapes. Many of us have a strong tendency to flop into a chair or sofa, throwing the back into a drooping "C" shape. This puts excessive stretch on the lumbar ligaments in the lower part of the back while compressing or impinging on the outlet of nerves through the spinal column and accentuating the pain. It is important to support the normal S-shaped curve that the back assumes when we stand or walk. One way to do this is to use a chair that allows you to place a small pillow two to three inches thick at the small of your back. By projecting your abdomen slightly forward, with the lower part of the S-curve supported, the neck and shoulders will usually fall into line so that the head is not thrust forward, but can be held farther back on the shoulders without strain. A very supportive pillow system is available from Aston Line Product Division, The Aston Training Center, P.O. Box 594, Mill Valley, CA 94941.

So much of the posture-related neck and shoulder pain that we see comes from hanging over a book or desk and forcing the muscles in the back of the neck and shoulders to continuously hold the head up. These muscles tend to contract more tightly as the day goes on until, usually by late afternoon or early evening, trigger points are filled with waste products and spasms begin. To avoid this holding pattern, the head should be balanced on the shoulders so that no muscle group is called upon to stay contracted for long periods of time. In other words, the head must be aligned in such a way that were it to fall, it could fall backwards just as easily as forward or to either side. Control and support of work and driving postures are an integral part of pain control.

Problems with seating apply doubly to driving an automobile, particularly if you have had an auto accident and drive with a lot more vigilance and tension than you used to. Many people drive with their heads thrust forward, especially when they are tired. This produces a painful strain on the neck and shoulder muscles. By placing a pillow at the small of your back as well as a wedge-shaped pillow under your buttocks, you can keep your pelvis rotated forward as you drive and thus maintain the S-shaped curve in your back. This will also help to keep your shoulders back and allow your head to rest comfortably balanced.

In addition, when you are driving to and from work, it may help to play quiet, soothing music on your radio or tape player instead of listening to news or to loud, jarring music. Listening to calming sounds can provide you with a transition period between home and work, and also give you a space in which to prepare for your next activity. If you usually drive in rush hour traffic, try waiting at work after the others leave. Listen to a relaxation tape and release the tensions of the day. When the traffic subsides, so will your tension, and a more relaxed, fun person will arrive home. Late, but a lot more pleasant to be with.

There are occupations that require various movements or postures that can lead to chronic pain in specific body areas. For example, people who work at supermarket check-out counters tend to have a particular shoulder, arm, and back pain syndrome that results from repeatedly moving merchandise over the check-out counter. People who have to lift heavy weights and twist their backs consistently in the same direction tend to develop mid- and low-back pain. Office workers who lean forward hour after hour to read their computer monitor screens develop a combination of shoulder, neck, and head pain. And people who sit at their desks all day without taking exercise breaks or varying their posture risk developing hip and leg pains along with lower back discomfort.

It's a good idea for each of us to check out our general movement patterns and postures to see if we're putting excessive strain on specific areas of the body. You should pay particular attention to the posture you habitually assume while reading. We have found that it is very important to keep the line of vision within a few degrees of parallel to the floor. In other words, when you read, try to look as straight ahead as possible. In order to do this, you may need to put reading material on a pedestal or book holder on top of your desk and raise or lower your chair to relieve neck strain. If you type or work at a computer console, be sure to give your arms adequate support. Forearm rests can be devised to fit your needs. You might also need to raise or lower the level of your typing instrument to prevent a stooped posture and poor angle of vision.

A special vision-related postural problem may occur for people who wear glasses, particularly those with bifocals. Often it is not eye disease and irregular eyeball shape that cause headache and neck pain, but simply inappropriate focal length adjustment in one's glasses. If you normally hold reading material eight to fifteen inches from your eyes, you probably have to thrust your head forward in order to read your monitor. Lenses for working at a monitor can be made so that the focal length is 18 to 22 inches, your probable distance from the monitor. This would allow you to straighten your head while working, relieving the strain on your neck muscles.

Having glasses made so that eye movements can remain relatively unimpeded is a major step in reducing neck muscle tension. If you wear glasses, also remember that your field of vision is restricted and you must track what you read with very minute and highly tuned neck movements. Bifocals provide an even smaller field of vision.

People who are partially deaf tend to thrust their heads forward and to the side in order to hear better. This is particularly true for those who are deaf in only one ear or for those who wear only one hearing aid. Ed, a physician, was partially deaf. He tended to listen to his patients with his head thrust forward in what appeared to be a posture of total interest and concern. By the end of each work day he developed a severe headache from the constant compression of the nerves and muscles at the base of his skull. When Ed finally realized his problem, he was able to control it by using two hearing aids and a pillow at the small of his back. Now he assumes a more balanced posture while working, and the headaches only come back when too much stress causes him to forget his body.

## SYMMETRY AND BALANCE

Each of us has a holding side of the body and a pivot side, usually determined by the dominant hemisphere of the brain. Right-handed people most often have a right-holding side of the body. That means they will tend to stand on the right foot when talking, pivoting and moving the left side, where most of the body's mobility is displayed.

Since trigger points occur in areas of least movement and maximum binding within muscles, usually more often in the body's holding side, it is important to determine if how you move is symmetrical enough to avoid muscle binding and the resulting trigger points.

A quick way to assess this is to look at the bottoms of any pair of tennis or Oxford-type shoes that you have worn for some time and inspect them for signs of wear. Notice where the shoe soles are worn the most and the least. When you hold both shoes together, are they wearing out symmetrically, or is there a distinct difference in the wear pattern? If the wear patterns are very uneven, chances are good that your movement pattern is asymmetrical and uneven, not only in your feet, but throughout your body.

It is very difficult to fit support mechanisms well. Rather than simply trying to correct "flat feet" or the tendency toward bunions, supports or orthotics should be fitted to assist in balancing an individual's movement pattern. When you purchase arch supports, heel lifts, orthotics, or other means of balancing your gait, test them by walking some distance. Notice not only how your feet feel, but if your walking movements are altered for better or worse. Pay real attention to what you see of your movement pattern in a mirror -- where your body holds tension and where it is free to move. If movements seem to stiffen, diminish, or become more one-sided, then it is likely that the new shoes or inserts are not right for you.

We can enjoy greater body symmetry by becoming more conscious of our movements and, as in the example of runners, lean upward and into our walking, feeling each muscle group expand and contract in a more balanced way. Whenever you move, become aware of the need that the body has to reach symmetry and balance. You can learn a lot, for example, by watching a mirror at home as you walk, or by following your reflection in store windows as you walk by.

We ask patients with extensive movement deformities to walk on a treadmill while we record their movements with a video camera. With a video monitor directly in front of the treadmill, they can watch their body movements while they walk. This provides immediate feedback so they can correct asymmetries of movement and gait almost instantaneously and without having to interpret a therapist's language.

If you have movement difficulties, you might find it helpful to view a slow-motion videotape of your walking and study those areas of your body you tend to hold most and move least, as well as those body areas that must compensate for these holding patterns. Areas of pain and spasm become clear as you observe these factors.

When one has a pattern of holding the shoulders too far back while walking, the heel strike is harder and sends a shock wave up the body.

The knee, which normally provides some shock absorption when flexed, is usually fully extended at heel strike when walking in this manner. The net effect is increased compression in the hip and low back. This is one of the major perpetuating factors of low back pain. It is a learned behavior, usually with such tried and untrue methods as teaching a child to walk with a book on his or her head. The combination of a shortened leg and "holding back" is a deadly combination usually found in chronic pain patients.

**EXERCISE**

Patients in our study complained of increasing pain both when they didn't get enough exercise and when they got too much. Each person had to learn to set limits and discover what type and amount of exercise was most beneficial.

Because unbalanced walking is so common, people with chronic pain are strongly advised not to walk long distances on sidewalks or road surfaces and not to do exercises in which absolute symmetry is required. When a person with movement asymmetry walks on a flat surface, the imbalance is accentuated rather than corrected. The connective tissue and muscle that are impacted progressively shorten and harden into a mass of pained tissue. Effective exercise can be had by walking over uneven terrain such as rocky slopes or the sides of hills. This increases the activity of the cerebellum and other balancing centers of the body. When you walk asymmetrically you must break old movement patterns in order to maintain balance. This means constantly shifting your holding pattern so as not to cause excessive strain or wear out any one group of muscles.

Similarly, jogging on paved roads or sidewalks or any hard surface usually makes this condition worse. If you enjoy jogging, it is better to run over nature trails and uneven, resilient surfaces where you must shift your weight. In this way you can avoid getting caught up in muscle holding patterns that produce pain.

Many people with chronic pain have long used exercise to get rid of a constant feeling of tension in their bodies. By exercising to the point of exhaustion they are finally able to rest. In fact, many become endorphin junkies, relying on the natural high of the body's own morphine, produced at the expense of overuse, stress and strain of intense exercise.

Susan was one of these. She was a 23 year old Olympic-class bicycle distance racer who trained by riding at least 40 miles a day, rain or shine. She neglected her feelings, nutrition and her relationships in order to maintain her competitive edge. One day, on a wet narrow road, she skidded into on oncoming car. Miraculously, there were no broken bones, but there was pain. The harder she tried to work the pain out of her stiff, sore muscles, the worse it got. Her efficiency and speed diminished as depression and pain grew worse. She felt like a total failure and a victim of the deadly trap of pain. Without the intense exercise, she lost the brief "highs" provided by endorphins. All that remained was depression. Only after coming to grips with her competitiveness and "type A" striving was Susan able to begin to care for herself as a person. Because she was born into a family that gave little reward for anything less than perfection, Susan had to find new ways to reward herself for feeling good.

Through the combination of biofeedback and psychotherapy, she gradually learned to recognize her tension and release it without guilt. With the help of Aston patterning, she learned to maintain effective movement and bicycling postures that would release the strain on the localized areas of tension/spasm in her body.

For many like Susan, it is hard to strike a balance between too little and too much exercise. The natural ebb and flow of energy cycles within the body are literally overwhelmed by the perfectionist and the "runner's high." As one loses touch with the need for rest and nutrition, this type A athlete also loses a competitive edge. The myth of "no pain, no gain" is no help in solving the problem.

By giving one's self permission to relax deeply as well as to perform with maximum effort, a balance can be achieved. Athletes who have lower muscle tension at rest have greater performance capacity while in competition. That is why such therapies as biofeedback and Aston Patterning have been used at the Olympic Training Center at Colorado Springs.

For those who do not aspire to athletic excellence, but must find a way to control their pain, exercise can be a real dilemma. Trying to find some method of avoiding extremes of prolonged tension, excessive stretching, or overuse of pained areas is a challenge in itself. Often we advise a graduated exercise program that ranges from swimming in a warm pool to soft, low impact physical conditioning, depending on the needs of each person.

**BAD WEATHER**

One of the most fundamental things pain patients forget is that muscles contracting at low levels of tension produce heat for the rest of the body. As a consequence, the muscles are likely to overdo their job when the body isn't adequately protected in cold weather. This results in spasms, tenderness, and trigger points. The obvious solution is to dress appropriately for the temperature. Let your clothes rather than muscle tension provide the warmth you need.

It is exceedingly important for people who have head and neck pain to wear a neck scarf in cold weather, since neck muscles have a tendency to become particularly tense when they are cold. Sweaters and jackets can be used to maintain warmth for the back and shoulders, and to provide additional protection for the neck.

One of the most effective means of breaking up severe muscle spasms is to take a hot bath or hot tub with water temperature around 98 to 103 degrees, immersing oneself right up to chin or ear level. This can be effective at the end of a stressful day, when pain is so distracting that you can think of nothing else. It can also help to prevent the build-up of tension in the neck and back when you anticipate great stress. Warmth penetrating the skin activates reflexes that release muscle tension to a great degree. Maintaining the muscles at this reduced level of tension for fifteen to thirty minutes enables them to drain their waste products and re-establish more normal reflex patterns, at least for a while. Some patients find gentle, flowing movements in the hot water are easier and less likely to produce rebound pain than exercise done "on dry land."

Very few patients have told me that warm baths were not helpful, once they decided it was alright to relax and enjoy themselves, made the effort to draw the water, take off their clothes, and get in. At one time I prescribed hot tubs and jacuzzis to patients and they could deduct the costs from their income tax. Unfortunately, the IRS no longer regards relaxation and hydrotherapy for pain as a legitimate deduction, even when prescribed by a physician. It is strange that the IRS and insurance carriers can now prescribe what one can do for good health.

## DIET

There are so many individual differences in pain patients that it is hard to give advice about diet. In general, one should try to get an adequate breakfast with sufficient protein to reduce the tendency for weakness during the day. Avoiding excessive sugar, caffeine and cola helps keep resting muscle tension down. As in all other aspects of self regulation and self awareness, you must learn to listen to your own internal nutritional needs -- not just feeding your fatigue as so many do.

## CHOOSING YOUR PROFESSIONAL HELPERS

### Finding The Right Physician

Some physicians are better at the diagnosis and management of chronic pain than others. Good physicians have a special empathy for problems they cannot cure, are good listeners, and give reasonable explanations and feedback. They do not believe everything that hurts should be either cut out or medicated to death. If your doctor doesn't have enough time to explain your problem to you or to listen long enough to understand you well, then you probably need to find someone with a less busy waiting room.

The long-term relationship with your primary treating physician should be one of mutual trust, empathy and advocacy. With certain specialists, however, there may be only a need for the answer to a specific diagnostic or therapeutic question. It is better not to expect this special relationship to extend very far. This is particularly true if the doctor you are seeing is employed by your insurance company rather than by you!

You can help minimize any problems you might have with your doctor by taking responsibility for insurance forms and legal matters as much as you can. He really wasn't trained to do these things and often resents being taken away from his/her role as a physician.

Dentists, like doctors are often highly specialized in dealing with chronic jaw pain problems. They are usually well known in your community for splinting the bite against bruxism or for specialized orthodontic procedures.

### Surgery

One must carefully balance the risks of surgery with any benefits that might be derived. Surgery may be a risky approach since any scar tissue left behind becomes a focus for more reflex pain and spasm. The physician recommending surgery should be certain that if a local anesthetic were to be injected into the proposed surgical site, it would result in complete, total remission of pain. Otherwise, surgery to that area is not likely to prove effective. If the surgeon promises a complete cure via surgery and cannot or will not answer your questions, but demands unquestioning loyalty and belief in his or her skill, forget surgery and try another route and/or doctor. Surgery for Thoracic Outlet Syndrome, Carpel Tunnel Syndrome and Laminectomies are rarely helpful in patients with fibrositis.

## Choosing A Body Therapist

If you find the problem of choosing a physician to be difficult, that of finding a helpful body therapist may be even more troublesome. Most body therapists we have encountered insist on stretching, strengthening, and stimulating painful areas involved in chronic muscle pain, usually producing rebound pain following such treatments. This is not because these people are sadistic. The usual course of therapy for rehabilitation following severe soft tissue injury involves mobilizing and strengthening the tissues that have sustained injury. This usually results in rapid recovery. Sadly, chronic muscle pain is often made worse by this approach. When the damaged segments of the spinal cord are stimulated by increased pain and stretch reflexes, the body may respond with a dramatic increase in muscle contracture that produces yet more pain and reflex (rebound) activity.

The body therapies that utilize gentle, functional approaches, such as those of Moshe Feldenkreis, Judith Aston, the Alexander Technique, and John Upledger's cranio-sacral therapy, have several features in common which are useful to those with fibrositis (fibromyalgia). These features should be helpful in guiding the pain patient to a more rational choice in body therapies. Many physical therapists have taken time to learn these techniques and apply them in their practice. Those who take the time and interest to learn effective management techniques for fibrositis must do so above and beyond their formal training in physical therapy. The following are some of the ways you can recognize therapeutic techniques utilized by advanced therapists. You may do well to discuss your prospective therapist's understanding and application of these approaches before beginning to work with them.

1. Treatments that are effective seldom produce rebound pain. Good therapists avoid vigorous stimulation of pained areas of the body.
2. Treatments are designed to reduce counterbalancing muscle tension between agonist and antagonist muscle groups.
3. Treatments address the postural support of pained areas of the body so that sitting, standing, sleeping and work postures are improved.
4. Effort is made to use graduated exercise to strengthen the body without causing excessive stretching, fatigue, or rebound pain.
5. Gentle massage is used to mobilized the body's support of pained areas. This avoids excessive stretching or premature deep-tissue mobilization.

6. Electrical stimulation or ultrasound treatments should never be the major focus of therapy. Tens units are often overused.
7. Finally, therapy should always be directed at giving the patient maximal education and an awareness of the bodily processes that perpetuate tension. Self-awareness of tension in everyday life is essential for long-term control.

**Acupuncture**

Acupuncture is a very old far eastern tradition for which we have no precedent in western medicine. We have only recently begun to recognize that temporary pain relief and biochemical changes, such as increases in endorphin levels within the brain, do occur with acupuncture treatment, even though we have little understanding of how or why. It is unfortunate that in the United States there are many relatively unskilled, but few well-trained practitioners capable of accurately assessing the problem and placing the needles in the appropriate meridians for pain relief.

Many of my patients can tell that there is a lot of "stuck energy" in their body, particularly around energy centers or chakras. These persons often find significant long term relief with acupuncture. For others, the use of acupuncture rarely provides lasting relief.

The key therapist for chronic pain is you! You can get totally lost in the jungle of pain management, letting others take control and lose yourself in the process. By understanding more about tension and its release you become the real expert. It's really a process of learning to listen to and love yourself. So many of us were not given consistent unconditional love and safety as children, thus we attuned ourselves to the expectations and needs of others to the exclusion of ourselves. We learned to define ourselves as caretakers; forgetting, or never learning to take care of ourselves. Thus, it is with a sense of guilt and shame that we come to acknowledge our own unmet needs and personal pain. It is not only possible, but necessary, that we forgive ourselves and others that made us "victims" of pain. Only then can we begin to feel our bodies as part of ourselves -- to be nourished, comforted and loved. Our inner child of the past must know that it is safe to feel good, to release tension, let the mind and it's concepts take a long break -- to come to know, accept and love ourselves. Only then can we really care for others.